MATHEMATICS

TOOLS AND MODELS

MATHEMATICS
TOOLS AND MODELS

Dalton R. Hunkins
Thomas L. Pirnot

Kutztown State College

Addison-Wesley Publishing Company
Reading, Massachusetts │ Menlo Park, California
London │ Amsterdam │ Don Mills, Ontario │ Sydney

ISBN 0-201-03046-2
EFGHIJKLMN-HA-89876543210

To Jeanette and
To Ann

Preface

The purpose of *Mathematics: Tools and Models* is to help the student develop an appreciation of mathematics through its applications. Often the approach used to stimulate an appreciation of mathematics is to emphasize its beauty and structure by covering such topics as logic, number theory, the real number system, etc. It is our belief that *Mathematics: Tools and Models* provides an appealing alternative to this more traditional approach by providing the student with an opportunity to see some of the diverse ways in which mathematics could be used. Our approach of presenting applications to students with interests outside of the natural sciences and engineering is supported by the fact that mathematics is finding a place in fields such as sociology, energy, political science, and economics.

Much of this material has been tried at a four-year state college over several semesters and we found that it works. We feel the topics have been successful because they are motivated by realistic applications, an approach that builds upon the attitude of many students that mathematics is primarily utilitarian. The following sample of comments obtained from our students seems to support this belief.

> The problems appeared relevant to contemporary life. The text seems very applicable to our world . . . I now realize some aspects of applied mathematics, therefore I respect it more today.

> I found this book quite helpful in (understanding) math because it was very relevant to problems in everyday life. It made it more enjoyable!

vii

In determining the depth of coverage, we felt that it need not be overly extensive and, therefore, we chose a level which is accessible to the student. For example, on linear programming, by restricting our attention to two-dimensional problems and graphical solutions, we can present this model without burdening the student with its more technical aspects. We have made a conscientious effort to make the material readable by the student. The dependency on first-year algebra is kept to a minimum although there are some exercises and examples in which simple equations must be solved. In addition, the format, examples, and exercises have been carefully chosen to provide motivation and reinforcement of the concepts being discussed.

There are numerous sets of short exercises labeled "Quiz Yourself." These questions are designed to reinforce the student's understanding of the major concept just presented before going on. In order to provide this immediate reinforcement, we have given the answers on the page on which the quiz occurs. These short exercises can be used by the instructor in the classroom for student participation and for a quick check of their understanding of the material. In addition, at the end of each part on "Tools," there is a mastery test which helps the student assess his comprehension of the material presented in that part. The answers for each of these questions are given in the appendix.

Each of the chapters, with the exception of the introduction and Chapter 10 on computers, consists of three parts. Part I is the statement of a problem which serves as motivation for the chapter. Part II is concerned with "Tools." In Part III, we construct a model of the problem and solve it using the tools developed for this purpose. In the table of contents we have starred those sections of Part II (if any) which are not essential for discussing the model in Part III.

In Chapter 10 on computers, we have varied from this three-part format and instead have used problems from earlier chapters to motivate and illustrate the discussion. Chapter 10 has no other dependency on the earlier chapters, except in the sense that examples and problems are related to topics previously discussed. We have restricted ourselves to considering algorithms from a flowcharting point-of-view. Since many students would not have access to a computer, we have completely avoided formal programming languages.

Our students have found Chapter 3, "Graph Theory," enjoyable and the discussion of PERT (Program Evaluation and Review Technique) provides an important application of graph theory. Chapter 4, "Legislative Apportionment and Inequalities," provides an interesting example of how extremely elementary mathematics can sometimes be used in solving important real-life problems.

Although some students may be quite familiar with solving systems of linear equations, we have provided depth and interest to the discussion in Chapter 5 by discussing topics such as population dynamics and simplified input-output models of an economy. Chapter 6, "Linear Programming," is written so as to be independent of Chapter 4. In this chapter, as throughout the book, we have avoided the temptation to rely on authority by simply quoting theorems; rather we give the student an intuitive feeling as to why two dimensional linear programming problems are solved in the described fashion.

Throughout the text the beginning and the end of an example are marked with the symbols ■ and □, respectively.

In keeping with the spirit of the book we have modeled the chapter dependency with the following directed graph. The solid arrows indicate strong dependency and the dotted ones show very slight dependency.

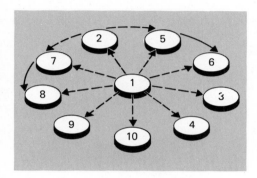

We are grateful to all those who have contributed to our writing of this text. In particular, we would like to thank our students who patiently endured the preliminary notes and our colleagues who were not only helpful with their comments and suggestions but who also gave encouragement through their good wishes. A special thank you goes to our typist Michele Readinger whose reliability was invaluable in meeting our deadlines and to Leroy Hart for his help in doing the exercises.

In addition, we thank the staff at Addison-Wesley for their enthusiasm in signing the text and their interest and assistance in developing and seeing our ideas to a successful conclusion.

Our gratitude is extended to the reviewers of the manuscript, Bill Bompart, John Brevit, Joseph T. Buckley, Bettye Ann Case, Joseph Cicero, Richard M. Crownover, Donald Freeburg, Richard Little, Nelson G. Rich, and Joseph Troccolo, whose thoughtful comments and constructive criticism improved it immeasurably.

Most of all we would like to thank our families, especially Jeanette and Ann, Matt and Tony, for their understanding and support throughout this project.

Kutztown, Pennsylvania   D.R.H.
November 1976   T.L.P.

Contents

xi

Introduction

While browsing through your library, would it surprise you to stumble across a book entitled *Formalized Music: Thought and Mathematics in Composition*?* Recently this book was reviewed in a major mathematics journal and to quote the reviewer, "This book is a creative chaotic potpourri of avant-garde musical composition based on stochastic processes, game theory, and algebraic structures. Xenakis' mathematics is as wild as the music it generates . . ."

In the most recent decades, mathematics has been used increasingly in areas where it had not previously been applied. It is not unusual now to see book with titles such as:

Mathematics in the Archaeological and Historical Sciences, Ed: F. R. Hodson.

The Theory of Graphs in Linguistics, Ernesto Zierer.

Mathematical Model Techniques for Learning Theories, Gustav Levine and C. J. Burke.

Mathematical Models in Marketing, Robert G. Murdick.

Elements of Mathematical Sociology, Murray A. Beauchamp.

The Foundations and Mathematical Models of Operations Research with Extensions to the Criminal Justice System, Haig Edward Bohigian.

Mathematical Models in Psychology, Frank Restle.

---

* This book, written by Iannis Xenakis, was reviewed in the April 1972 volume of the *American Mathematical Monthly*, p. 432.

The above list of books points out that we should begin to broaden our view of mathematics and recognize its applications outside the physical sciences. The above book list also points out that mathematics is being applied in such areas as sociology, biology, political science, economics, and management. *Mathematics: Tools and Models* should help you to appreciate the potential use of mathematics in your own area of interest.

In order to recognize this potential, you will be doing some computational work as you study the material in this book. Although you will encounter a limited number of the existent tools of mathematics, this exposure will be helpful since you will be learning some of the basic concepts and language of the subject.

In applying mathematics to a particular problem, certain steps, which comprise the modeling process, are performed. The first step is to determine the important characteristics that should be considered. This interpretation of the real-world situation is an extremely important aspect of the process since it determines the model to be used in solving the problem. However, at times we make some simplifying assumptions or ignore some features in order to give a description which is not overly complicated.

Once interrelationships between the components have been described, the next step in the process is to represent this description symbolically. This symbolic representation is called the **mathematical model**. A model can take many forms; for instance, it could be a simple equation, a set of points and lines, or a collection of possibilities with a probability assigned to each. We solve the problem in this model and then test the solution to see whether or not it is acceptable.

To exemplify these remarks, put yourself in the following situation. Suppose we are planning an auto trip from Kutztown, Pennsylvania to Boston, Massachusetts and we want to estimate the amount of time it would take. We must first select a road map for planning our trip.

Assumption 1:   We will travel on major highways as much as possible.

Thus, we will use an interstate highway-system map of northeastern United States. Unfortunately, though Boston is on our map, Kutztown is not. However, the distance from Kutztown to Interstate 78 at Allentown, Pennsylvania is small in relation to the total distance we will probably be traveling.

Assumption 2:   For the sake of computing the distance and time of travel, our trip begins on I–78.

There are several routes which interconnect with I–78 and which will get us to Boston. The designations for each of these possible routes appear on the map as being of the same type.

Assumption 3:   We can travel at the same average speed on any one of the possible routes. Therefore, the best route with respect to time is the shortest one.

There is a chart of distances between major cities included with the map.

It seems that the shortest route takes us through New York City, New Haven and Hartford, Connecticut and then to Boston.

Assumption 4:   We can use the chart to determine the distance of our trip.

Adding these distances, we get 316 miles. The map does not indicate the speed limit but since most of the roads on our route are major highways the speed limit should be 55 mph for all of them. However, we cannot expect to maintain the speed limit over the entire trip.

Assumption 5:   The speed we can average is 50 mph.

Now that we have described our distance and speed we can model the situation with the equation $t = d/r$. Computing t, we get 6 hours and 20 minutes.

We have a solution for our problem. Unfortunately, based on the authors' past experience, we would arrive in Boston about 1 hour and 25 minutes later than our solution tells us. It is clear that we solved the equation for t accurately. Why then do we get an incorrect answer?

The problem is not in the mathematical model $t = d/r$; it correctly reflects the relation between speed, distance, and time. The difficulty arises because our assumptions led us to use this formula with one speed for the entire trip. Of course, if we think about it, it would be hard to determine an average speed in advance.

In order to construct a better model, we may wish to consider the following characteristics in our description of the situation:

1) travel will be done on secondary roads;

2) speed will be reduced in congested areas;

3) time will be spent in refueling, at rest stops, etc.

Based upon these considerations, we can give a more refined model,

$$t = \frac{d_1}{r_1} + \frac{d_2}{r_2} + \frac{d_3}{r_3} + T,$$

where d_1 and r_1 represent the distance and speed on secondary roads; d_2 and r_2 represent the distance and speed on major highways in congested areas; d_3 and r_3 represent the distance and speed on major highways in open areas; and T represents time spent in refueling, etc. Breaking the trip into segments and using this model would allow us to obtain a better estimate of the travel time.

Mathematical models can take a variety of forms. Let us now look at a different type of model taken from Chapter 3, "Graph Theory."

A lobbyist for a large company wants a committee in the state senate to introduce a bill which would favor his company. He feels that his best strategy would be to obtain the support of the most influential committee member,

not to contact all the committee members. To this end, the lobbyist has made
the following observations:

Committee member	Influences
Adams	Collins, Fisher, and Evans
Baker	Fisher
Collins	Evans and Baker
Davis	Baker, Collins, and Evans
Evans	Baker
Fisher	Collins and Evans

In addition to the described relationship between the committee members,
the following assumption is made.

Assumption: All influence is exerted with equal intensity.

With this interpretation, we can construct the diagram in Fig. 1.1 as
a model of the committee. The committee members are represented by points
which are connected with arrows to indicate the direction of influence. For
example, an arrow is drawn from *E* to *B*, corresponding with Evans' influence
over Baker.

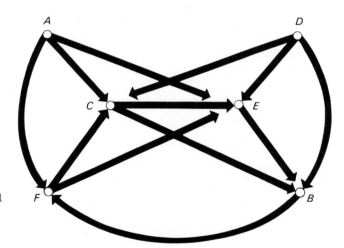

Figure 1.1

In Chapter 3, we will work with this model to answer questions regarding
committee influence.

In later chapters, we will construct models to solve problems such as,

- project-scheduling
- disease-communication

- how the seats in the United States House of Representatives are apportioned
- urban redevelopment
- fish-management
- the validity of a pharmaceutical company's claim
- determining the amount of stock a merchant should order
- curving students' grades

In constructing models and solving the various problems, it will be necessary to develop new mathematical concepts and techniques which we call "tools." The principal "tools" that we will encounter in this book are set theory, graph theory, linear equations, inequalities and linear programming, counting techniques, probability theory, and descriptive statistics.

It is hoped that the models of real-life situations that are presented herein will "whet your appetite." As you continue your education, you may wish to select other courses which build upon this first exposure to mathematical models.

2 Set theory

In 1851, the Italian mathematician Bolzano wrote a work entitled *Paradoxes of the Infinite* in which he began a development of set theory. In doing so, he took a first step toward answering perplexing questions regarding infinity which had troubled mathematicians and philosophers for over 2000 years. During the latter part of the nineteenth century, Bolzano's work was further developed by the German mathematician Georg Cantor, who today is generally considered to be the father of set theory.

At the University of Berlin, Cantor began his preparation for the important role he would play in mathematics. In addition to philosophy and physics, he studied mathematics with Kummer, Weirstrass, and Kronecker, three of the leading mathematicians of that time. In 1879, he started writing a series of papers which were devoted to some of the more difficult aspects of set theory and which would begin one of the fiercest intellectual battles in the history of mathematics. Using ingenious methods he obtained results so startling and revolutionary that they were not immediately accepted by his contemporaries. Cantor anticipated the controversy his work would generate when he wrote, "I am well aware that . . . I am putting myself in opposition to widespread views regarding . . . mathematics . . ."

His most devasting critic was his former teacher, Kronecker, who concluded that these new methods were "a dangerous type of mathematical insanity." Kronecker's opposition not only caused Cantor difficulty in getting his work published but it also kept him from obtaining a coveted professorship

Georg Cantor (1845–1918)

at the University of Berlin. Others, such as the Dutch mathematician Brouwer (1882–1966) and the German mathematician Hermann Weyl (1885–1955), agreed that Cantor's methods and results were not allowable in mathematics. At the Fourth International Congress of Mathematicians, the French mathematician Henri Poincaré proclaimed, "Later generations will regard (Cantor's set theory) as a disease from which one has recovered."

To make matters worse, towards the end of the nineteenth century several contradictions were found in Cantor's set theory. These "paradoxes," as they were called, prompted two of the foremost mathematicians of the time, David Hilbert of Germany and Bertrand Russell of England, to propose modifications which would eliminate the paradoxes in Cantor's theory. These modifications, however, did not impress Brouwer, who was still unsatisfied with Cantor's work. He wrote, "Nothing of mathematical value will be attained in this manner; a false theory which is not stopped by a contradiction is none the less false . . ."

Although the battle still continues among some mathematicians, most have accepted the validity of Cantor's results. Today set theory is recognized as a convenient language for expressing mathematical ideas and it is known that it provides a foundation from which virtually all mathematics can be developed. It is perhaps, in the words of Hilbert, "the most admirable fruit of the mathematical mind and indeed one of the highest achievements of man's intellectual processes."

I. A SURVEY PROBLEM

Senator Goodheart while running for reelection hired a pollster to make a survey of the voters in his district. The pollster interviewed 100 people and then presented Senator Goodheart with the following information:

1. There are 34 blue-collar workers with an income of at least $10,000 a year.

2. Twenty-six of the 34 blue-collar workers are conservative.

3. Twenty-three of those who earn at least $10,000 a year are not blue-collar workers.

4. There are 19 conservative, blue-collar workers who earn at least $10,000 a year.

5. Twenty-four of the conservative people earn at least $10,000 a year.

6. The blue-collar workers and those earning at least $10,000 a year together total 60.

7. The conservatives who earned less than $10,000 total 27.

After studying the results of the survey, the senator realized that something was wrong and immediately fired the pollster. Can you find out what is wrong with the above information?

The language of set theory allows us to express mathematical ideas precisely. We will see that when a problem is formulated using set theory, the solution may be more readily seen. (Photo by Laurence Lowry.)

Do not be disturbed if the above situation seems very confusing to you at this point. Unless you've encountered this type of problem before, it is unlikely that you will organize your thinking well enough to find what is wrong with the pollster's results. By the time you have finished studying this chapter you will have learned enough about set theory to easily analyze the results of this survey and determine what is wrong. Also, the set theory we develop will be an extremely useful tool throughout the book.

II. THE TOOLS OF SET THEORY

THE LANGUAGE OF SETS

We begin our discussion with the notion of a **set**. Intuitively, a set is a collection of objects. For example, several persons are brought together to form a committee. That single thing, the committee, is a set. The persons serving on this committee are the members of this set. Whenever we consider objects together in a single collection, this collection is called a set and the individual objects in this collection are called the **members** of the set. Sometimes the members of a set are called **elements**.

The names given to sets will generally be capital letters, A, B, C, etc. For example, we may label a committee of people with the letter C. Sometimes, we may want to use the same letter for several sets; however, by doing so, we would not be able to distinguish between the sets. Therefore we may subscript the capital letter with numbers or letters in order to make the distinction. For instance, if there are three committees which we would like to label C, then we would use C_1, C_2, C_3 for the names of the three sets.

Two sets, A and B, are said to be **equal** if they have identical members. In this case we write $A = B$. If A and B are not equal we write $A \neq B$.

A particular set can often be given by **listing members** within braces. To illustrate let us consider a set composed of seasons of the year. If we call this set S, then we may write $S = \{$spring, summer, fall, winter$\}$.

Even though it may be possible to list all the elements of a set, we sometimes find it inconvenient to do so. Suppose for instance that we wanted to express a set A consisting of all counting numbers from 1 to 1000 inclusively. Clearly, listing all the members of this set would be cumbersome, so instead, we might write $A = \{1,2,3, \ldots ,1000\}$. The first few members of A were written in order to establish a pattern. The dots indicate that the list continues in the same manner up to the last number in the set, which is 1000. A set C, consisting of all counting numbers, could be written in a similar fashion as $C = \{1,2,3, \ldots \}$. Not putting a number after the dots indicates that the list does not end.

A second method for giving a set is to use **set builder notation**. To use set builder notation we must have a set whose members all share some common characteristic which is satisfied by no other object. We can then describe this set by using this common characteristic. For example, suppose that P is the set of all past presidents of the United States. Using set builder notation we write

$$P = \{x: x \text{ is a past president of the United States}\}$$

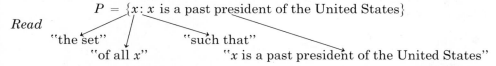

Read

"the set"   "such that"

"of all x"   "x is a past president of the United States"

Translating these symbols back into words (as we have indicated above), we read "P is the set of all x such that x is a past president of the United States." In set builder notation the sentence following the colon allows us to test whether or not a particular object belongs to the set. This comment becomes clear if we consider the sentence "x is a past president of the United States." If we replace x with "George Washington," we obtain a true sentence and so George Washington is a member of P. On the other hand, if we replace x with "Benjamin Franklin," we get a false sentence which means that Benjamin Franklin is not a member of P. Such a sentence, which uses a variable name like x, is called an **open sentence**.

When a set is described using set builder notation, it is important that it be **well-defined**—that is, we must be able to tell whether or not any particular

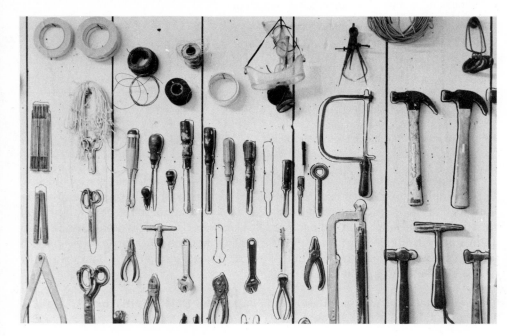

It is important that all sets we consider be well-defined; that is, we must clearly be able to tell which elements belong to the set and which do not. Do the tools that belong to this workshop wall constitute a well-defined set? (Photo by Jeff Albertson, Stock Boston.)

object is a member of the set. For example, $\{x: x$ is a person who is over 6 feet tall$\}$ is a well-defined set, since for any person it is always possible to tell whether or not he or she belongs to this set. On the other hand, $\{x: x$ is a tall person$\}$ is not well-defined since the word "tall" is somewhat ambiguous. Whether or not a person belongs to this set is a matter of how we interpret "tall."

A set which is given by the listing method can often be given by set builder notation, or vice versa.

EXAMPLES   ■ Suppose we define the set T to be

$$T = \{A, B, AB, O\}.$$

Then we can write T using set builder notation as follows:

$$T = \{x: x \text{ is a blood type}\}. \quad \square$$

■ If

$$B = \{y: y \text{ is a color of the American flag}\},$$

then B can be written using the listing method as

$$B = \{\text{red, white, blue}\}. \quad \square$$

QUIZ YOURSELF*   In the following examples each set has been given by either listing its members or by set builder notation. Read each statement aloud and identify the method used. Then express the set using the alternate type of set description.

1. $A = \{y: y$ is a day of the week$\}$
2. $B = \{1, 2, 3, \ldots, 60\}$
3. $C = \{$Canada, United States, Mexico$\}$
4. $D = \{z: z$ is one of the first six months of the year$\}$

---

*ANSWERS
1. $\{$Monday, Tuesday, . . . ,Sunday$\}$
2. $\{x: x$ is a counting number less than or equal to 60$\}$
3. $\{x: x$ is a country in North America$\}$
4. $\{$January, February, March, April, May, June$\}$

It is possible to write an open sentence that is not satisfied by any objects. For example, consider the sentence "x is a woman and a past president of the United States." Since no object makes this sentence true, you might feel that being a woman and a past president of the United States cannot be a defining characteristic of a set. However, we will find it more convenient to allow a set with no members. Thus $W = \{x: x$ is a woman and a past president of the United States$\}$ is a set.

DEFINITION   **The set which contains no members is called the EMPTY SET or NULL SET. This set is labeled by the symbol $\varnothing$.**

The set W mentioned above is the null set—that is,

$\{x: x$ is a woman and a past president of the United States$\} = \varnothing$.

Another set that is frequently needed is the **universal set**. The universal set is simply the set of all members under consideration in a given discussion. For example, we may agree to use only the counting numbers from 1 to 10 in a certain problem involving sets. For this discussion then, the universal set U would be $U = \{1,2,3, \ldots ,10\}$. In another situation we might want to consider only the female consumers in the United States. In this case the universal set U would be $U = \{x: x$ is a female consumer living in the United States$\}$.

Since we will often refer to a member of a set, we would like to have a special symbol for the phrase "is a member of." The symbol that is used is $\in$. The notation "$3 \in A$" expresses that 3 is a member of the set A. If we wish to state that 3 is not an element of the set A, we write $3 \notin A$.

Below are some examples to illustrate the notation which was just introduced.

■ a) $3 \in \{2,3,4,5\}$ **EXAMPLES**

   b) anthropology $\in$ {psychology, sociology, anthropology, archeology}

   c) timber wolf $\in \{y: y$ is a predatory animal of North America$\}$

   d) $7 \in \{1,2,3, \ldots \}$

   e) Benjamin Franklin $\notin \{x: x$ is a past president of the United States$\}$ □

Read aloud each of the following statements and decide whether it is true or false. Be   **QUIZ YOURSELF***
prepared to support each of your answers with an explanation.

  1. Westinghouse $\in$ {RCA, Philco, GE, Westinghouse}

  2. Alchemy $\in \{x: x$ is a modern day science$\}$

  3. $2 \notin \varnothing$

  4. $0 \in \varnothing$

  5. Denver $\notin \{x: x$ is a state of the Union$\}$

---

1. true   2. false   3. true   4. false   5. true   ***ANSWERS**

EXERCISES

  1. Determine which of the following sets are well-defined.

    a) $\{x: x$ is an American citizen$\}$

    b) $\{2,4,6,8,10, \ldots \}$

    c) $\{y: y$ is intelligent$\}$

    d) $\{t: t$ is a number greater than 5$\}$

    e) {Bach, Brahms, Beethoven}

2. Use the alternative method to express the following sets.
   a) {$x: x$ is an integer which is greater than 1 and less than 10}
   b) {2,4,6,8}   c) {$a: a$ is a letter in the word *set*}
   d) {s,i,n,g}   e) {$y: y$ is an integer between 6 and 11}

3. Use the alternative method to express the following sets.
   a) {$t: t$ is a digit in the number 173,268}
   b) {$z: z$ is an integer between 1 and 999 inclusive}
   c) {l,o,g,i,c}
   d) {$x: x$ is a positive even integer}
   e) {$y: y$ is a letter of the word *set* and y is a letter of the word *ring*}

4. Use the alternative method to express the following sets.
   a) {Aries, Taurus, Gemini, Cancer, . . . ,Aquarius, Pisces}
   b) {$y: y$ is a month of the year}
   c) {Alabama, Alaska, Arizona, Arkansas, California, . . . ,Wyoming}
   d) {$a: a$ is a position on a baseball team}

5. For each of the following, replace the # with either ∈ or ∉ to express a true statement.
   a) Woodrow Wilson # {$x: x$ is a past president of the United States}
   b) John Audubon # {$x: x$ is an American ornithologist}
   c) Albert Einstein # {$y: y$ is a living American poet}
   d) Vancouver # {$p: p$ is a Canadian province}
   e) IBM # {$w: w$ is a manufacturer of computers}

6. Consider the following universal set U.
   U = {Voltaire, Archimedes, Beethoven, Bach, Leonardo da Vinci, Ernest Heming-way, Golda Meir, Napoleon, Louis Pasteur, Rembrandt, Winston Churchill, Julius Caesar, Shakespeare, Harry Truman, Madame Curie, Leonard Bernstein}.
   From U we can choose a set of elements which all share some common characteristic and can give this set using both the listing method and set builder notation. For example, {Louis Pasteur, Leonardo da Vinci, Archimedes, Madame Curie} can be written as {$x: x$ is a famous scientist}. Another example is {Shakespeare, Winston Churchill} which can be written as {$y: y$ is British}. Find as many such sets as you can and give them using both the listing method and set builder notation.

7. Consider the following universal set U.
   U = {apple, TV set, hat, radio, fish, sofa, washing machine, shoe, dog, automobile, potato chip, bread, banana, vacuum cleaner, hammer, bed}.
   From U we can choose a set of elements which all share some common characteristic and can give this set using both the listing method and set builder notation. For example,

   {TV set, radio, washing machine, vacuum cleaner}

   $$= \{x: x \text{ is an electrical appliance}\}.$$

   Find as many such sets as you can and give them using both the listing method and set builder notation.

RELATIONS AND OPERATIONS WITH SETS

From your past experiences you know that mathematical objects are frequently compared with one another. We say that "2 is less than 7" or "$\frac{6}{4}$ equals $\frac{3}{2}$" or "$x^2 + 1$ is greater than zero." In a somewhat similar way we frequently compare sets. The most fundamental relationship that might occur between two sets is given in the following definition.

The set A is a SUBSET of the set B provided every member of A is also a member   DEFINITION
of B. This relationship is symbolized by writing $A \subset B$. If A is not a subset of
B, then we write $A \not\subset B$.

An important relationship that can exist between two sets is that one is a subset of the other. Frequently a subset will arise naturally in a discussion because all of its members share some characteristic which identifies them within the set. (Photo by Ira Kirschenbaum, Stock Boston.)

In order to show that $A \subset B$ we must convince ourselves that every member of A also occurs as a member of B.

EXAMPLES

■ Suppose $A = \{1,2,3\}$ and $B = \{1,2,3,4\}$. Since every member of A also occurs in B we can say $A \subset B$. Because there is a member of B, namely 4, which is not a member of A, we can write $B \not\subset A$. □

■ If $A = \{x: x$ is a predatory bird$\}$ and $B = \{y: y$ is a flesh eating animal$\}$, then $A \subset B$. However, B is not a subset of A since lion is a member of B but is not a member of A. □

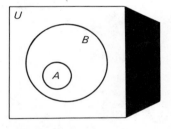

If C and D are sets, the only way C can fail to be a subset of D is if there is a member of C which is not a member of D. We can realize from this remark that $\varnothing$ is a subset of any set since it has no members.

Given that A is a subset of B we can indicate this by means of the following diagram, which is called a **Venn diagram.** The rectangular region which has been labeled U represents the universal set for our discussion. Note that the region labeled by A is completely contained in the B region. This shows us that all of A's members are found in B and hence A is a subset of B. We will frequently use Venn diagrams to visualize abstract situations.

Now that we have defined subset let us take another look at the important notion of equality of sets. Recall that sets A and B are equal if they have identically the same members. Therefore every member of A must be a member of B and every member of B must be a member of A. Another way of saying this is, for the two sets A and B to be equal we must have A as a subset of B and B as a subset of A.

EXAMPLES

■ a) $\{$crude oil, sulfur, mercury$\} = \{$mercury, sulfur, crude oil$\}$. Note that according to our definition of set equality, a set remains the same even though we change the order in which its elements are listed.

b) $\{$lion, eagle$\} \neq \{y: y$ is a flesh eating animal$\}$.

c) $\{z: z$ is a letter in the word *ecology*$\} = \{e,c,o,l,g,y\}$.

d) $\{1,2,1,3,1,4\} = \{1,2,3,4\}$. If you do not see this, recall again our definition of set equality. Certainly every member of the left set is also found in the right set and so $\{1,2,1,3,1,4\} \subset \{1,2,3,4\}$. Furthermore, $\{1,2,3,4\} \subset \{1,2,1,3,1,4\}$. The two sets are therefore equal.

e) If $A = \{x: x$ is a past president of the United States$\}$ and $B = \{y: y$ was born in the United States$\}$, then $A \neq B$. Even though $A \subset B$, $B \not\subset A$. □

You have often seen how numbers are combined to make other numbers. For instance, we add 2 and 3 to get 5 or divide 12 by 4 to get 3. Sets can also be combined in various ways to form other sets and we now turn our attention to this important aspect of set theory.

Let A and B be sets. The UNION of A and B is that set of elements which are   **DEFINITION**
members of either A or B or both. This set is denoted by $A \cup B$. Using set builder notation,

$$A \cup B = \{x: x \text{ is a member of } A \text{ OR } x \text{ is a member of } B\}.$$

In forming the union of two sets we are essentially combining the members of the two sets to form another set.
(Photo by Mike Mazzaschi, Stock Boston.)

We may also form the union of more than two sets. This union is the set of all those members that belong to at least one of the sets.

■ Let

$$A = \{1,3,4,5\}; \qquad B = \{2,4,6\}; \qquad \text{and} \qquad C = \{3,5,7\}.$$

Then

$$A \cup B = \{1,2,3,4,5,6\} \qquad \text{and} \qquad A \cup B \cup C = \{1,2,3,4,5,6,7\}.$$

Since 4 is a member of both A and B you might have been tempted to list 4 twice in $A \cup B$. However, it is unnecessary to list an element more than once in a set. □

■ Let

A = {x: x is letter of the English alphabet},
V = {x: x is a vowel},
C = {x: x is a consonant},
W = {x: x is a letter in the word *Mississippi*}.

We shall obtain the sets $V \cup C$, $A \cup V$, and $W \cup V$.

The set $V \cup C$ is the set of letters of the alphabet which are either vowels or consonants, and so we see that $V \cup C$ is the entire set of letters of the alphabet. That is, $V \cup C = A$.

We also have that $A \cup V = A$. This becomes clear if we write this statement out using the listing method,

$$\{a,b,c, \ldots ,z\} \cup \{a,e,i,o,u\} = \{a,b,c, \ldots ,z\}.$$

Lastly, $W \cup V = \{m,i,s,p\} \cup \{a,e,i,o,u\} = \{m,i,s,p,a,e,o,u\}$. □

It is a good idea in these examples to practice reading the statements aloud in order to become more comfortable with the notation and terminology we are using.

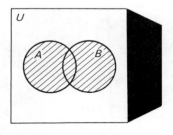

We can visualize the union of two sets in a Venn diagram as shown at the left. Again the region U represents the universal set. The entire shaded region represents $A \cup B$.

Let A and B be sets. The INTERSECTION of A and B is that set of elements which are members of both A and B. This set is denoted by $A \cap B$. Using set builder notation,

$$A \cap B = \{x : x \text{ is a member of } A \text{ AND } x \text{ is a member of } B\}.$$

We may also form the intersection of more than two sets. This intersection is the set of all those elements which belong to each of the sets.

■ If we let $A = \{1,3,4,5\}$, $B = \{2,4,6\}$, and $C = \{3,5,7\}$, then

EXAMPLE

$$A \cap B = \{4\},$$
$$A \cap B \cap C = \varnothing. \ \square$$

If $A \cap B = \varnothing$, then we say that A and B are DISJOINT.

DEFINITION

■ Let

EXAMPLE

$A = \{x: x \text{ is a letter of the English alphabet}\}$,
$V = \{x: x \text{ is a vowel}\}$,
$C = \{x: x \text{ is a consonant}\}$.

The meaning of a mathematical term can sometimes be remembered if one thinks of its everyday usage. Just as the intersection of two tracks can be described as the region common to both, the intersection of two sets is that set of elements contained in both sets. (Photo by Harry Wilks, Stock Boston.)

Then

$$A \cap C = \{x: x \text{ is a letter of the alphabet and is a consonant}\} = C.$$

The set $V \cap C = \varnothing$ since there is no letter of the alphabet which is both a consonant and a vowel. Thus V and C are disjoint. $\square$

The set of elements excluded from a given set make up its complement. (Photo by Peter Menzel, Stock Boston.)

The intersection of two sets can be pictured as shown in the Venn diagram at the right. As usual U represents the universal set and the shaded region represents $A \cap B$.

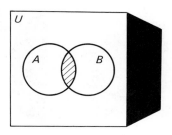

The third operation that we shall consider is complementation.

Let A be a subset of the universal set U. The COMPLEMENT of A is the set of elements of U which are not members of A. This set is denoted by A'. Using set builder notation

$$A' = \{x : x \in U \text{ but } x \notin A\}.$$

DEFINITION

If $U = \{1,2,3, \ldots ,10\}$ and $A = \{1,3,5,7,9\}$, then $A' = \{2,4,6,8,10\}$. $\square$

EXAMPLE

The complement of A can be pictured as shown in the next Venn diagram.

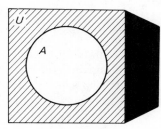

Let A and B be sets. The COMPLEMENT OF A RELATIVE TO B is that set of elements which are members of B but which are not members of A. This set is denoted by $B - A$. Using set builder notation,

$$B - A = \{x : x \text{ is a member of } B \text{ AND } x \text{ is NOT a member of } A\}.$$

DEFINITION

■ Let

$A = \{x : x$ is a letter in the English alphabet$\}$,

$V = \{x : x$ is a vowel$\}$,

$C = \{x : x$ is a consonant$\}$.

EXAMPLE

Since every member of V is also a member of A, when we remove the members of V which are members of A we obtain the empty set, that is, $V - A = \varnothing$. Removing the vowels from the letters of the alphabet, we are left with the consonants; therefore

$$A - V = C.$$

Finally, since no vowel is a consonant, we have

$$V - C = V. \square$$

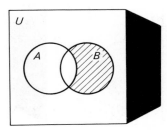

The complement of A relative to B is shown as the shaded region of the Venn diagram at the left.

QUIZ YOURSELF*

1. Let $A = \{1,2,3,4,5\}$ and $B = \{2,4,6,8,10\}$. Find $A \cup B$ and $A \cap B$.
2. Let $U = \{a,c,e,g,i,k,m,q,o\}$ and $C = \{x: x$ is a letter in the word *game*$\}$. Find C'.
3. Let $A = \{x: x$ was born after the end of the American Revolution$\}$ and $B = \{x: x$ was born before the end of World War I$\}$. Find $A - B$ and $B - A$.

***ANSWERS**

1. $A \cup B = \{1,2,3,4,5,6,8,10\}$; $A \cap B = \{2,4\}$
2. $C' = \{c,i,k,o,q\}$
3. $A - B = \{x: x$ was born after the end of World War I$\}$;
   $B - A = \{x: x$ was born before the end of the American Revolution$\}$.

At this point we have defined three operations on sets, namely, union, intersection, and complement. Also, we saw that a Venn diagram could be used to give a pictorial representation of a set obtained from two sets by one of the three set operations. We will find that a Venn diagram is very helpful in picturing a set which is obtained from more than two sets. Let us consider the following examples.

EXAMPLE

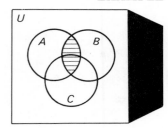

■ Let A, B, and C be sets in a universal set U. We will draw Venn diagrams to compare the set $(A \cap B) \cup C$ with the set $(A \cup C) \cap (B \cup C)$.

To shade the region corresponding to the set $(A \cap B) \cup C$, we proceed as follows: First, let us shade with horizontal lines the region which represents the set $A \cap B$ as shown at the left. Thus any point in the shaded region of this diagram is in the set $A \cap B$.

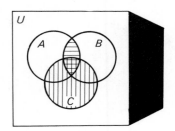

To this diagram we add shading with vertical lines to the region corresponding to the set C. Our Venn diagram will now appear as shown at the left.

Since we know that for a point to be in the set $(A \cap B) \cup C$ it must either be in the region $A \cap B$ shaded with horizontal lines or in the region C shaded with vertical lines or both, we can now display the region $(A \cap B) \cup C$ with one type of shading as we do at the right.

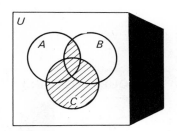

Let us now draw a Venn diagram for the set $(A \cup C) \cap (B \cup C)$. In our diagram we first shade the region $A \cup C$ with horizontal lines. We then shade the region $B \cup C$ with vertical lines. $(A \cup C) \cap (B \cup C)$ will be those elements which are common to both $A \cup C$ and $B \cup C$. Therefore this region will have been shaded twice. To simplify this diagram, we will now indicate this region by diagonal shading.

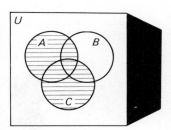
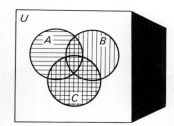
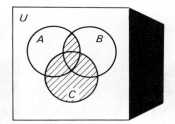

Comparing the diagram for $(A \cap B) \cup C$ with that of $(A \cup C) \cap (B \cup C)$, we see that both of these sets are the same. □

With our understanding of the definitions of the set operations, we may be able to directly shade the desired region in a Venn diagram without doing the intermediate diagrams. To illustrate, consider the next example.

■ Let A and B be sets in the universal set U. We will draw Venn diagrams to compare the set $(A \cap B)'$ with the set $A' \cap B'$.   **EXAMPLE**

In the diagram on the left below we have shaded the region $(A \cap B)'$, which is the set of all elements in U which are not in $A \cap B$.

We now shade $A' \cap B'$ in another diagram. For an element to be in this set, it must be outside of A and at the same time outside of B.

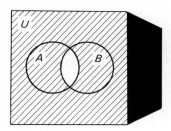
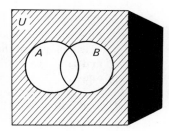

Comparing the above two diagrams, we see that $(A \cap B)'$ and $A' \cap B'$ are not the same sets. However, it can be shown using Venn diagrams that $(A \cap B)'$ and $A' \cup B'$ are equal; you might want to try to show this as an exercise. □

QUIZ YOURSELF*

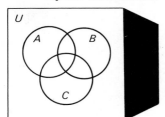

Let A, B, and C be sets in U. In the diagram, shade the region corresponding to each of the following sets.

1. $A \cap B \cap C$

2. $A \cup B \cup C$

3. $(A \cup B \cup C)'$

4. $(A \cup B) - C$

---

*ANSWERS

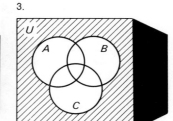

1.

2.

3.

4.

We will need a notation for the number of elements in a set. Therefore, let us denote the number of elements in a set A by $n(A)$.

EXAMPLE   ■ $n(\{a,b,c\}) = 3$.

$n(\{x \colon x \text{ is a state in the Union}\}) = 50$. □

EXERCISES

1. Determine which of the following pairs of sets are equal.
    a) {CBS,ABC} and {$y \colon y$ is a major television broadcasting company}
    b) {$x \colon x$ is a season of the year} and {winter, spring, summer, fall, July}
    c) {gold, silver, brass} and {brass, gold, silver, gold}
    d) {Abraham, Martin, John} and {John, Abraham, Martin}

2. Determine which of the following pairs of sets are equal.
    a) {$z \colon z$ is a month of the year} and {January, February, March, . . . ,December}
    b) $\varnothing$ and {$x \colon x$ is a living American born before 1800}
    c) $\varnothing$ and {$\varnothing$}
    d) {$x \colon x$ is a member of the United States Congress} and {$y \colon y$ is a member of the United States Senate}

3. In exercises 1 and 2 above determine whether either set is a subset of the other.

4. For this question, let the universal set U be the same as in problem 6 of the previous exercise set.

   Let

   $A = \{x : x \text{ is an American}\}, \qquad F = \{t : t \text{ is French}\},$

   $B = \{y : y \text{ is British}\}, \qquad W = \{x : x \text{ is a woman}\},$

   $D = \{z : z \text{ is deceased}\}, \qquad P = \{y : y \text{ is a political leader}\}.$

   Describe each of the following sets first using set builder notation and then give the set by the listing method.

   a) $A \cap P$ b) $F \cup B$ c) $B - D$

   d) $P \cap F'$ e) $(F \cup B)'$ f) A'

   g) $A - W$ h) $W - (F \cup B)$

5. Let $U = \{x : x \text{ is a letter of the alphabet}\}$,

   $A = \{x : x \text{ is a letter in the word } rose\}$,

   $B = \{x : x \text{ is a letter in the word } flowers\}$.

   Give each of the following sets by either the listing method or by set builder notation.

   a) $A \cup B$ b) $A \cap B$ c) $(A \cap B)'$

   d) $B - A$ e) $A - B$

6. For each part draw a Venn diagram similar to the one given and shade the appropriate region corresponding to the given set.

   a) $A \cup B$

   b) $(A \cap B) - C$

   c) $(A \cup C)'$

   d) $(A \cup B \cup C)'$

   e) $(A \cup B \cup C) - (A \cap B \cap C)$

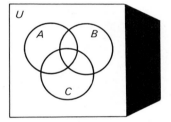

7. For each of the following Venn diagrams, describe the shaded region using set theory notation.

a)

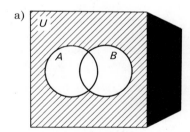

b)

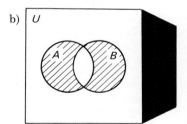

c)

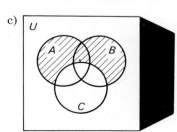

d)

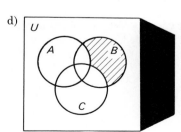

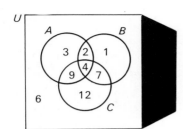

8. In the Venn diagram, we have indicated the number of elements in each region. Find each of the following:
   a) $n(A)$
   b) $n(C')$
   c) $n(A \cap C)$
   d) $n(A - B)$
   e) $n((A \cup C) - (B \cup C))$

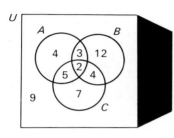

9. In the Venn diagram, we have indicated the number of elements in each region. Find each of the following:
   a) $n(A \cup B)$
   b) $n(A \cap C)$
   c) $n((A \cup B) \cap C)$
   d) $n(A \cap B \cap C)$
   e) $n((A \cup B \cup C)')$

10. Suppose A and B are subsets of a universal set U. Draw Venn diagrams for $(A \cup B)'$ and $A' \cap B'$. Do you notice any relationship between these sets?

11. Assume $A \subset B$. Express the indicated set in each part in a simpler way.
    a) $A \cap B$ b) $A \cup B$
    c) $A - B$ d) $A' \cap B'$

12. Find all subsets of each of the following sets.
    a) $\varnothing$ b) $\{1\}$ c) $\{1,2\}$ d) $\{1,2,3\}$
    e) Without computing, can you guess how many subsets the set $\{1,2,3,4\}$ has?
    f) How many subsets does the set $\{1,2,3, \ldots ,n\}$ have?

13. A certain committee consists of senators Allen, Brown, Cianci, Devlin, and Eastman. Each senator's vote has been weighted so that Allen's and Cianci's votes will be counted twice, Brown's will be counted three times, Devlin's four times, and Eastman's once. In order for the committee to pass a given bill, a vote of 9 or more is required. How many different subsets of this committee have the property that the sum of the voting strengths of its members is 9 or more?

14. Let us suppose that Iran, Iraq, Kuwait, Libya, and Saudi Arabia are meeting to plan future strategy for controlling world petroleum prices. Each country's vote has been weighted according to their current oil producing ability, so that Saudi Arabia's vote will count as 4 votes, Iran's and Kuwait's votes will each count as 3, Libya's vote will count as 2, and Iraq's vote will count only once. In order for any policy to be accepted by this group, it has been agreed that a vote of 10 or more is needed. How many different subsets of these 5 countries have a total voting strength of 10 or more?

MASTERY TEST: TOOLS OF SET THEORY

1. For each of the following pairs of sets, determine whether either set is a subset of the other and whether the two sets are equal.
   a) {France, England, Germany} and {France, England, France, Germany, England}
   b) {x: x is a letter in the word *middle*} and {x: x is a letter in the word *mild*}
   c) {y: y is a manufactured food product} and {bread, cheese, ice cream}.

2. For each of the following Venn diagrams, describe the shaded region using set theory notation.

   a) 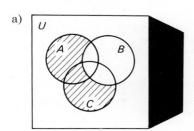 b)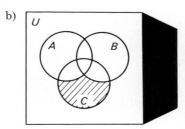

3. Let A, B, and C be sets contained in a universal set U. Draw a Venn diagram for each of the following sets:
   a) $A \cup (B \cap C)$ b) $(A \cap B) - C$ c) $A \cap C'$

4. Consider the universal set U = {bread, apple, automobile, shirt, cheese, banana}. Let E = {x: x may be eaten}, F = {y: y is a fruit}, and M = {z: z is manufactured}. Find:
   a) $E - M$ b) $F \cap M$ c) $E' \cup M'$ d) $(E \cup M) \cap F$.

5. Let W = {x: x is a woman}, C = {y: y is a United States citizen}, and K = {z: z is a past king of England}. Are any two of the sets W, C, and K disjoint? Explain.

6. a) How many distinct subsets can be formed from a set of five elements?
   b) How many committees can be formed using five people where a committee must contain at least one member?

7. In the Venn diagram, we have given the number of elements in each region. Find each of the following.

   a) $n(B)$ b) $n((A \cup B) - C)$ c) $n(A \cap C)$ d) $n((A \cup B \cup C)')$

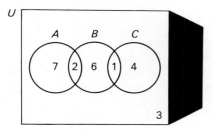

III. THE SURVEY PROBLEM: A MODEL AND A SOLUTION

We have now developed enough set theory to model and solve the survey problem posed at the beginning of this chapter. Recall that 100 people were surveyed and Senator Goodheart had been presented with the following information.

1. There are 34 blue-collar workers with an income of at least $10,000 a year.
2. Twenty-six of the 34 blue-collar workers are conservative.
3. Twenty-three of those who earn at least $10,000 a year are not blue-collar workers.
4. There are 19 conservative, blue-collar workers who earn at least $10,000 a year.
5. Twenty-four of the conservative people earn at least $10,000 a year.
6. The blue-collar workers and those earning at least $10,000 a year together total 60.
7. The conservatives who earned less than $10,000 total 27.

Our first step in modeling this problem will be to translate the above information into the mathematical language of sets.

We will consider our universal set for this problem to be the 100 people surveyed and will define the sets A, B, and C as follows:

$$A = \{x : x \text{ earns at least } \$10,000 \text{ a year}\},$$
$$B = \{x : x \text{ is a blue-collar worker}\},$$
$$C = \{x : x \text{ is a conservative}\}.$$

Let us consider condition (1.) for a moment. Since 34 people are blue-collar workers who also earn at least $10,000 a year, this means there are 34 people who are in both sets B and A. This can be expressed using set theory notation as $n(B \cap A) = 34$. Similarly, the other 7 conditions can be restated symbolically as:

2. $n(B \cap C) = 26$ 3. $n(A - B) = 23$
4. $n(A \cap B \cap C) = 19$ 5. $n(A \cap C) = 24$
6. $n(B \cup A) = 60$ 7. $n(C - A) = 27$

It is helpful at this stage to draw a Venn diagram as shown at the left.

Let us focus our attention on the set A. Note that it is the union of two disjoint sets, namely, $A \cap B$ and $A - B$. From condition (1.) we know that $n(A \cap B) = 34$, and from condition (3.) we have $n(A - B) = 23$. This means that $n(A) = 34 + 23 = 57$.

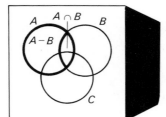

Now let us consider the set B. We know that $n(B \cap C) = 26$. This information together with the fact that $n(A \cap B \cap C) = 19$ forces the region S in the diagram at the right to contain 7 elements.

The 57 elements of A plus the 7 elements of S force $B \cup A$ to have at least 64 members. However, in condition (6.), the pollster said $n(B \cup A) = 60$. In other words, when we systematically analyze our Venn diagram model of the pollster's information, we see that it is inconsistent. This is the reason why the pollster was fired.

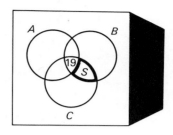

■ A United States Senator has polled 100 of his constituents to determine which energy policy he should pursue. Partial results from his poll are as follows:

EXAMPLE

1. 12 people favor the increased construction of nuclear plants only,
2. 20 people think that both solar energy research and increased nuclear plant construction are desirable,
3. 22 people favor both tax credits for oil companies and increased nuclear plant construction,
4. 14 people would like to see all three areas pursued.

From this information we would like to determine the total number of people who favored increased nuclear plant construction.

As in our last example, we can rephrase the above information in terms of sets. First, let

$T = \{x: x \text{ favors tax credits for oil companies}\}$,

$N = \{x: x \text{ favors increased nuclear plant construction}\}$,

$S = \{x: x \text{ favors federal aid for solar energy research}\}$.

Then conditions (1.)–(4.) can be rewritten as:

1. $n(N \cap T' \cap S') = 12$, 2. $n(S \cap N) = 20$,
3. $n(T \cap N) = 22$, 4. $n(T \cap N \cap S) = 14$.

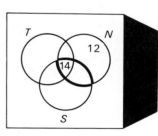

It is helpful to display the above information in a Venn diagram. For example, the fact that $n(T \cap N \cap S) = 14$ and $n(N \cap T' \cap S') = 12$ can be represented graphically as shown in the upper diagram at the right.

In our Venn diagram we have outlined the set $S \cap N$ with heavy lines. From condition (2.) we see that there are 20 in this set. Since 14 of these are already accounted for in the subset $T \cap N \cap S$, there must be 6 in the set $S \cap N \cap T'$. In a similar manner we can use the fact that $n(T \cap N) = 22$ to account for 8 more in the set N. Adding these numbers to our Venn diagram, we have the result shown in the lower diagram at the right.

We see that N contains $8 + 14 + 6 + 12 = 40$ people. □

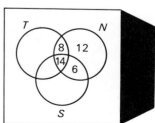

We should observe that if sufficient information had been given in the previous example, we could also have found $n(T)$ and $n(S)$ in a manner similar to the one in which we calculated $n(N)$.

EXERCISES

For problems 1 through 5, find, if possible, the number of elements in sets A, B, and C using the given information. If there is an inconsistency in the information, state where it occurs.

1. $n(A \cap B) = 5$, $n(A \cap B \cap C) = 2$, $n(B \cap C) = 6$, $n(B - A) = 10$, $n(B \cup C) = 23$, $n(A \cap C) = 7$, $n(A \cup B \cup C) = 31$.

2. $n(B - C) = 4$, $n(C - B) = 9$, $n(A \cap B \cap C) = 3$, $n(B \cup C) = 22$, $n(A - C) = 7$, $n(A \cap B) = 7$, $n(A \cap C) = 5$.

3. $n(B - C) = 8$, $n(C - B) = 9$, $n(A \cap B \cap C) = 3$, $n(B \cup C) = 17$, $n(A - C) = 6$, $n(A \cap B) = 7$, $n(A \cap C) = 5$.

4. Suppose $A \cap C = \varnothing$, $n(A \cap B) = 3$, $n(C - B) = 2$, $n(B - C) = 7$, $n(B \cup C) = 11$, $n(A \cup B) = 16$.

5. Suppose $A \subset B$, $A \cap C = \varnothing$, $n(C - A) = 8$, $n(A \cup C) = 12$, $n(C - B) = 3$, $n(B \cup C) = 17$.

6. There are 82 people watching a movie in the Strand Theater. If there are 47 males and 28 children, 13 of which are girls, how many men are there in the theater?

7. A survey is taken of 100 magazine subscribers. The following information is obtained:
   a) 17 read all three of the magazines *Time*, *TV Guide*, and *Reader's Digest*.
   b) 28 read both *Time* and *TV Guide*.
   c) 24 read both *Time* and *Reader's Digest*.
   d) 42 read *Time*.
   e) 68 read either *Time* or *TV Guide*.
   f) 14 read only *TV Guide*.
   g) 18 read none of the three magazines.
   How many read *TV Guide*? How many read only *Reader's Digest*?

8. Repeat exercise 7, given that
   a) 17 read all three of the magazines *Time*, *TV Guide*, and *Reader's Digest*;
   b) 39 read both *Time* and *TV Guide*;
   c) 26 read both *Time* and *Reader's Digest*;
   d) 9 read *Time* only;
   e) 65 read either *Time* or *TV Guide*;
   f) 14 read *TV Guide* but do not read *Reader's Digest*.

9. A survey of 100 persons was taken to determine which of the various media they regularly use to obtain the news. The following information was obtained:
   a) Of the 36 people who read the newspaper, 13 use only the newspaper to learn the news.
   b) Of the 48 people who listen to the radio, 11 use only the radio to learn the news.
   c) All three of these media are used by 23 people to learn the news.
   How many people use both the radio and television regularly to learn the news?

10. A survey was made of 200 people to study their use of mass transit facilities. It was found that:
    a) 83 did not use mass transit,
    b) 68 used the bus,
    c) 44 used only the subway,
    d) 28 people used both the bus and subway,
    e) 59 people used the train.
    Explain how you can use the above information to deduce that some people must use both the bus and train.

SUGGESTED READINGS

GARDNER, M., "Mathematical Games: Boolean Algebra, Venn Diagrams, and the Propositional Calculus." *Sci. Am.*, February 1969, p. 110.

An explanation on using Venn diagrams to test the validity of a statement with a brief introduction to Boolean Algebra as related to sets.

HAHN, H., "Is There An Infinity?". *Sci. Am.*, November 1952, pp. 76–84.

A nontechnical description of the work of Georg Cantor on infinite sets.

KLINE, M., *Mathematical Thought from Ancient to Modern Times.* New York: Oxford University Press, 1972.

Chapter 41 presents a historical development of set theory.

MESCHKOWSKI, H., *Evolution of Mathematical Thought.* San Francisco: Holden-Day, 1965.

Gives a low-level introduction to the questions of infinity, Cantor's set theory, and some of the paradoxes that arose.

WHITESITT, J. E., *Boolean Algebra and Its Applications.* Reading, Mass.: Addison-Wesley, 1961.

An introduction at the beginning level to Boolean algebra as related to sets with applications to switching circuits. The presentation of switching is not dependent on symbolic logic.

3 Graph theory

The Swiss mathematician Leonhard Euler, who wrote a paper in 1736 solving a popular puzzle known as the Koenigsberg Bridge Problem,* was the first mathematician to work in graph theory. Euler was a man of many and extraordinary talents. Although he prepared for the clergy at his father's insistence, at the age of 20 he chose instead to accept an appointment in medicine at St. Petersburg Academy in Russia. However, his main love was mathematics and six years later he was named the chief mathematician at the academy.

Euler produced mathematics at an unbelievable rate, which led the biographer Arago to say that he could calculate without effort "as men breathe, or as eagles sustain themselves in the wind." During his life he authored almost 900 papers touching on virtually all areas of mathematics. Euler was probably the most prolific mathematician of all time. This fantastic output seems even more incredible when we consider that about 400 of these papers were written the last 17 years of his life, during which time he was totally blind.

After Euler's work, the progress of graph theory was rather erratic. The English mathematician Arthur Cayley (1821–1895) became briefly interested in it through another puzzle called the Four-Color Problem.† Subsequently he wrote several papers concerning the application of graph theory to chemistry. In the nineteenth and twentieth centuries many contributions were

---

* We will discuss this famous problem in detail later in this chapter.
† We will also discuss this problem later in the chapter.

LEONARD EULER
Analytical Theory of the
Circular Functions

Leonard Euler (1707–1783)

made by scientists working in various disciplines who needed new mathematical structures with which to model their ideas. The area known as operations research has had a particularly great influence on the development of graph theory. Operations research is concerned with applying mathematical techniques to study the operation of a system—here system could mean a cancer cell, an industrial plant, or even the United States government. Graph theory is especially useful because it provides a means for describing the interrelationships of the various components of intricate, complex systems such as those we have mentioned.

One particularly important application of graph theory which we will study in this chapter is an organizational technique known as PERT (Program Evaluation and Review Technique). This procedure, which is widely used, was originally invented to aid in the scheduling of the construction of the Polaris submarine.

Graph theory has now become an important area of mathematics in which a great deal of research is being done. The current interest in graph theory among mathematicians is perhaps due to the fact that it has a large number of applications in fields such as sociology, biology, psychology, political science, and urban studies.

I. SCHEDULING THE CONSTRUCTION OF A SPACE COLONY

The United States Congress has asked a committee of scientists and engineers to develop a timetable for constructing and populating an orbiting space colony. It has been decided that this project can be subdivided into ten major tasks.

Task	Time required, months
1. Train construction workers to work in space	6
2. Build the shell of the colony in sections on earth	8
3. Build life support systems on earth	14
4. Recruit colonists to live in the colony	12
5. Assemble the shell of the colony in space	10
6. Train the colonists in the operation of the colony's systems	10
7. Install the life support systems	4
8. Install the solar energy systems	3
9. Test the life support and energy systems	4
10. Bring colonists to the colony and orient them to living in space	4

One of the important applications of graph theory which we will see in this chapter involves the scheduling of the interrelated jobs needed to complete a complex construction project. (Photo by Laurence Lowry.)

We are interested in developing a schedule so that this project can be completed in as short a time period as possible. Also, since building the life support systems requires the most time, can the length of the whole project be decreased by devoting extra effort to reducing the time it takes to do this task?

After completing the next part of this chapter, we will have enough tools of **graph theory** available to model and solve this problem. In addition to modeling this scheduling problem, we will see how graph theory can be applied to other situations.

II. THE TOOLS OF GRAPH THEORY

GRAPHS, PUZZLES, AND MAP COLORING

We begin our development of the tools of graph theory by considering an example of a well-known puzzle (Fig. 3.1). Before reading further, you might want to try the puzzle. Do you think it can be done? We will soon learn a very simple test which can be applied to any such puzzle to determine whether or not it can be accomplished. But first, we need some terminology.

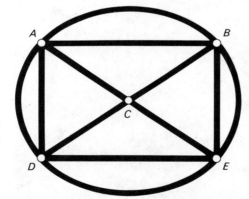

Fig. 3.1

Puzzle tracing: Trace the figure completely without lifting the pencil from the paper and without tracing any line twice.

DEFINITION Consider a finite set of points called VERTICES. An EDGE is a line joining two different vertices. Two vertices which are joined by an edge are called the ENDPOINTS of that edge. A GRAPH is a finite set of vertices together with a finite number of edges.

The drawing in Fig. 3.1 illustrates a graph; the points labeled A, B, C, D, and E are the vertices of the graph and the 12 connecting lines are the edges.

Generally, capital letters will be used to designate vertices. An edge can be labeled by the endpoints it connects provided there is no confusion.

For example, the edge joining vertices A and C can be called the edge AC or even the edge CA. However, to speak of the edge BE would be confusing since two different edges join B and E. As an alternative for naming edges, we shall sometimes use small letters such as e, or e_1, e_2, etc.

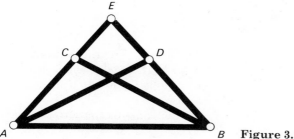

Figure 3.2

The graph of Fig. 3.2 has five vertices, A, B, C, D, and E. We see that the edges BC and AD intersect. Since we do not want to consider this point of intersection as a vertex of the graph, we do not label it. Any point of intersection of a pair of edges will not be considered as a vertex *unless* designated and labeled as one.

Although the origin of graph theory can be traced to mathematicians' attempts to solve a popular 18th century puzzle, it now has many important real-life applications. (Photo by Peter Southwick, Stock Boston.)

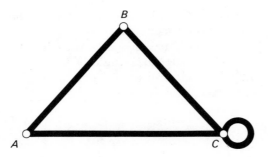

Figure 3.3

Our definition does not allow Fig. 3.3 to be a graph. The line which starts at C and loops back to C is not an edge. Recall that an edge must join two *different* vertices. We wish to eliminate these loops from our graphs since they usually contribute nothing to the discussion.

We will now give several examples that show some of the various ways in which the use of graphs may arise.

One of the most famous and yet elementary examples where graph theory is applied had its origins in the city of Koenigsberg, Prussia, during the eighteenth century. The Pregelarme River divided the village of Koenigsberg into four distinct sections as shown on the map of Fig. 3.4.

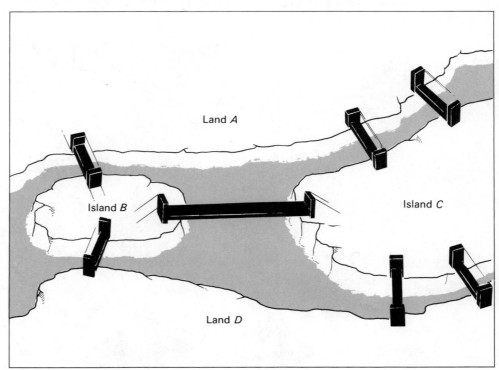

Fig. 3.4

Map of Koenigsberg.

The four portions of Koenigsberg were connected by seven bridges. It was a popular pastime for the citizens of Koenigsberg to start in one section of the city and try to take a walk visiting all sections of the city, crossing each bridge exactly once and returning to the original starting point. This problem is generally called the **Koenigsberg bridge problem**.

You may wish to take a moment and attempt to trace such a walk on the map of Koenigsberg. However, you would be wise to use the eraser end of your pencil since there is an excellent chance that your first attempt will fail!

At first, it might not be clear that this is a graph theory problem. However, we should realize that any two bodies of land which are joined by a bridge can be represented in a graph as two vertices connected by an edge. Therefore the four pieces of land which make up Koenigsberg can be modeled by four vertices, A, B, C, and D, and the seven bridges can then be represented by seven edges in the graph. In other words, the map in Fig. 3.4 can be modeled by the graph in Fig. 3.5.

We see that our problem can now be stated as follows. Place your pencil at some vertex of the graph. Then without lifting your pencil, trace the graph so that each edge is traced exactly once and the tracing finishes at the starting vertex. When stated this way, it is clear that the Koenigsberg bridge problem is very similar to the puzzle mentioned at the beginning of this section.

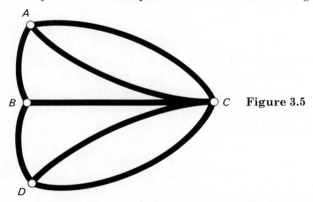

Figure 3.5

In 1735, Leonhard Euler discovered a brilliant, yet simple, way to determine when a graph can be traced.* First, to **trace** a graph you must always

1) place your pencil at some vertex;
2) draw the entire graph without lifting the pencil; and
3) do not trace any edge more than once.

In order to give the solution of this graph-tracing problem, we need a few more definitions.

---

* A very readable translation of Euler's original paper is "Leonhard Euler and the Koenigsberg Bridges" by L. Euler, J. R. Newman (ed.), *Scientific American*, July, 1953, pp. 66–70.

DEFINITION   **A graph is CONNECTED if it is possible to travel from any vertex of the graph to any other vertex of the graph by moving along successive edges.**

In Fig. 3.6 we see examples of a connected graph and a nonconnected graph.

Clearly, if a graph is not connected then it is not possible to draw the entire graph without lifting the pencil—that is, it is not possible to trace the graph.

Fig. 3.6

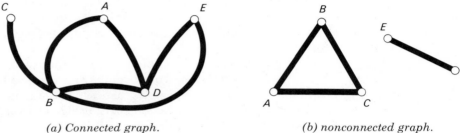

(a) Connected graph.   *(b) nonconnected graph.*

DEFINITION   **A vertex of a graph is said to be ODD if it is an endpoint of an odd number of edges of the graph. Similarly, a vertex is EVEN if it is an endpoint of an even number of edges.**

EXAMPLE   ■ Consider the graph given below. The vertices A, B, and C are even whereas D and E are odd. □

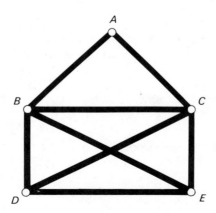

For each of the graphs given below answer the following questions:

a) Is the graph connected?

b) Which vertices are odd?

c) Which vertices are even?

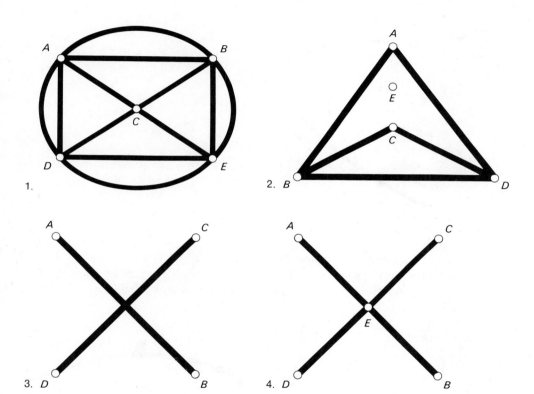

1.   2.

3.   4.

---

1. This graph is connected. Vertices A, B, D, and E are odd. Vertex C is even.   ***ANSWERS**
2. This graph is not connected. Vertices B and D are odd. Vertices A, C, and E are even. We realize that vertex E is even since it is the endpoint of 0 vertices and 0 is an even number.
3. This graph is not connected. All vertices are odd.
4. This graph is connected. A, B, C, and D are odd. E is even.

Suppose we have a graph G which can be traced; then we would like to make the following two observations.

In tracing a graph G, if we neither begin nor end at a vertex A, then A must be an even vertex.   **OBSERVATION 1**

This fact is easily seen. Suppose A is a vertex of graph G and assume that we neither begin nor end our tracing at A. Since we are tracing the whole graph G, we must eventually come into A by means of some edge e_1 and, since we are not ending at A, we must leave by another edge e_2.

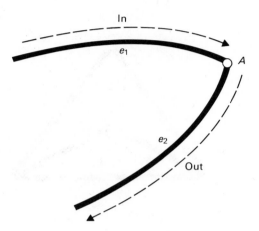

If no other edges are joined to A, then clearly A is even because it is the endpoint of exactly two edges. On the other hand, if more edges are joined to A, then we will come into A again using a third edge e_3 and leave by a fourth edge e_4.

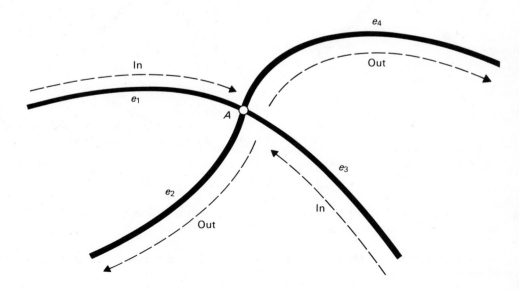

If there are no other edges joined to A, then we have seen that A is an endpoint of four different edges and is therefore even.

Should A be the endpoint of more than four edges, we could continue this analysis, realizing that each time we enter by one edge we must also leave by another edge. Since we can never finish at A, it is clear that A must be the endpoint of an even number of edges and so A is even.

It is clear from our first observation that an odd vertex can be used only as a starting or ending point.

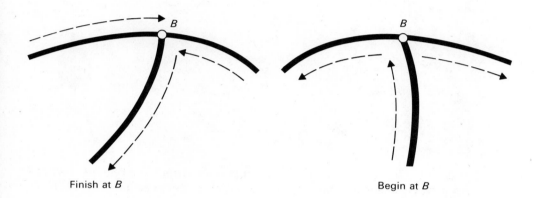

Finish at *B*   Begin at *B*

Continuing the above line of thinking we make the next observation.

If a graph can be traced, then it can have at most two odd vertices.   OBSERVATION 2

In tracing a graph, one vertex will be used as a starting point and possibly another will be used as an ending point. Since no vertices other than these two will be used as starting or ending points, all other vertices must be even. Thus if a graph can be traced, it can have at most two odd vertices.

The net result of our two observations is that we can now agree that *if a graph has more than two odd vertices, then it cannot be traced.*

For the complete answer to the question of when a graph can be traced, we paraphrase Euler's theorem.

A graph can be traced provided: EULER'S THEOREM

1) **It is connected.**

2) **It has no odd vertices or two odd vertices.**

3) **If it has two odd vertices, the tracing must begin at one of these and end at the other.**

4) **If all the vertices are even, then the tracing must begin and end at the same vertex; it does not matter at which vertex this occurs.**

Let us consider again the puzzle graph given earlier. We realize now that we cannot trace this graph because it has four odd vertices, namely, *A*, *B*, *D*, and *E*.

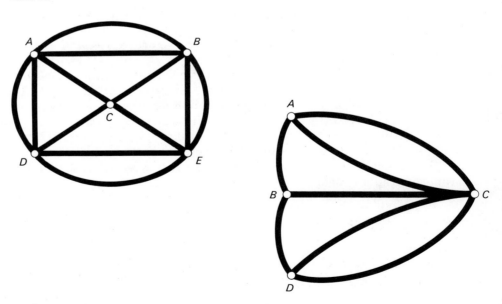

Similarly, if we take another look at the Koenigsberg bridge problem, we see that this graph cannot be traced because all four of its vertices are odd.

Using Euler's theorem, state which of the following graphs can be traced. For a graph which cannot be traced, state which part of Euler's theorem fails.

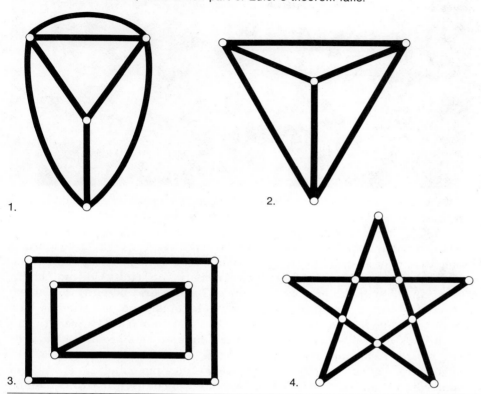

1.

2.

3.

4.

1. It can be traced.
2. It cannot be traced because it has more than two odd vertices.
3. It cannot be traced because it is not connected.
4. It can be traced.

The situation we encounter in the Koenigsberg bridge problem is typical of those which can be modeled with a graph. In the statement of this problem we have:

1) a set of objects (the four bodies of land), and
2) a relationship among these objects (two bodies of land are "related" if they are connected by a bridge).

To draw the graph we first represent the set of objects with a set of vertices. Then, if two objects of the set are related, we join the two corresponding vertices with an edge.

State a relationship that could be considered between the eyeglasses in this photo. This set with the relationship that you have in mind can now be modeled by a graph. (Photo by Mike Mazzaschi, Stock Boston.)

We will also find that a graph can be used as a model when we consider a most interesting question, **the four-color problem**.

In 1852, Francis Guthrie, a student at University College, London, first posed this famous question to his mathematics professor, Augustus de Morgan. Unable to find an answer, de Morgan communicated the problem to his friend Sir William R. Hamilton, a professor of mathematics at Trinity College, Dublin, Ireland. In the more than 100 years since Guthrie first posed this problem, neither de Morgan, Hamilton, nor many subsequent mathematicians have been able to solve it.* In 1967, Professor Oystein Ore of Yale University published a book entirely devoted to the work done on this problem (*The Four Color Problem*, Academic Press, 1967).

The statement of the four-color problem is so simple and direct that you will immediately understand it. Using four or fewer colors, is it always possible to color a map so that any two regions which share a common border receive different colors?

For an illustration of the statement of the problem, see Fig. 3.7, a map of South America. Using at most four colors, try to color this map so that we use a different color for any two countries having a common border. For example, we cannot use the same color for Chile and Peru, however, we can use the same color for Paraguay and Colombia.

For our four-color problem, we have a set of countries with the relation of sharing a common border. Thus we can use a graph as a model of this situation. We can represent each country with a vertex. If two countries share a common border, we then can draw an edge between the corresponding vertices. In this manner the map can be modeled by the graph in Fig. 3.8.

---

* Kenneth Appel and Wolfgang Haken, professors of mathematics at University of Illinois, Urbana-Champaign, have recently announced a proof of the four color theorem. They utilized some 1200 hours on high-speed computers to obtain their proof. Appel and Haken stated in the announcement that it would have been impossible without the aid of computers.

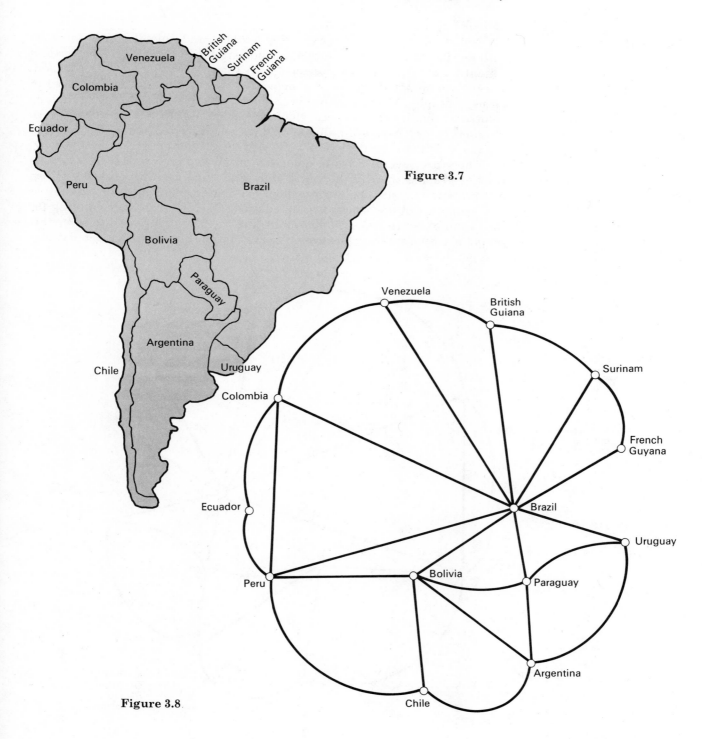

Figure 3.7

Figure 3.8

Note that the vertices representing Ecuador and Colombia are connected with an edge because they share a common stretch of boundary. The vertices Argentina and Peru are not connected since they have no boundary in common.

We see now that the original map-coloring question can be rephrased in terms of graphs. Using four or fewer colors, can the vertices of the graph be colored so that no two vertices which are endpoints of the same edge receive the same color? You will probably find it easier now to think about coloring the graph rather than it was working with the original map.

There are several ways to color the graph in Fig. 3.8. However, unlike tracing graphs, there is no particular procedure for this problem. Keen observation and trial and error make it possible to color the graph so that no two endpoints of the same edge are colored alike. One possible coloring for the graph of South America appears in Fig. 3.9. You may wish to color this graph in a different way; however, observe that the minimum number of colors that can be used for this graph is four.

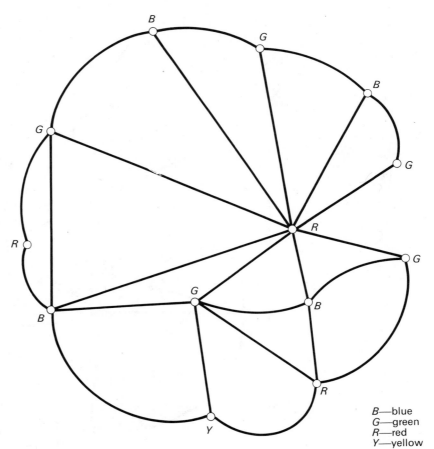

Fig. 3.9

*Graph coloring of
South America*

B—blue
G—green
R—red
Y—yellow

Before leaving the topic of map coloring, we should reiterate that an answer to the four-color problem is not known. In 1890, P. J. Heawood proved that at most five colors are needed to color any map, but no one has yet produced a map which requires five colors.

EXERCISES

1. Answer the following questions for each of the given graphs.
   Is the graph connected?
   Which are the odd vertices of the graph?
   Which are the even vertices of the graph?

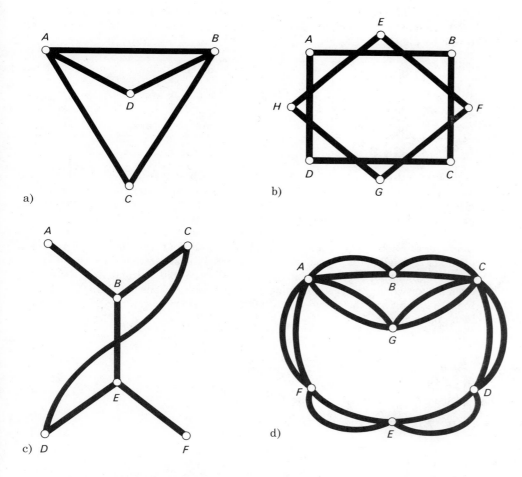

a)

b)

c)

d)

2. Use Euler's theorem to decide which graphs in exercise 1 can be traced. If a graph cannot be traced, tell which conditions of the theorem fail.

3. An ice cream vendor wishes to travel over each of the streets indicated in the map below, but he does not want to travel over any part of the route more than once. Can this be done? Explain your answer.

Figure 3.10

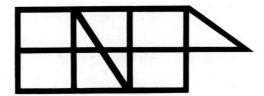

4. Different notes are obtained on a trumpet by moving its three valves up and down. We have indicated the eight possible positions for these valves and a note which would be sounded.

Valve									
	1	up	up	up	up	down	down	down	down
	2	up	up	down	down	up	up	down	down
	3	up	down	up	down	up	down	up	down
Note		C	F♯	B	E♭	F	D	E	C♯

Is it possible to play each of these eight notes once in some sequence—that is, to start with one note and then by moving one valve at a time obtain each of the other notes? (*Hint*: Consider the notes as the objects of a set and that two notes are related if one can be obtained from the other by moving one valve.)

5. Below is the floor plan of a suite of offices. Assume all the doors indicated are open. Is it possible for a security guard to enter the suite from the hallway, pass through each door locking it behind him and then exit without ever having to open a door which has been previously locked? (*Hint*: Consider the set of objects to be the rooms and the hallway and that any two are related if they are connected by an open door.)

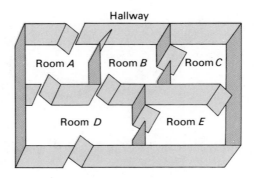

6. Below is another floor plan for a suite of offices. The situation is the same as in exercise 5, except there is now only one door by which the security guard can enter the suite. Place an exit door in one of the rooms *A*, *B*, or *C* so that the security guard can enter by the door marked "enter," pass through each door and lock it behind him, and then exit by the door you've placed in the plan. Explain why this is the only possible place to locate the door.

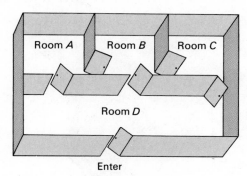

7. Using the map of the continental United States (Fig. 3.10), represent each of the following groups of states with a graph; consider that the relation is "sharing-a-common-border."
   a) Washington, Oregon, Idaho, Montana, and Wyoming.
   b) Ohio, Kentucky, Tennessee, Arkansas, Missouri, Illinois, Indiana, Iowa, Minnesota, and Wisconsin.
   c) Texas, Oklahoma, Arkansas, Louisiana, Mississippi, Alabama, Tennessee, North Carolina, South Carolina, Georgia, and Florida.

8. Use as many of the colors red, green, blue, and yellow as needed to color the vertices of the graphs you obtained in exercise 7. As usual, two vertices which are joined by an edge cannot be the same color.

9. The Chief of Protocol is arranging a State Department luncheon for the ambassadors of 12 countries. It is very important that any two ambassadors who are not on friendly terms be seated at different tables.

Ambassador from	Is not friendly with the ambassador from
Russia	United States, Poland, Canada, Latvia, Spain, France, Britain, Portugal
Germany	Latvia, Canada
Mexico	United States, Britain
Italy	Spain, Portugal
Poland	Canada, France, Russia
Britain	France, Russia, United States, Mexico
Latvia	Spain, Canada, Russia, Germany
Portugal	United States, Spain, Italy, Russia

Determine a satisfactory seating arrangement for the luncheon using as few tables as possible.

Fig. 3.10

10. The designer of a new entertainment complex called "Jungle World" wants to include several large enclosures in which wild animals can roam freely about. Obviously if one of the animals can harm another, then those two animals cannot be placed in the same enclosure.

Animal	Cannot be placed with
tiger	zebra, leopard, rhinoceros, giraffe, antelope, ostrich
leopard	zebra, boar, antelope, giraffe, ostrich
crocodile	ostrich, heron
boar	tiger, crocodile, zebra

Using the information given in the above table, determine the smallest number of enclosures necessary to contain the animals. Also tell how you would assign the animals to these enclosures.

DIRECTED GRAPHS

In using graphs to model various situations it is often necessary to assign directions to the edges of the graphs. When an edge is given a direction it is called a **directed edge**. A graph in which all edges are directed is called a **directed graph**.

If the relationship between the objects in a set is exerted in only one direction, then we could model it with a directed graph. (Photo by Jack Prelutsky, Stock Boston.)

■ Suppose we wish to model how rumors are spread among four friends Al, **EXAMPLE** Barbara, Charlie, and Donna. Let us assume that we have gathered the following information:

1) If Al hears a rumor, he will communicate it to Barbara, but Barbara will not communicate the rumors she hears to Al.

2) Barbara and Donna will tell any rumors they hear to each other.

3) Barbara tells Charlie every rumor she hears, but Charlie never relays the rumors he hears back to Barbara.

A model for this situation is given in the following directed graph.

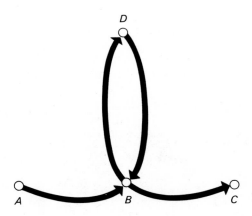

We have drawn arrows on the edges of this graph to indicate the way in which rumors travel. For instance, the directed edge from A to B reflects the fact that Al communicates all rumors he hears to Barbara. Likewise, the absence of a directed edge from B to A models the fact that Barbara never relays rumors to Al. □

A directed graph can be used to model a relationship between objects in a set whenever that relationship is not necessarily exerted in both directions; that is, it is possible for X to be related to Y without Y having the same relationship to X.

In our above example, the relation is "tells a rumor." For this relationship, we see that Al is "related" to Barbara but Barbara is not "related" in the same manner to Al.

DEFINITION   **Let us suppose that X and Y are vertices in a directed graph. If it is possible to begin at X, follow a sequence of edges in the direction indicated, and end at Y, then we call this sequence of directed edges a DIRECTED PATH FROM X TO Y. We will denote a directed path by the sequence of endpoints encountered. The LENGTH of a directed path is the number of edges along that path.**

EXAMPLE   ■ Consider the following directed graph.

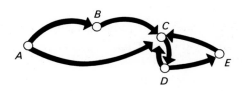

The sequence $ACDE$ indicates a directed path of length 3 from A to E. The sequence $BCDEC$ is a directed path of length 4 from B to C. Note that the sequence $ABCE$ is not a directed path from A to E since the last edge CE has the wrong direction. ☐

Use the graph given below to answer the following questions.

QUIZ YOURSELF*

1. Can you find two directed paths from A to C?

2. Can you find two directed paths from C to B?

3. Are there any directed paths from C to A?

4. What is the length of the directed path $ABCDB$?

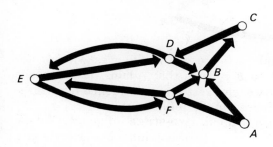

---

1. yes; ABC and $AFBC$ 2. yes; CDB and $CDEDB$ 3. no   4. 4

*ANSWERS

■ A lobbyist for a large company wants a committee in the state senate to introduce a bill which would favor his company. He feels that on this issue his best strategy would be to obtain the support of the most influential member of the committee and, from past experience, he knows that some committee members have influence over others.

EXAMPLE

Committee member	Other committee members he influences
Adams	Collins, Evans, Fisher
Baker	Fisher
Collins	Baker, Evans
Davis	Baker, Collins, Evans
Evans	Baker
Fisher	Collins, Evans

The relationship we have on this set of people is influence. Since influence is not exerted in both directions (Adams influences Collins but Collins does not influence Adams), we can use a directed graph to model this relationship between the committee members.

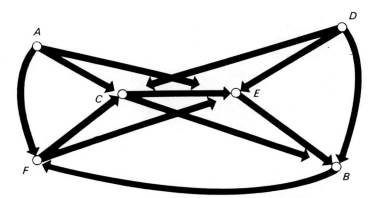

The directed edges in the graph indicate who influences whom. For instance, the directed edge from A to E reflects that Adams has influence over Evans. Since Adams and Davis each exert direct influence over three people, we may wish to say that both are equally influential.

However, let us consider what we will call **two-stage** influence. We observe that Adams influences Collins, who in turn has influence over Evans. Therefore we will say that Adams has two-stage influence over Evans. Likewise, Davis has influence over Baker, who in turn influences Fisher. Thus Davis has two-stage influence over Fisher. In terms of our model, if there is a directed path of length 2 from vertex X to vertex Y, then committee member X exerts two-stage influence over Y.

Let us determine now the number of ways that Adams can influence Evans either directly or in two stages. In our graph, in addition to the directed edge AE, there are also the directed paths ACE and AFE. Therefore there are three ways in which Adams can influence Evans in one or two stages.

We would like to determine this number for every other pair of committee members. For convenience, we have recorded in the following table the number of directed paths of length 1 or 2 between each pair of vertices.

	A	B	C	D	E	F
A	0	2	2	0	3	1
B	0	0	1	0	1	1
C	0	2	0	0	1	1
D	0	3	1	0	2	1
E	0	1	0	0	0	1
F	0	2	1	0	2	0

We have labeled the columns and the rows of our table A, B, C, D, E, and F. We saw that there are three paths of length 1 or 2 from A to E, we therefore put a 3 in row A, column E, of our table. The entry of row C, column B, is 2 since there are two paths of length 1 or 2 from C to B, namely, CB and CEB. In general, the entry made in row X, column Y, is the number of paths of length 1 or 2 from vertex X to vertex Y in the graph.

Now that we have determined the number of one- or two-stage influences, let us use this information to rank the committee members. Even though both Adams and Davis can influence Evans directly and by way of Collins, we see that Adams can also influence Fisher who can influence Evans. There are three ways in which Adams can exert influence over Evans but only two ways in which Davis can. Therefore it would be reasonable to say that Adams can exert more influence over Evans than Davis can. Thus we will rank the committee members according to the number of ways they can exert their influence in one or two stages.

If we add the entries in row A of our table, we obtain $0 + 2 + 2 + 0 + 3 + 1 = 8$. Thus there are eight ways in which Adams can exert his influence over other committee members in one or two stages. If we add the entries for each of the other rows, we obtain

Row	A	B	C	D	E	F
Sum of entries	8	3	4	7	2	5

Then according to our measure of influence, the committee members would be ranked as follows:

Adams	Most influential
Davis	
Fisher	
Collins	
Baker	
Evans	Least influential

□

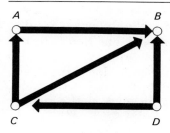

QUIZ YOURSELF*

Give a table which displays the number of directed paths of length 1 or 2 between each pair of vertices in the graph at the left.

***ANSWER**

$$\begin{array}{c}\\A\\B\\C\\D\end{array}\begin{array}{cccc}A & B & C & D\\\left[\begin{array}{cccc}0 & 1 & 0 & 0\\0 & 0 & 0 & 0\\1 & 2 & 0 & 0\\1 & 2 & 1 & 0\end{array}\right]\end{array}$$

A directed graph can model the way members of a set dominate each other. Model the situation depicted in this photo. (Photo by Malcolm Perkins, Stock Boston.)

As in the last example, we are often interested in ranking two or more objects with respect to some property. For example, we could rank football teams by using their number of victories. If in the same number of games played, Ohio State has four victories and Oklahoma has three, then we would rank Ohio State above Oklahoma. Suppose instead that each of these two teams has four victories. In order to break a tie in the ranking order, we could consider the notion of two-stage dominance; that is, if Ohio State has beaten Michigan and Michigan has defeated Indiana, then Ohio State has demonstrated two-stage dominance over Indiana. If Ohio State has more two-stage dominances over other teams than Oklahoma has, then we would rank Ohio State above Oklahoma. If the two teams are still tied with regard to the number of two-stage dominances, we could then consider the number of three-stage dominances and so forth.

In using this method of ranking, it may happen that it would be impossible to break a tie in the ranking order. For example, suppose we want to rank Penn State, Alabama, and Southern Cal using only the following results: Penn State beats Alabama, Alabama beats Southern Cal, and Southern Cal beats Penn State. All three teams will be tied in victories; they will also have the same number of two-stage dominances, three-stage dominances, etc. Using this method of ranking, all three teams should be ranked the same. You could also consider some other method of ranking, such as the number of points scored by each team.

You should keep the above comments in mind while performing the exercises, given at the end of this section, which involve ranking the objects of a set.

In our next example we use a directed graph to help us answer a different type of question.

■ Collegeville University is faced with a crisis. During the past week, eight students reported to the University Health Center suffering from infectious mononucleosis. Health Center officials have isolated these students and hope that none of the other students on campus have contracted this disease.

After interviewing each of the infected students, it was determined how the disease was contracted within this group. Health Center officials believe that the disease was introduced on campus by one person and then communicated by people on campus to others. If we accept this assumption, then we can use the information given in the table below to determine whether or not all the infected people on campus have been isolated.

Student	Others within the group who could have contracted the disease from this student
Al	Dave, Joe
Barbara	Ellen, Fred, Ivy
Dave	Ellen
Ellen	Fred
Fred	Kathy
Ivy	Barbara, Fred
Joe	Al, Ellen, Fred
Kathy	Ellen

It is easy to see that we can model this situation by using a directed graph. First, we represent each of the eight students with a vertex. We then draw a directed edge from vertex X to vertex Y, if X could have transmitted the disease to Y. Using this interpretation of vertices and directed edges, we have the following directed graph.

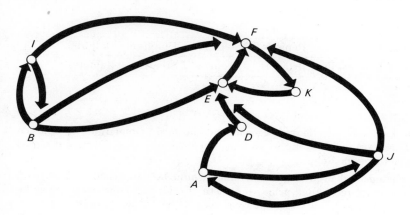

Since Al could have given mononucleosis to Joe, we draw a directed edge from *A* to *J*. Similarly, since Barbara could have transmitted the disease to Fred, there is a directed edge drawn from *B* to *F*. The directed paths in the graph show how the infection might have spread within the group. For example, the directed path *ADEF* shows that it was possible for the mononucleosis to travel from Al to Fred; first from Al to Dave, then from Dave to Ellen, and finally from Ellen to Fred.

It is important to realize that if the disease spread from student *X* within the group to student *Y*, then there must be a directed path in the graph from vertex *X* to vertex *Y*.

By examining our graph, we can easily see that it was impossible for the disease to have started with Al and spread within the group to Barbara since there is no directed path from *A* to *B*. Similarly, the absence of a directed path from *F* to *J* means that the disease could not have started with Fred and then spread within the group to Joe. In fact, if we check each of the eight students, we will see that it was impossible for the disease to start from any one of them and then spread within the group to all the others. Therefore according to our information and assumption, there is at least one other person on campus who has the disease but who has not yet been identified. □

In our examples, we have touched lightly on modeling with graphs. A book devoted entirely to applications of directed graphs to sociology is *Structural Models: an Introduction to the Theory of Directed Graphs*, F. Harary, et al., New York: John Wiley, 1965.

EXERCISES

1. Let us assume that in the National Football League over a period of several weeks the following happens:

Team	Steelers	Bengals	Steelers	Dolphins	Dolphins
Beat	Redskins	Broncos	Broncos	Steelers	Raiders

   Draw a directed graph to model these results.

2. The African hare and gazelle are vegetarians who feed primarily on grass. Gazelles and hares are eaten by lions, jackals, cheetas, and man. Draw a directed graph to model this food chain.

3. When purchasing food products, a consumer would be concerned with ease-of-preparation, nutritional value, price, and taste. Conduct a consumer survey by asking five people you know to complete a copy of the following ballot.

> **For each of the following pairs, circle which quality is most important to you**
>
> ease-of-preparation   nutritional value
>
> ease-of-preparation   price
>
> ease-of-preparation   taste
>
> nutritional value   price
>
> nutritional value   taste
>
> price   taste

On the third line of the ballot, if more people chose taste over ease-of-preparation, then we can say that taste is preferred over ease-of-preparation. Use the results of your survey to determine which quality is preferred for each pair and then summarize your results by means of a directed graph.

4. Consider the directed graph at the left below.
   a) Find two directed paths of length 3 from *A* to *E*.
   b) Find a directed path from *C* to *B* and give its length.

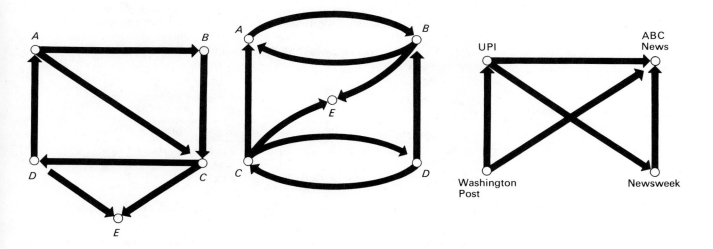

5. Give a table which displays the number of directed paths of length 1 or 2 between each pair of verticles in the graph in the center above.

6. Four news organizations made public the same bit of classified information. The directed graph on the right above indicates how this information could have passed among the four news organizations. From this graph, determine which news organization first obtained this bit of classified information.

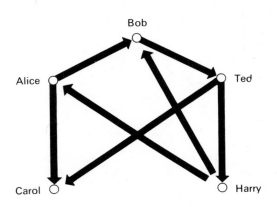

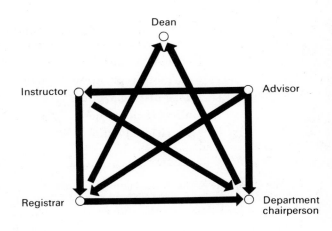

7. The directed graph on the left above indicates how a rumor could spread among five neighbors, Alice, Bob, Ted, Carol, and Harry. Which neighbors could start a rumor that would eventually spread to the others?

8. In a particular college bureaucracy, five signatures are needed on a form. It is known that certain officials will not give their approval before others. This situation is depicted in the directed graph on the right above. A directed edge from vertex A to vertex B indicates that A must sign the form before B. If a student must hand-carry the form from office to office, in which sequence can all the signatures be obtained?

9. The following influence has been observed among five committee members:

Member	Pratt	Quigley	Ross	Stickle	Thums
Influences	no one	Stickle and Thums	Pratt and Stickle	Pratt	Pratt and Ross

Draw a directed graph to model this situation. Determine a ranking order for the committee members using one- and two-stage influence.

10. In a round-robin singles handball tournament, each of six competitors play each other once. The results of the tournament are tabulated below.

Player	Defeated
Charlie	Bob, Fred, Sam, and Ed
Bob	Tom, Sam, and Ed
Tom	Charlie, Fred, and Ed
Fred	Bob and Ed
Sam	Tom and Fred
Ed	Sam

Draw a directed graph to model this situation. Determine a ranking order for the players using one- and two-stage dominance.

11. In a "paired-comparison" test, an individual is asked to compare five brands, two at a time, and for each pair, the individual is to indicate a preference. The following results are obtained:

Brand	A	B	C	D	E
Preferred over	B	E	B	A	A
	C		D	B	
			E	E	

Model this situation with a directed graph then determine a ranking order for the five brands using one- and two-stage preference.

12. When shopping for an automobile, a buyer may consider each of the following factors: economy, safety, comfort, style, manufacturer, and availability of service. Consider yourself as a car buyer and compare each possible pair of factors indicating which of the two is the most important to you. After completing this paired-comparison test, determine a ranking order by importance of the six factors.

MATRIX MODELS FOR DIRECTED GRAPHS

In the previous section we used a table in our influence example to display the number of directed paths of length 1 or 2 between each pair of vertices. Each table entry was determined by directly counting the number of paths within the graph. In a complicated graph, this procedure can be tedious, and we could also overlook some directed paths of interest. You will see that it is possible to avoid these difficulties if we first represent a directed graph by a table of numbers called an **incidence matrix**. By defining multiplication and addition for matrices, we then obtain the desired information about the graph.

In the next example, we describe how to obtain the incidence matrix for a directed graph.

Directed graph

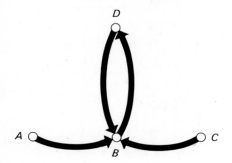

Incidence matrix

$$\begin{array}{c c} & \begin{array}{c c c c} A & B & C & D \end{array} \\ \begin{array}{c} A \\ B \\ C \\ D \end{array} & \left[\begin{array}{c c c c} 0 & 1 & 0 & 0 \\ 0 & 0 & 0 & 1 \\ 0 & 1 & 0 & 0 \\ 0 & 1 & 0 & 0 \end{array} \right] \end{array}$$

EXAMPLE

■ Our matrix has four rows, one for each of the vertices A, B, C, and D, and four columns which are also labeled A, B, C, and D. We have entered 0's and 1's in the incidence matrix according to the following rules.

1) If there is a directed edge from vertex X to vertex Y in the directed graph, we place a 1 in row X, column Y.

2) If there is not a directed edge from vertex X to vertex Y, we place a 0 in row X, column Y.

We can see that there is a directed edge from A to B in the graph, and thus we enter a 1 in row A, column B, of the matrix. We do not enter a 1 in row B, column A, since there is no directed edge from B to A. Corresponding to the other directed edges, we put 1's in the appropriate places in the matrix. After we account for all directed edges in the graph, 0's are entered in the 12 remaining places in the matrix. □

The following terminology will aid in the discussion of matrices.

DEFINITION   **The number in row X, column Y, of a matrix is called the X, Y-ENTRY of the matrix.**

If the rows and columns of a matrix have not been labeled, it is more common to speak of the i, j-entry of the matrix. For example, the 2, 3-entry indicates the number which is located in the second row, third column of the matrix. Therefore in referring to either an X, Y-entry or an i, j-entry, the first character gives us the row and the second the column.

DEFINITION   **The size of a matrix is determined by the number of rows and columns it contains. An m by n matrix has m rows and n columns. If $m = n$, the matrix is said to be SQUARE.**

The incidence matrix of our previous example is a square (4 by 4) matrix. In fact, any incidence matrix will be square.

EXAMPLE   ■ Below is the incidence matrix for the directed graph of the committee-influence example given in the previous section.

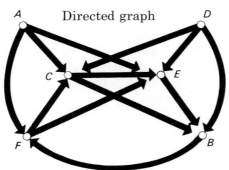

Directed graph

Incidence matrix

$$
\begin{array}{c c c c c c c}
 & A & B & C & D & E & F \\
A & 0 & 0 & 1 & 0 & 1 & 1 \\
B & 0 & 0 & 0 & 0 & 0 & 1 \\
C & 0 & 1 & 0 & 0 & 1 & 0 \\
D & 0 & 1 & 1 & 0 & 1 & 0 \\
E & 0 & 1 & 0 & 0 & 0 & 0 \\
F & 0 & 0 & 1 & 0 & 1 & 0 \\
\end{array}
$$

Our matrix has 6 rows and 6 columns because there are 6 vertices in the directed graph; that is, it's size is 6 by 6. Therefore the matrix has 36 entries. The A, C-entry is 1 because there is a directed edge from A to C. The D, F-entry is 0 because there is no directed edge from D to F. Note that there are 12 entries of 1 since the directed graph has 12 edges. □

Give the incidence matrix for the two following directed graphs.

a)

b)

a)

$$\begin{array}{c c} & \begin{array}{ccc} A & B & C \end{array} \\ \begin{array}{c} A \\ B \\ C \end{array} & \left[\begin{array}{ccc} 0 & 1 & 1 \\ 1 & 0 & 1 \\ 0 & 0 & 0 \end{array}\right] \end{array}$$

b)

$$\begin{array}{c c} & \begin{array}{cccc} A & B & C & D \end{array} \\ \begin{array}{c} A \\ B \\ C \\ D \end{array} & \left[\begin{array}{cccc} 0 & 0 & 0 & 0 \\ 0 & 0 & 0 & 1 \\ 1 & 1 & 0 & 1 \\ 0 & 0 & 1 & 0 \end{array}\right] \end{array}$$

In addition to incidence matrices, we often encounter square matrices having entries other than 0 or 1. These matrices, of course, cannot be called incidence matrices.

In order to use matrices to determine information about the directed paths in a graph, we need to know how to multiply them. Let us begin with a not too complicated example.

■ Let us find the product

$$\begin{bmatrix} 2 & 4 \\ 5 & 3 \end{bmatrix} \times \begin{bmatrix} 9 & 7 \\ 8 & 6 \end{bmatrix}$$

The product is a matrix with two rows and two columns. To obtain the 1, 1-entry of the product matrix, we do the following:

a) As we move from left to right in row 1 of the first matrix, we multiply the entries by those of column 1 of the second matrix, moving down the column.

$$\begin{bmatrix} 2 & 4 \\ 5 & 3 \end{bmatrix} \times \begin{bmatrix} 9 & 7 \\ 8 & 6 \end{bmatrix}$$

In particular, we obtain the two numbers $2 \cdot 9$ and $4 \cdot 8$.

b) We then add the two products obtained in Step (a). Therefore the 1, 1-entry of the product matrix is the number $2 \cdot 9 + 4 \cdot 8 = 50$.

The 1, 2-entry of the product is found by repeating the above steps, using row 1 of the first matrix and column 2 of the second matrix.

$$\begin{bmatrix} 2 & 4 \\ 5 & 3 \end{bmatrix} \times \begin{bmatrix} 9 & 7 \\ 8 & 6 \end{bmatrix}$$

Thus the 1, 2-entry is $2 \cdot 7 + 4 \cdot 6 = 38$.

Similarly, we see that the 2, 1-entry is $5 \cdot 9 + 3 \cdot 8 = 69$.

$$\begin{bmatrix} 2 & 4 \\ 5 & 3 \end{bmatrix} \times \begin{bmatrix} 9 & 7 \\ 8 & 6 \end{bmatrix}$$

Finally, the 2, 2-entry is $5 \cdot 7 + 3 \cdot 6 = 53$.

$$\begin{bmatrix} 2 & 4 \\ 5 & 3 \end{bmatrix} \times \begin{bmatrix} 9 & 7 \\ 8 & 6 \end{bmatrix}$$

We can summarize our discussion by writing the following:

$$\begin{bmatrix} 2 & 4 \\ 5 & 3 \end{bmatrix} \times \begin{bmatrix} 9 & 7 \\ 8 & 6 \end{bmatrix} = \begin{bmatrix} 50 & 38 \\ 69 & 53 \end{bmatrix} \quad \square$$

DEFINITION   Assume that M and N are two square matrices of the same size. The PRODUCT of these matrices is a matrix of the same size, which we denote by $M \times N$. To obtain the i, j-entry of $M \times N$, multiply the entries of the ith row of M, moving left to right, by the entries of the jth column of N, going down the column, and add these products together.

EXAMPLE

$$\begin{bmatrix} 5 & 0 & 2 \\ 0 & 3 & 1 \\ 2 & 0 & 3 \end{bmatrix} \times \begin{bmatrix} 6 & 0 & 2 \\ 4 & 5 & 0 \\ 1 & 3 & 4 \end{bmatrix} = \begin{bmatrix} 32 & 6 & 18 \\ 13 & 18 & 4 \\ \textcircled{15} & 9 & 16 \end{bmatrix}$$

■ For example, the 3, 1-entry of the above product matrix was obtained by using the third row of the first matrix and the first column of the second matrix. In particular, the 3, 1-entry is $2 \cdot 6 + 0 \cdot 4 + 3 \cdot 1 = 15$.

In a similar way you can verify that the other eight entries of the product matrix are correct. $\square$

QUIZ YOURSELF*

$$\begin{bmatrix} 1 & 0 & 2 \\ 3 & 4 & 0 \\ 0 & 1 & 5 \end{bmatrix} \times \begin{bmatrix} 0 & 0 & 3 \\ 2 & 1 & 4 \\ 5 & 1 & 2 \end{bmatrix} = ?$$

***ANSWER**

$$\begin{bmatrix} 10 & 2 & 7 \\ 8 & 4 & 25 \\ 27 & 6 & 14 \end{bmatrix}$$

With the next two examples we shall see that the matrix, which displays the number of directed paths of length 2 in a graph, can be obtained from the product of the incidence matrix with itself.

■ Consider the following directed graph.

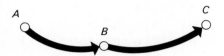

By counting directly in the graph, we tabulate in matrix P the number of directed paths of length 2.

$$P = \begin{array}{c} \\ A \\ B \\ C \end{array} \begin{array}{ccc} A & B & C \\ \left[\begin{array}{ccc} 0 & 0 & 1 \\ 0 & 0 & 0 \\ 0 & 0 & 0 \end{array}\right] \end{array}$$

Let us now obtain the product of the incidence matrix for the graph with itself, that is, the square of the incidence matrix.

$$\begin{bmatrix} 0 & ① & 0 \\ 0 & 0 & 1 \\ 0 & 0 & 0 \end{bmatrix} \times \begin{bmatrix} 0 & 1 & 0 \\ 0 & 0 & ① \\ 0 & 0 & 0 \end{bmatrix} = \begin{bmatrix} 0 & 0 & 1 \\ 0 & 0 & 0 \\ 0 & 0 & 0 \end{bmatrix}$$

We see that the square of M, $M \times M = M^2$, is the same as P. We can satisfy ourselves that this is no coincidence by studying the manner in which the product was obtained. For example, the A, C-entry of M^2 is $0 \cdot 0 + 1 \cdot 1 + 0 \cdot 0 = 1$. We see that this entry 1 is a result of the multiplication of the 1's that are circled in the incidence matrix. These 1's represent a directed edge from A to B and another from B to C; therefore there is a directed path of length 2 from A to C, namely, ABC.

Considering another entry, we see that the A, B-entry of M^2 is $0 \cdot 1 + 1 \cdot 0 + 0 \cdot 0 = 0$. Each of the three terms of this sum is 0 indicating that a directed path of length 2 from A to B by way of A, B, or C does not exist in the graph.

In like manner, we can prove that the other entries of M^2 and P should agree. □

Let us look at a slightly more complicated example.

■ Consider the directed graph of our committee-influence example given in Fig. 3.9. Given below is the square of its incidence matrix:

$$
\begin{array}{c|cccccc}
 & A & B & C & D & E & F \\
\hline
A & 0 & 0 & 1 & 0 & 1 & 1 \\
B & 0 & 0 & 0 & 0 & 0 & 1 \\
C & 0 & 1 & 0 & 0 & 1 & 0 \\
D & 0 & 1 & 1 & 0 & 1 & 0 \\
E & 0 & 1 & 0 & 0 & 0 & 0 \\
F & 0 & 0 & 1 & 0 & 1 & 0 \\
\end{array}
\times
\begin{array}{c|cccccc}
 & A & B & C & D & E & F \\
\hline
A & 0 & 0 & 1 & 0 & 1 & 1 \\
B & 0 & 0 & 0 & 0 & 0 & 1 \\
C & 0 & 1 & 0 & 0 & 1 & 0 \\
D & 0 & 1 & 1 & 0 & 1 & 0 \\
E & 0 & 1 & 0 & 0 & 0 & 0 \\
F & 0 & 0 & 1 & 0 & 1 & 0 \\
\end{array}
=
\begin{array}{c|cccccc}
 & A & B & C & D & E & F \\
\hline
A & 0 & 2 & 1 & 0 & 2 & 0 \\
B & 0 & 0 & 1 & 0 & 1 & 0 \\
C & 0 & 1 & 0 & 0 & 0 & 1 \\
D & 0 & 2 & 0 & 0 & 1 & 1 \\
E & 0 & 0 & 0 & 0 & 0 & 1 \\
F & 0 & 2 & 0 & 0 & 1 & 0 \\
\end{array}
$$

Let us focus our attention on the A, B-entry of M^2 which is 2. In computing this entry we used row A and column B of the incidence matrix.

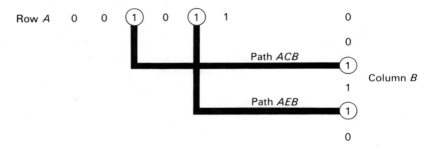

We have matched the pairs of 1's, which are multiplied together and contribute to the A, B-entry of M^2. Note that these two pairs of 1's represent two distinct directed paths of length 2 from A to B, namely, ACB and AEB. Thus the 2 in the A, B-entry of M^2 reflects the fact that there are two directed paths of length 2 from A to B in the graph.

Considering another entry, the B, E-entry of M^2 is 1, which means that there is one directed path of length 2 from B to E. Verify this from the graph. □

The incidence matrix M is a model for the directed graph in that it depicts the directed edges—that is, the directed paths of length 1. We now see that M^2 provides the number of directed paths of length 2 between each pair of vertices. Similarly, it can be shown that the matrix $M^3 = M \times M \times M$ indicates the number of directed paths of length 3. In general, we have the following principle.

PRINCIPLE   **Suppose M is the incidence matrix for some directed graph. Then M^n gives us information about the directed paths of length n in the graph. In particular, the i, j-entry of M^n is the number of directed paths of length n from the ith vertex to the jth vertex in the graph.**

Suppose M is the incidence matrix for some directed graph. We also give M^2 and M^3:

$$
M \quad\quad\quad M^2 \quad\quad\quad M^3
$$

	A	B	C	D
A	0	1	1	1
B	0	0	1	1
C	1	1	0	1
D	0	1	1	0

	A	B	C	D
A	1	2	2	2
B	1	2	1	1
C	0	2	3	2
D	1	1	1	2

	A	B	C	D
A	2	5	5	5
B	1	3	4	4
C	3	5	4	5
D	1	4	4	3

1. How many directed paths of length 2 are there from A to D?
2. How many directed paths of length 2 are there from C to A?
3. Are there any directed paths of length 3 from D to C? If so, how many?
4. How many directed paths of length 3 are there from A to D?

Let us again return to our committee influence example. Instead of counting the number of directed paths of length 1 or 2 in the graph, we use M and M^2 to obtain this same information:

$$
M = \begin{array}{c|cccccc} & A & B & C & D & E & F \\ \hline A & 0 & 0 & 1 & 0 & 1 & 1 \\ B & 0 & 0 & 0 & 0 & 0 & 1 \\ C & 0 & 1 & 0 & 0 & 1 & 0 \\ D & 0 & 1 & 1 & 0 & 1 & 0 \\ E & 0 & 1 & 0 & 0 & 0 & 0 \\ F & 0 & 0 & 1 & 0 & 1 & 0 \end{array}
\qquad
M^2 = \begin{array}{c|cccccc} & A & B & C & D & E & F \\ \hline A & 0 & 2 & 1 & 0 & 2 & 0 \\ B & 0 & 0 & 1 & 0 & 1 & 0 \\ C & 0 & 1 & 0 & 0 & 0 & 1 \\ D & 0 & 2 & 0 & 0 & 1 & 1 \\ E & 0 & 0 & 0 & 0 & 0 & 1 \\ F & 0 & 2 & 0 & 0 & 1 & 0 \end{array}
$$

For example, by adding the A, E-entries of M and M^2, we get 3, which means there are three directed paths of length 1 or 2 from A to E. If we add each of the entries of M to the corresponding entries of M^2, we obtain the matrix

$$
T = \begin{array}{c|cccccc} & A & B & C & D & E & F \\ \hline A & 0 & 2 & 2 & 0 & 3 & 1 \\ B & 0 & 0 & 1 & 0 & 1 & 1 \\ C & 0 & 2 & 0 & 0 & 1 & 1 \\ D & 0 & 3 & 1 & 0 & 2 & 1 \\ E & 0 & 1 & 0 & 0 & 0 & 1 \\ F & 0 & 2 & 1 & 0 & 2 & 0 \end{array}
$$

The entries of T are the number of paths of length 1 or 2 that exist between pairs of vertices in the graph. You should compare T to the matrix we obtained in the previous section, where we first did the committee-influence example, to see that they are the same.

The matrix T is the sum of M and M^2. Let us now make this matrix-addition notion precise.

DEFINITION   **Suppose M and N are two matrices of the identical size. The SUM of these matrices is a matrix of the same size, which we denote by $M + N$. Each entry of the sum is found by adding the corresponding entry of M to the corresponding entry of N. If two matrices are not of the same size, they cannot be added.**

EXAMPLE   ■ $\begin{bmatrix} 1 & 2 & 3 \\ 0 & 4 & 6 \\ 5 & 1 & 1 \end{bmatrix} + \begin{bmatrix} 8 & 2 & 2 \\ 4 & 5 & 0 \\ 3 & 4 & 5 \end{bmatrix} = \begin{bmatrix} 9 & 4 & 5 \\ 4 & 9 & 6 \\ 8 & 5 & 6 \end{bmatrix}$ □

QUIZ YOURSELF*   Perform the following additions if possible.

a) $\begin{bmatrix} 3 & 0 & 1 \\ 0 & 2 & 2 \\ 0 & 3 & 0 \end{bmatrix} + \begin{bmatrix} 6 & 9 & 0 \\ 1 & 3 & 5 \\ 8 & 8 & 7 \end{bmatrix}$ b) $\begin{bmatrix} 1 & 2 & 3 \\ 6 & 0 & 2 \\ 1 & 0 & 0 \end{bmatrix} + \begin{bmatrix} 1 & 2 \\ 3 & 4 \end{bmatrix}$

***ANSWERS**   a) $\begin{bmatrix} 9 & 9 & 1 \\ 1 & 5 & 7 \\ 8 & 11 & 7 \end{bmatrix}$ b)   This addition cannot be done because the matrices are not the same size.

EXAMPLE   ■ Consider the directed graph at the left.

Given below are the incidence matrix M for the graph and the computed matrices M^2 and M^3.

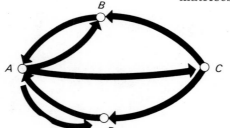

$$\begin{array}{c} M \\ \begin{array}{c c} & \begin{array}{cccc} A & B & C & D \end{array} \\ \begin{array}{c} A \\ B \\ C \\ D \end{array} & \begin{bmatrix} 0 & 1 & 1 & 1 \\ 1 & 0 & 0 & 0 \\ 0 & 1 & 0 & 1 \\ 1 & 0 & 0 & 0 \end{bmatrix} \end{array} \end{array} \quad \begin{array}{c} M^2 \\ \begin{array}{c c} & \begin{array}{cccc} A & B & C & D \end{array} \\ \begin{array}{c} A \\ B \\ C \\ D \end{array} & \begin{bmatrix} 2 & 1 & 0 & 1 \\ 0 & 1 & 1 & 1 \\ 2 & 0 & 0 & 0 \\ 0 & 1 & 1 & 1 \end{bmatrix} \end{array} \end{array} \quad \begin{array}{c} M^3 \\ \begin{array}{c c} & \begin{array}{cccc} A & B & C & D \end{array} \\ \begin{array}{c} A \\ B \\ C \\ D \end{array} & \begin{bmatrix} 2 & 2 & 2 & 2 \\ 2 & 1 & 0 & 1 \\ 0 & 2 & 2 & 2 \\ 2 & 1 & 0 & 1 \end{bmatrix} \end{array} \end{array}$$

If we form the sum $M + M^2 + M^3$, we get a matrix which we will call T:

$$T = M + M^2 + M^3 = \begin{array}{c c} & \begin{array}{cccc} A & B & C & D \end{array} \\ \begin{array}{c} A \\ B \\ C \\ D \end{array} & \begin{bmatrix} 4 & 4 & 3 & 4 \\ 3 & 2 & 1 & 2 \\ 2 & 3 & 2 & 3 \\ 3 & 2 & 1 & 2 \end{bmatrix} \end{array}$$

Consider, for example, the A, B-entry of T, which is the sum of the A, B-entries of M, M^2, and M^3. This entry being 4 tells us there are four directed paths from A to B having lengths of 1, 2, or 3. Similarly, the C, B-entry of T is 3, which means there are three directed paths from C to A of length 1, 2, or 3. $\square$

Besides offering a means for avoiding the tedious chore of counting directed paths in the graph, a matrix model for the graph offers a neat, computational procedure for obtaining the same information. Furthermore, this matrix model can be placed into a computer and the computer could then carry out the computations for us. Thus a problem that could be extremely time consuming if done by hand could be solved with the aid of a computer in a matter of seconds.

EXERCISES

1. Give the incidence matrix for each of the directed graphs.

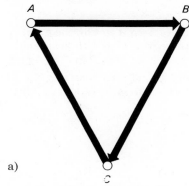
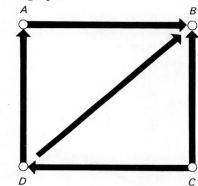

a)

b)

2. Draw a directed graph such that the given matrix is the incidence matrix for the graph.

a)

$$\begin{array}{c} \\ A \\ B \\ C \end{array}\begin{array}{ccc} A & B & C \\ \left[\begin{array}{ccc} 0 & 1 & 1 \\ 1 & 0 & 1 \\ 1 & 1 & 0 \end{array}\right] \end{array}$$

b)

$$\begin{array}{c} \\ A \\ B \\ C \\ D \end{array}\begin{array}{cccc} A & B & C & D \\ \left[\begin{array}{cccc} 0 & 1 & 1 & 0 \\ 0 & 0 & 1 & 1 \\ 0 & 0 & 0 & 1 \\ 1 & 0 & 0 & 0 \end{array}\right] \end{array}$$

3. For each of the following, perform the indicated matrix computation.

a) $\begin{bmatrix} 2 & 0 \\ 3 & 5 \end{bmatrix} + \begin{bmatrix} 1 & 1 \\ 0 & 2 \end{bmatrix}$

b) $\begin{bmatrix} 1 & 0 & 1 \\ 0 & 2 & 3 \\ 1 & 3 & 0 \end{bmatrix} + \begin{bmatrix} 2 & 3 & 1 \\ 4 & 0 & 0 \\ 0 & 5 & 1 \end{bmatrix}$

c) $\begin{bmatrix} 3 & 0 & 2 \\ 4 & 1 & 6 \\ 3 & 0 & 0 \end{bmatrix} \times \begin{bmatrix} 1 & 0 & 0 \\ 2 & 0 & 0 \\ 0 & 0 & 3 \end{bmatrix}$

d) $\begin{bmatrix} 0 & 0 & 0 & 3 \\ 1 & 0 & 2 & 1 \\ 5 & 0 & 0 & 3 \\ 6 & 3 & 1 & 4 \end{bmatrix}^2$

e) $\begin{bmatrix} 0 & 1 & 1 \\ 0 & 0 & 1 \\ 0 & 0 & 0 \end{bmatrix}^3$

4. Find a matrix M such that

$$M + \begin{bmatrix} 2 & 0 \\ 3 & 4 \end{bmatrix} = \begin{bmatrix} 4 & 6 \\ 5 & 4 \end{bmatrix}.$$

5. Suppose that

$$M = \begin{bmatrix} a & 3 \\ 2 & b \end{bmatrix}.$$

Find values for a and b so that

$$\begin{bmatrix} 3 & 1 \\ 0 & 1 \end{bmatrix} \times \begin{bmatrix} a & 3 \\ 2 & b \end{bmatrix} = \begin{bmatrix} 14 & 10 \\ 2 & 1 \end{bmatrix}.$$

6. Suppose that

$$M^3 = \begin{array}{c} \\ A \\ B \\ C \\ D \end{array} \begin{array}{c} \begin{array}{cccc} A & B & C & D \end{array} \\ \begin{bmatrix} 0 & 0 & 2 & 0 \\ 1 & 4 & 3 & 2 \\ 0 & 1 & 1 & 4 \\ 1 & 2 & 3 & 0 \end{bmatrix} \end{array}.$$

   a) How many directed paths of length 3 are there from A to C in the graph?
   b) How many directed paths of length 3 are there from C to A in the graph?
   c) How many directed paths of length 3 are there from D to D in the graph?

7. For the given graph, give the incidence matrix M and then compute M^2. Verify from the graph that M^2 gives the number of directed paths of length 2 between the pairs of vertices.

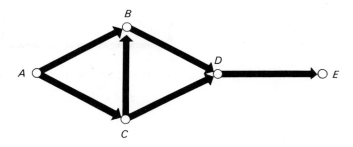

8. Repeat exercise 7 for the graph given below.

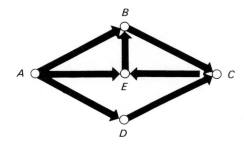

9. Suppose M is the incidence matrix for some directed graph:

$$M = \begin{array}{c} \\ A \\ B \\ C \\ D \end{array} \begin{array}{cccc} A & B & C & D \\ \begin{bmatrix} 0 & 1 & 0 & 0 \\ 1 & 0 & 1 & 1 \\ 0 & 1 & 0 & 1 \\ 1 & 1 & 0 & 0 \end{bmatrix} \end{array}$$

Compute $T = M + M^2 + M^3$. Explain what matrix T tells you about the graph.

10. Redo exercise 9 of the previous section. However, this time determine the number of one- or two-stage influence(s), and thus the ranking order of the committee members, by doing matrix computations.

11. In a round-robin singles ping-pong tournament, each of six competitors play each other once. The results of the tournament are given below:

Player	Defeated
Bruce	Max, Ron, John, Charlie
Max	Ron
Charlie	Max, John
Ron	John, Charlie
Tom	Bruce, Max, Ron, Charlie
John	Tom, Max

Give the incidence matrix which corresponds to the above table. Using this incidence matrix, compute M^2. Based upon your computed results, determine the ranking order of the six players.

12. A paired-comparison test was done for five qualities a consumer would consider when buying a household appliance. The following results were obtained for a particular consumer:

Quality	Preferred over
A	B, C, and E
B	E
C	B and D
D	A and B
E	C

Give the incidence matrix which corresponds to the above table. Use matrix computations to determine one- and two-stage preference. Based upon your computed results, give a ranking order for the five qualities.

NEW TERMS

Vertex

Edge

Graph

Tracing a graph

Connected graph

Even (odd) vertex

Koenigsberg bridge problem

Four-color problem

Directed edge

Directed graph

Directed path

Length of directed path

One- and two-stage influence

Incidence matrix

X, Y-entry

Size of a matrix

Square matrix

Product of matrices

Sum of matrices

MASTERY TEST: TOOLS OF GRAPH THEORY

1. Use Euler's theorem to determine which of the graphs shown in Fig. 3.11 can be traced?

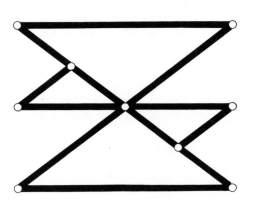 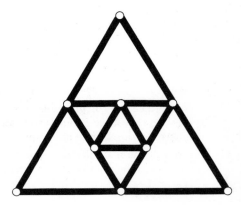

Figure 3.11

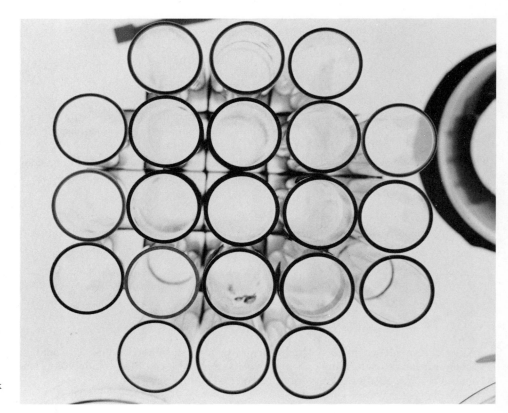

Is it possible to fill all of these test tubes in the following fashion: begin by filling one tube, then move either horizontally or vertically to an adjacent, unfilled tube, continuing in this manner until all tubes are filled? (Photo by Fredrik D. Bodin, Stock Boston.)

2. Model the map in Fig. 3.12 with a graph. With the aid of this graph, solve the four-color problem related to the map.

Figure 3.12

3. The Civil Air Patrol is searching a forest area for some lost hikers. Seven planes will each be assigned to search one of the seven regions indicated on the map in Fig. 3.13. If the planes in adjacent regions are to fly at different altitudes, what is the least number of altitudes that can be used?

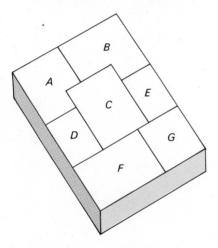

Figure 3.13

4. Six amateur radio hobbyists are located throughout the world. Because of variations in their equipment, not all can receive and broadcast messages equally well. Draw a directed graph to model the given information. Determine which persons can initiate a message that will eventually reach everyone.

Hobbyist	Can receive messages from
Alice	Carol
Bill	Frank
Carol	Frank
Dick	Bill
Eve	Dick, Carol, Bill
Frank	Alice

5. With reference to the auto, steel, coal, and railroad service industries, suppose that Fig. 3.14 depicts how the slackening demand for each product causes a corresponding reduction in demand for the other products.
    a) Give the incidence matrix for the graph.
    b) Verify through matrix computations that each industry is directly or indirectly dependent upon the demand for automobiles.

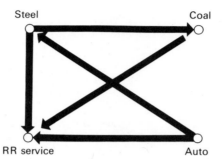

Figure 3.14

6. a) Give the incidence matrix for the directed graph in Fig. 3.15.
    b) Compute M^2.
    c) How many directed paths of length 2 are there from A to B? From B to C?

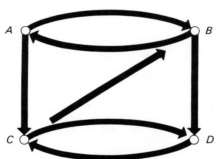

Figure 3.15

III. PERT: A MODEL USED IN SCHEDULING THE CONSTRUCTION OF THE SPACE COLONY

We can now use our "tools" of graph theory to schedule the completion of the space colony mentioned in the first part of this chapter. Recall that it has been decided that this project can be subdivided into ten major tasks.

Task	Time required, months
1. Train construction workers	6
2. Build shell	8
3. Build life-support systems	14
4. Recruit colonists	12
5. Assemble shell	10
6. Train colonists	10
7. Install life-support systems	4
8. Install solar-energy systems	3
9. Test life-support and energy systems	4
10. Bring colonists to the colony	4

It is our goal to schedule these tasks efficiently and also to determine how much of the time allotted for the project can be shortened by reducing the amount of time needed to build the life-support systems.

The model we will use in determining the schedule is a special type of directed graph called a **PERT diagram.**

In determining the total time required for the project, you may be tempted to simply add the time required for each task and say that 75 months are needed to construct and populate the colony. However, we should realize that the tasks need not be done one at a time—some can be worked on simultaneously. Therefore to devise a schedule, we must know which tasks must wait for others to be completed. In the following table, we have indicated which tasks must precede others.

Task	Preceding tasks
1. Train construction workers	None
2. Build shell	None
3. Build life-support systems	None
4. Recruit colonists	None
5. Assemble shell	1, 2
6. Train colonists	2, 3, 4
7. Install life-support systems	1, 2, 3, 5
8. Install solar-energy systems	1, 2, 5
9. Test life-support and energy systems	1, 2, 3, 5, 7, 8
10. Bring colonists to the colony	1, 2, 3, 4, 5, 6, 7, 8, 9

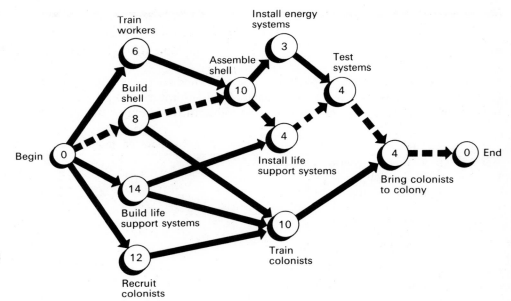

Train
workers

Install energy
systems

Assemble
shell

Test
systems

Build
shell

Begin

Install life
support systems

Build life
support systems

Bring colonists
to colony

End

Train
colonists

Figure 3.16

Recruit
colonists

Now the two tables of information can be displayed by the PERT diagram given in Fig. 3.16. Each task is represented by a small circular vertex containing the number of months necessary for completion of that task. The beginning and end of the project are shown in our diagram and we assume that these require no time.

We have drawn a directed edge from vertex X to vertex Y if the task represented by vertex X immediately precedes the task represented by Y. For example, we agreed that training the workers to work in space precedes assembling the shell in space, so we have drawn a directed edge from the vertex "train workers" to the vertex "assemble shell."

Although building the shell on earth does come before installing the life-support systems, we did not draw a directed edge from "build shell" to "install life-support systems." Drawing such an edge is not necessary because we have indicated in the diagram that "building the shell" must be done before "assembling the shell in space," which in turn must precede "installing the life-support systems." Hence this sequence of tasks already shows that "building the shell" must come before "installing the life-support systems."

A project consists of nine separate tasks *A, B, C, D, E, F, G, H,* and *I*. The table indicates the time required for each task and also the dependency among the various tasks. Draw a PERT diagram for this project.

Task	Time required, days	Preceding tasks
A	3	None
B	4	*A*
C	6	*A*
D	2	*A, B, C*
E	5	*A, B, C, D*
F	7	*A, B, C, D*
G	3	*A, B, C, D*
H	8	*A, B, C, D, F, G*
I	3	*A, B, C, D, E, F, G, H*

*ANSWER

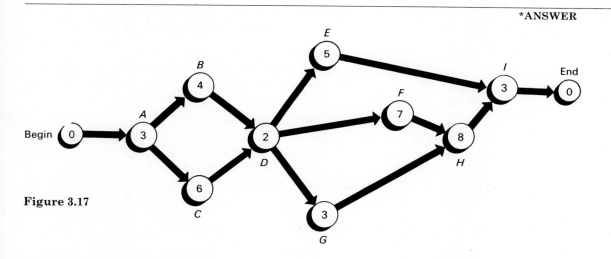

Figure 3.17

Returning to our space-colony scheduling problem, let us use our PERT diagram to determine the earliest time that the life-support systems could be installed. We see from Fig. 3.16 that there are several different sequences of tasks that must precede installing the life-support systems.

For example, the sequence "begin," "train workers," and "assemble shell" comes before "install life-support systems." This sequence requires $0 + 6 + 10 = 16$ months to complete. Another sequence is "begin" and "build life-support systems," which takes 14 months to complete. By examining the diagram, it is not difficult to see that the most time-consuming sequence of tasks preceding "install life-support systems" is the sequence "begin," "build shell," and "assemble shell" and this requires $0 + 8 + 10 = 18$ months. Thus 18 months will allow enough time to complete *all* those tasks which must precede installing the life-support systems.

Observe that an efficient schedule would have construction on the shell and on the life-support systems going on simultaneously. Hence we should schedule the installation of the life-support systems to begin during the nineteenth month of the project. We can determine a schedule for the other tasks using similar reasoning. In order to simplify our discussion, we introduce the following.

DEFINITION   **Suppose T is a task in a PERT diagram. Let us consider all directed paths from "Begin" to T. If we add the time along each of these paths, the one requiring the most time to complete is called a CRITICAL PATH for T.**

Using this new terminology, we can say that the path "begin," "build shell," "assemble shell," and "install life-support systems" is a critical path for the task "install life-support systems."

To determine when to schedule a task T, we do the following:

1) Find a critical path for the task.
2) Add up all the time along this critical path with the exception of the time required for the task T. The sum we obtain gives us the time to be allowed before scheduling T.

Let us use the above procedure to determine when to schedule the testing of the colony's systems. From our PERT diagram in Fig. 3.16, we see that a critical path for this task is "begin," "build shell," "assemble shell," "install life-support systems," and "test systems." Considering this path we realize that $0 + 8 + 10 + 4 = 22$ months must be allowed before the testing of the colony's systems can begin. Therefore this task should be scheduled to begin in the twenty-third month of the project.

Time schedules for the remaining tasks are found in a similar manner. This information is summarized in the next table.

Task	Task begins, month
1. Train construction workers	1st
2. Build shell	1st
3. Build life-support systems	1st
4. Recruit colonists	1st
5. Assemble shell	9th
6. Train colonists	15th
7. Install life-support systems	19th
8. Install solar-energy systems	19th
9. Test life-support and energy systems	23rd
10. Bring colonists to the colony	27th

In order to determine the time needed for the entire project, we must find a critical path for the vertex "End." Such a path is indicated by the dotted edges in our diagram. From this it is clear that $0 + 8 + 10 + 4 + 4 + 4 + 0 = 30$ months is needed for the project.

Let us now consider whether it is possible to shorten the total length of the project by reducing the time spent in building the life-support systems. Referring to Fig. 3.16, we see that the task of building the life-support systems does not lie on a critical path for the vertex "End." Therefore reducing this time would in no way decrease the total length of the project.

The PERT diagram shown in Fig. 3.18 is for a project consisting of tasks A, B, C, D, E, F, G, and H.

QUIZ YOURSELF*

1. Find a critical path for E in the PERT diagram.
2. Find a critical path for G.
3. What is a critical path for "End"?
4. Assuming that the numbers in the vertices represent days, when should task E be scheduled?
5. When should H be scheduled?
6. How long will the whole project take?

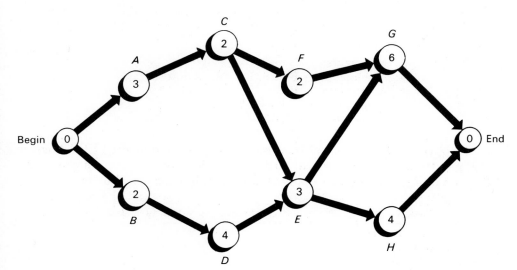

Figure 3.18

EXAMPLE ■ Let us suppose that you are chairperson of a committee which is organizing a spring arts festival on your campus. At the committee's first meeting it is agreed that the following individual tasks must be accomplished in order for the festival to be successful.

Task	Time required, weeks
1. Poll the student body to determine their interest	2
2. Obtain funds from the Student Government Association	1
3. Canvass local merchants for financial support	4
4. Hire performers	3
5. Rent an auditorium	2
6. Print the program	1
7. Advertise the festival	2

Your job as chairperson of the committee is to develop a schedule so that the project can be completed in the shortest possible time.

As in our space-colony example, we first need to know the dependency among the various tasks. We give this information in the following table:

Task	Tasks which must precede this one
1. Poll	None
2. S.G.A.	None
3. Merchants	None
4. Performers	Poll, S.G.A., Merchants
5. Auditorium	Poll
6. Program	All of the above
7. Advertise	All of the above except "Program"

Now, the two tables of information can be displayed by the PERT diagram in Fig. 3.19. By finding critical paths for each of the vertices in the PERT diagram of Fig. 3.19, we can easily obtain the following schedule of the tasks:

Task	Week during which task should begin
1. Poll students	1
2. Obtain funds from S.G.A.	1
3. Canvass merchants	1
4. Rent auditorium	3
5. Hire performers	5
6. Advertise	8
7. Print program	8

Using critical paths, we can also determine the shortest time period in which the project can be completed. The most time-consuming sequence of

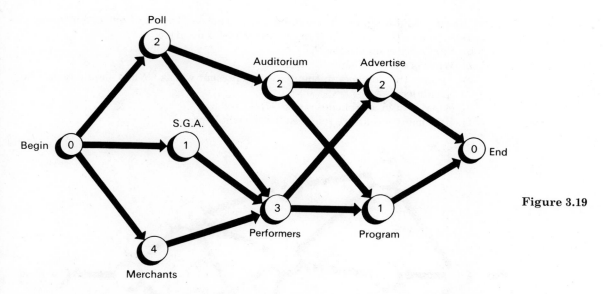

Figure 3.19

tasks in a project is found along a critical path for the vertex "End." For the festival project, such a path is "Begin," "Merchants," "Performers," "Advertise," and "End." The sum of time along this path is nine weeks. Therefore the entire festival project can be scheduled so as to be completed in nine weeks. □

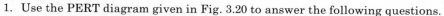

EXERCISES

1. Use the PERT diagram given in Fig. 3.20 to answer the following questions.
   a) Find a critical path for task I.
   b) Find a critical path for task E.
   c) On what day will task H begin?
   d) When will task I be completed?
   e) What is the least number of days needed to complete this project?
   f) How many critical paths are there for the vertex "End"?

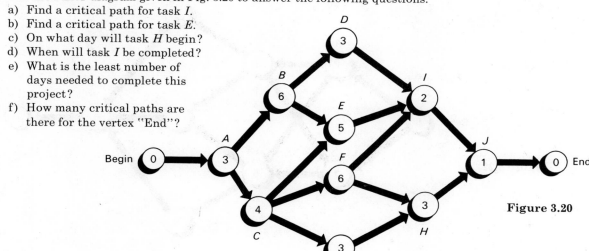

Figure 3.20

2. Use the PERT diagram shown in Fig. 3.21 to answer the questions given below:
   a) Find a critical path for G in the above PERT diagram.
   b) Find a critical path for H.
   c) What is a critical path for "End"?
   d) Assume the numbers in the vertices represent weeks. When should task F be scheduled?
   e) When should we schedule task G?
   f) How many weeks are required for the whole project?

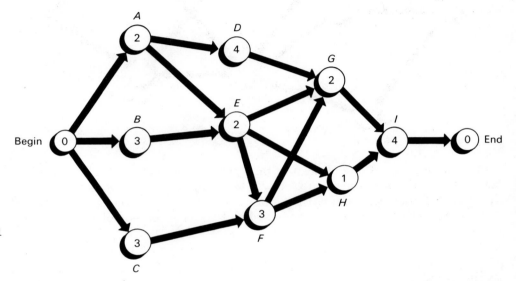

Figure 3.21

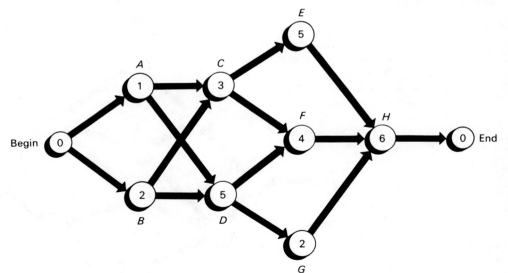

Figure 3.22

3. Use the PERT diagram shown in Fig. 3.22 to schedule the tasks so the project is completed in the least possible amount of time. Assume the numbers in the vertices refer to days.

4. Repeat exercise 3 using the PERT diagram shown in Fig. 3.23.

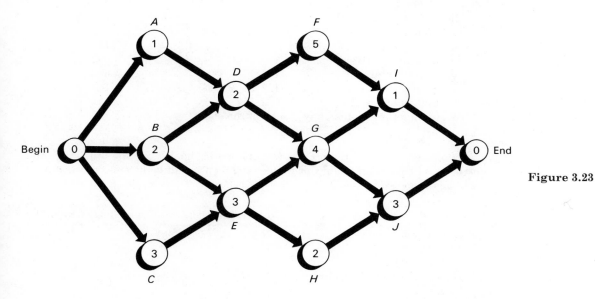

Figure 3.23

5. A contractor is planning to build a new development consisting of houses and commercial buildings. The contractor has divided the project into nine tasks. Using the following table, draw a PERT diagram for this project and give a schedule for the tasks.

Task	Tasks which precede this task	Time needed for this task, months
1. Buy land	None	3
2. Draw plans	None	4
3. Clear land	1	2
4. Install streets, etc.	1, 2, 3	5
5. Build commercial buildings	1, 2, 3, 4	12
6. Build houses	1, 2, 3, 4	16
7. Rent commercial buildings	1, 2, 3, 4, 5	8
8. Sell houses	1, 2, 3, 4, 6	6
9. Landscape	1, 2, 3, 4, 5, 6	3

6. An emerging nation plans to improve the health of its citizens. Use the following table to draw a PERT diagram for this project and then schedule the tasks so the project can be completed as efficiently as possible.

Task	Preceding tasks	Time required, months
1. Appropriate money	None	3
2. Build health clinics	1	8
3. Construct hospitals	1	18
4. Educate citizens in health practices	1	12
5. Recruit students	1	8
6. Train doctors	1, 3, 5	30
7. Train aids	1, 2, 5	15
8. Innoculate	1, 2, 5, 7	4
9. Conduct follow-up study	All of the above	2

7. A publisher plans to produce a new mathematics book. Use the following table to draw a PERT diagram for this project and then schedule the tasks so the project can be completed as efficiently as possible.

Task	Preceding tasks	Time required, months
1. Conduct survey	None	3
2. Develop budget	1	1
3. Hire author	1	6
4. Set-up production schedule	1, 2, 3	1
5. Write book	1, 2, 3	8
6. Advertise book	1, 2, 3, 5	2
7. Produce book	1, 2, 3, 4, 5	6

8. A local electric company wants to build an experimental house that will conserve energy. Company officials feel the project breaks down naturally into the following tasks (the time required for each task is given in parentheses).

   1. Draw plans for house (2 months)
   2. Design energy systems for house (6 months)
   3. Develop new insulation techniques (3 months)
   4. Purchase land (4 months)
   5. Build conventional shell of house (6 months)
   6. Install new energy systems (2 months)
   7. Install new insulation (1 month)
   8. Finish the remaining conventional parts of the house (3 months)
   9. Landscape (1 month)
   10. Perform tests to determine actual energy usage in house (8 months)

   Devise a dependency table for the tasks; draw a PERT diagram based upon your dependency table; and give a schedule for the tasks.

SUGGESTED READINGS

GARDNER, M., "Mathematical Games: Map Coloring." *Sci. Am.*, September 1960.
A history of the four-color problem with a discussion on related problems.

GARDNER, M., "Mathematical Games: Plotting the Crossing Numbers of Graphs."
Sci. Am., June 1973, pp. 106–109.
*Determining the number of edge crossings that will occur in a graph with a given
number of nodes; includes an answer to the "utilities problem."*

GRECOS, A. P., AND R. W. GIBBERD, "A Diagrammatic Solution to the 'Instant Insanity'
Problem." *Math Mag.*, May 1971, pp. 119–124.
Gives a graph theoretic solution to the instant insanity puzzle.

MIZRAHI, A., AND M. SULLIVAN, *Finite Mathematics with Applications for Business and
Social Sciences* (2nd ed.). New York: John Wiley, 1976.
*Chapter 10 contains applications of directed graphs to three types of structures with
easy to understand examples and exercises.*

ORE, O., *Graphs and Their Uses.* Westminster, Md.: Random House, 1963.
A readable introduction to the subject.

WIEST, J., AND M. LEVY, *A Management Guide to PERT/CPM.* Englewood Cliffs, N.J.:
Prentice-Hall, 1969.
A comprehensive study of PERT and the critical path method.

The founding fathers felt that the apportionment of the members of the House of Representatives was so important that it was discussed at the very beginning of the United States Constitution. Article I, Section 2, states:

> Representatives and direct taxes shall be apportioned among the several states which may be included within this Union, according to their respective numbers . . . The actual Enumeration shall be made within three years after the first Meeting of the Congress of the United States, and, within every subsequent Term of ten years, in such Manner as they shall by Law direct. The number of Representatives shall not exceed one for every thirty thousand, but each State shall have at least one representative . . .

Although the constitution originally set the number of representatives at 65, Congress does have the power to vary the size of the House, provided, of course that there is no more than one representative for every 30,000 people in the United States. After the country's first census in 1790, Congress passed an apportionment act in 1792, which allotted 120 seats to the then 15 states. Secretary of State Thomas Jefferson felt that the bill was not in keeping with the spirit of the founding fathers and wrote in opposition to the bill, "No invasions of the Constitution are fundamentally so dangerous as the tricks played on their own numbers . . ."* George Washington, acting on Jefferson's advice, turned down the bill in what was to be the very first presidential veto.

---

* *The Works of Thomas Jefferson*, Vol. VI, Paul Leicester Ford, editor, New York: G. P. Putnam's Sons, 1904, pp. 460–471.

George Washington turned down the apportionment bill of 1792 in what was to be the very first presidential veto.

Another apportionment bill was drafted fixing the House at 103 members for a population of then roughly 4 million. After each succeeding census a further apportionment was enacted, usually with a corresponding increase in the size of the House. In 1910, the number of representatives was set at 435, which is the size of the present House. It is interesting to note that if the principle of approximately one representative for every 30,000 people were applied with our present population, there would be over 6000 members in the House!

The constitution does not expressly state what system is to be used in allocating representatives to states except that the apportionment is to be made "according to their respective numbers." This vagueness has caused debate over methods of apportionment from the time of Washington to the present. Daniel Webster felt that it is not possible to assign representatives so that each one would have the same number of constituents. Speaking before the House of Representatives on April 5, 1832, he said,

> The Constitution . . . must be understood, not as enjoining an absolute relative equality, because that would be demanding an impossibility, but as requiring of Congress to make the apportionment of Representatives among the several States according to their respective numbers, as *near as may be*. That which cannot be done perfectly must be done in a manner as near perfection as can be.*

---

* *The Works of Daniel Webster*, Vol. III, 16th Ed. Boston: Little, Brown and Company, 1872.

I. THE ALABAMA PARADOX

In 1881 while considering reapportionment of the members of the House of Representatives, Congress discovered the following startling fact. Using the then current apportionment system known as the Vinton method, Alabama would be entitled to 8 representatives in a House having 299 members, but it would receive only 7 representatives in a 300-member House. What we are saying is that without any change in population, Alabama would receive fewer representatives in a larger House. This strange occurrence, known as the Alabama paradox, was no isolated incidence. Following the 1890 census, using the same Vinton method, it was learned that Arkansas would deserve fewer representatives in a House having 360 members than in a House of 359.

Under an apportionment plan proposed following the census of 1890, both Colorado and Maine were subject to this same paradoxical situation. In fact in the case of Maine, the number of representatives fluctuated up and down

Surprisingly, sometimes very elementary mathematics can be used to solve important, real-life problems. As we will see in this chapter, the "tools" used in resolving the Alabama paradox are not difficult. (Photo by Owen Franken, Stock Boston.)

so rapidly as different sized Houses were considered that representative Littlefield of Maine remarked:

> . . . it does seem as though mathematics and science had combined to make a shuttlecock and battledoor of the State of Maine in connection with the scientific basis upon which this bill is presented to this House.
>
> Now you see it and now you don't. In Maine comes and out Maine goes. The House increases in size and still she is out. It increases a little more in size, and then, forsooth, in she comes. A little further increase, and out she goes, and then a little further increase and in she comes. God help the State of Maine when mathematics reach for her and undertake to strike her down in this manner in connection with her representation on this floor . . .*

It is not our intention to do the extensive calculations needed to explain why the Alabama paradox took place. Instead we will discuss a hypothetical and much simplified example in which the same type of situation arises.

Collegeville University is organized into three main divisions: humanities, science, and education. There are presently 103 faculty members in the humanities division, 63 in the science division, and 34 in the education division. The 200-member faculty governs its affairs by means of a 20-member council.

Since the humanities division has $\frac{103}{200}$ or 51.5 percent of the faculty, common sense tells us that it is entitled to 51.5 percent of the 20 members on the council, which is 10.3 representatives. In the following table, similar calculations are made for the other divisions.

Division	Number of faculty	Percent of faculty	Exact number of representatives deserved
Humanities	103	51.5	10.3
Science	63	31.5	6.3
Education	34	17.0	3.4
Totals	200	100.0	20.0

Obviously the humanities division cannot elect 10.3 representatives; they must choose either 10 or 11. Similarly the science division will select either 6 or 7 representatives, and the education division will select either 3 or 4.

We realize that if the humanities division is given 10 representatives, the science division 6, and the education division 3, then we have only 19 representatives on the council. In order to reach the required 20, it is necessary to decide which division is deserving of the 1 additional representative. Since the education division has the highest fractional part of a representative, namely .4, it seems that the fairest thing would be to give the extra representative to that division. In fact, this is exactly the way the **Vinton method** of apportionment would allocate this one extra faculty representative.

---

*   The full text of representative Littlefield's remarks can be found in the Congressional Record of the 56th Congress, Second Session, Vol. 34, beginning on p. 590.

Suppose the council decides to increase its size to 21 members. Let us consider how to use the same method to apportion the 21-member faculty council.

Division	Number of seats allotted in 20-member council	Exact number of seats in 21-member council	Number of seats allotted in 21-member council
Humanities	10	10.815	11
Science	6	6.615	7
Education	4	3.570	3
Totals	20	21.000	21

The last column of the above table shows the apportionment according to our "common-sense" system. We start by assigning 10 seats to the humanities division, 6 to the science division, and 3 to the education division, for a total of 19 seats. In the 21-member council, the two additional seats are given to the divisions with the highest fractional parts. We see that one more seat should go to the humanities division and the second to the science division. Thus this method gives 11 seats to the humanities division, 7 to the science division, but only 3 to the education division. Even though the number of faculty members in each division remains the same, the education division lost one seat when the council increased in size. In particular, by the Vinton method of apportionment, the education division deserves 4 seats on a 20-member council but only 3 seats on a 21-member council.

One of the main reasons the "paradox" occurs in our example when using this method is that every time the size of the council changes we have to completely reassign the seats. We saw that even though 20 representatives had already been apportioned, when the council size increased to 21, we had to reassign the original 20 seats. This presented the possibility of some of the divisions losing representatives which had already been assigned to them.

Certainly one way to prevent an Alabama paradox is to develop an apportionment method based on the principle that once a seat has been assigned, it cannot be reassigned if there is an increase in the size of the council. The method we use to do this does not depend on the size of the council; specifically we assign the seats one at a time until all seats are apportioned. If we had proceeded in this manner in our example, after the 20 seats were assigned, we would then only have to decide which division would receive the extra seat on the 21-member council. Thus no division can lose any representatives and the paradox is avoided.

You should realize that we have not *fully* described an apportionment method. In order to assign the seats one at a time, we must have some criterion for assigning the first seat, then the second, etc. In general, at any stage in the apportionment process, we must be able to decide which division should be given the next seat on the council. In Part II of this chapter, our goal will be to develop such a criterion.

II. APPORTIONMENT AND THE TOOLS OF INEQUALITIES

MEASURES OF THE UNFAIRNESS OF AN APPORTIONMENT

Before we can develop the desired criterion for assigning representatives to the faculty council, we must look at some general concepts related to the apportionment process. Let us consider the hypothetical representation of two states in the United States House of Representatives.

Suppose state A has 2,000,000 people and 8 representatives while state B has 800,000 people and 4 representatives. Each representative from state A will have an average of 2,000,000/8 = 250,000 constituents in his district, whereas each representative from state B averages 800,000/4 = 200,000. Since each legislator from state A must speak on behalf of more people, we would say that state A is more poorly represented in the House than state B.

In general, for an apportionment of representatives to the two states A and B, state A would be more poorly represented than state B if

$$\frac{\text{Population of } A}{\text{Number of representatives from } A} > \frac{\text{Population of } B}{\text{Number of representatives from } B}.$$

It would be ideal, of course, to have an apportionment in which the average number of constituents per representative would be the same for both states. This would be consistent with the "one-man, one-vote" concept; however, in dealing with an actual apportionment, this may not be a realistic goal. If we cannot have equality, then it is natural that we try to assign representatives so that the average number of constituents per representative for both states is as close as possible. One measure of an apportionment is therefore the difference between these two numbers.

DEFINITION   **Let us suppose that for some particular apportionment of representatives to states A and B, it happens that state A is more poorly represented in comparison to state B. We define the ABSOLUTE UNFAIRNESS for this apportionment as the difference**

$$\frac{\text{Population of } A}{\text{Number of representatives from } A} - \frac{\text{Population of } B}{\text{Number of representatives from } B}$$

EXAMPLE   ■ Suppose state C has a population of 1542 and 6 representatives while state D has a population of 1445 and 5 representatives. Determine which state is more poorly represented and calculate the absolute unfairness for this assignment of representatives.

Each representative from state C has $\frac{1542}{6} = 257$ constituents while each representative from state D has $\frac{1445}{5} = 289$ constituents. Thus state D is more poorly represented in comparison to C and the absolute unfairness for this apportionment is $289 - 257 = 32$. □

Calculate the average number of constituents per representative for each of the following pairs of states to determine which state is more poorly represented. Then compute the absolute unfairness for the given apportionment.

QUIZ YOURSELF*

1. State *A* has a population of 11,710 and 5 representatives; state *B* has a population of 16,457 and 7 representatives.

2. State *C* has a population of 1,000,000 and 10 representatives; state *D* has a population of 360,000 and 3 representatives.

3. State *E* has a population of 10,443 and 3 representatives; state *F* has a population of 38,291 and 11 representatives.

---

1. In state *A* there are 2342 constituents per representative, while in state *B* there are 2351. So state *B* is more poorly represented; 2351 − 2342 = 9.

2. In state *C* there are 100,000 constituents per representative, while in state *D* there are 120,000. Thus state *D* is more poorly represented; 120,000 − 100,000 = 20,000.

3. Both states *E* and *F* on the average have 3481 constituents per representative; they are equally well represented; 3481 − 3481 = 0.

*ANSWERS

Although absolute unfairness does provide a way of calculating the inequality of an apportionment between two states, it is inadequate as the following discussion indicates.

Suppose a state *X* has a population of 974,116 and 4 representatives, while a state *Y* has a population of 730,779 and 3 representatives. The absolute unfairness of this apportionment is

$$\frac{730{,}779}{3} - \frac{974{,}116}{4} = 243{,}593 - 243{,}529 = 64.$$

According to the way we have been measuring unfairness, it seems as though this apportionment is worse than the apportionment involving states *C* and *D*, which has an absolute unfairness of 32. An analogy will help you see that we should consider a different measure of unfairness.

Suppose in a state election Smith receives 5,000,050 votes, while Jones gets 4,999,950. Smith wins the election by 100 votes and it is likely that newspapers would proclaim:

"SMITH AND JONES FINISH IN VIRTUAL TIE."

On the other hand, if Smith and Jones were competing for a borough council seat in a small town and Smith got 115 votes while Jones received 35, the town paper might read:

"SMITH BURIES JONES IN LANDSLIDE."

Although the difference in the number of votes is important, the significance of that difference depends upon the number of votes they received. A difference of 100 votes is perceived to be small when one receives 4,999,950, whereas an 80-vote difference seems large if one of them receives only 35 votes.

In a similar way, we may feel that the difference between the average number of constituents per representative for the two states X and Y is *relatively* small because of the size of the two numbers. On the other hand, since the average number of constituents for state C is 257 and the average number for state D is 289, we may feel the difference of 32 is *relatively* large.

These remarks point out that in measuring the unfairness of an apportionment we should take into consideration not only the difference in the average number of constituents per representative but also the *size* of these numbers. We give one way of doing this in our next definition.

DEFINITION   **Let us assume that for a given apportionment of representatives, state A is more poorly represented in comparison with state B. We define the RELATIVE UNFAIRNESS of this apportionment as**

$$\frac{\text{Absolute unfairness of the apportionment}}{\left(\dfrac{\text{The population of } B}{\text{Number of representatives from } B}\right)}.$$

In computing relative unfairness, we should observe that the denominator of the quotient is the smaller average number of constituents per representative for the two states.

Let us compute the relative unfairness for each of our two earlier examples. For states X and Y, the absolute unfairness of the apportionment is 64. State Y has the smaller number of constituents per representative so the relative unfairness of this apportionment is $64/243{,}529 = .0003$. The absolute unfairness of the apportionment for states C and D is 32. Since state C has the smaller number of constituents per representative, namely 257, the relative unfairness is $\frac{32}{257} = .1245$. Using relative unfairness to measure the inequality of the apportionment, we see that the representatives have been more equally apportioned between states X and Y than between C and D.

QUIZ YOURSELF*   Use the information given below to calculate the relative unfairness of the apportionment for each of the following pairs of states.

1. State C has a population of 1,000,000 and 10 representatives; state D has a population of 360,000 and 3 representatives.

2. State A has a population of 11,710 and 5 representatives; state B has a population of 16,457 and 7 representatives.

3. State E on the average has 3485 constituents per representative from the state; and, state F on the average has 3000 constituents per representative.

---

*ANSWERS   a) $\dfrac{\left(\dfrac{360{,}000}{3}\right) - \left(\dfrac{1{,}000{,}000}{10}\right)}{\left(\dfrac{1{,}000{,}000}{10}\right)} = \dfrac{120{,}000 - 100{,}000}{100{,}000} = .2$

b) $\dfrac{2351 - 2342}{2342} = .0038$

c) $\dfrac{485}{3000} = .1617$

Handwritten margin notes:

✗ (Example)

absolute unfairness

$\begin{array}{cc} x & y \\ .9\text{ million} & 4\text{ million} \\ 2 & 10 \end{array}$

$\dfrac{.9}{2} = .45 \qquad \dfrac{4}{10} = .4$

$.45 - .4 = .05$

Relative unfairness

$\dfrac{.05}{.4} = .125$

In order to avoid an Alabama paradox, we recall that the seats on the faculty council were to be assigned one at a time. The procedure used to do this must meet some criterion. Thus the following criterion, based on relative unfairness, will be used.

When assigning a representative to one of two divisions, assign the representative to the division which results in a smaller relative unfairness.

APPORTIONMENT
CRITERION

The next section on inequalities will provide the tools needed for developing an apportionment principle. This principle will enable us to describe a method of apportioning the seats on the council one at a time, so that the above criterion is satisfied.

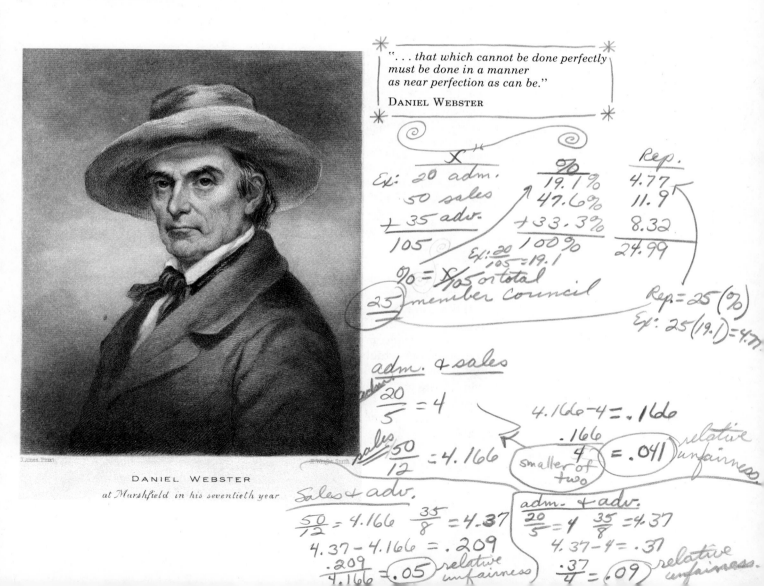

DANIEL WEBSTER

at Marshfield in his seventieth year

"*. . . that which cannot be done perfectly
must be done in a manner
as near perfection as can be.*"

DANIEL WEBSTER

EXERCISES

1. Recall from Part I of this chapter, that from a 20-member council at Collegeville University, the 103-member humanities division received 10 representatives, while the 63 member science division received 6. Calculate the absolute and relative unfairness for this apportionment.

2. Redo exercise 1 for the humanities and education divisions. Recall that the 34-member education division gets 4 representatives on a 20-member council.

3. Use the information contained in exercises 1 and 2 to compute the absolute and relative unfairness of the apportionment of representatives to the science and education divisions on a 20-member council.

4. According to the 1970 census, Colorado had a population of approximately 2.2 million people, while Delaware had .6 million people. Colorado was allotted 5 seats in the House and Delaware was allotted 1. Calculate the absolute and relative unfairness of this apportionment.

5. According to the 1970 census, Iowa had a population of approximately 2.8 million people while Utah had 1 million people. Iowa was allotted 6 seats and Utah was allotted 2. Calculate the absolute and relative unfairness for this apportionment.

6. Assume that of a 21-member faculty council at Collegeville University, the science division receives 7 representatives and the education division receives 3. Suppose one additional representative can be given to either the science or education division, but not to both.
   a) Calculate the relative unfairness of the apportionment if the additional representative is given to the science division.
   b) Calculate the relative unfairness of the apportionment if the additional representative is given to the education division.

INEQUALITIES

The techniques we use in working with inequalities are quite similar to those you have encountered in your previous experiences with equations. You may recall that there are several basic manipulations which can be performed upon equations that enable us to write them in an equivalent form. For example, adding or subtracting the same expression to both sides of an equation results in an equation which is equivalent to the original one. It is this principle which allows us to subtract $2x$ from both sides of the equation $2x + 20 = 6x$ to obtain the equivalent equation $20 = 4x$.

You may also remember that multiplying or dividing both sides of an equation by the same nonzero expression also yields an equation equivalent to the original one. For instance, we can multiply the equation

$$\frac{x}{5} = -\frac{7}{2} + \frac{3x}{10}$$

by 10 to obtain the equivalent equation $2x = -35 + 3x$.

The rules for manipulating inequalities are quite similar to the rules used in working with equations, with one important exception—when an inequality is multiplied by a negative number, the inequality sign must be reversed. (Photo by Patricia Hollander Gross, Stock Boston.)

QUIZ YOURSELF*   Determine which of the following pairs of equations are equivalent. If a pair of equations are equivalent, explain how the second equation can be obtained from the first using the principles given earlier.

a) $2x + 3 = 7x - 13$
   $16 = 5x$

b) $7x + 4 = x$
   $8x = -4$

c) $\dfrac{x}{4} + 3y = 7$
   $x + 12y = 28$

d) $\dfrac{x}{3} = -\dfrac{1}{4} + \dfrac{2x}{3}$
   $x = \frac{3}{4}$

---

*ANSWERS   a) Yes, add $13 - 2x$ to both sides
b) No
c) Yes, multiply both sides by 4
d) Yes, multiply both sides by 12, add $3 - 4x$ to both sides, and then divide by 4

Let us now turn our attention to inequalities. The material that we consider will be useful not only in discussing apportionment later in this chapter but also in Chapter 6 on linear programming. An inequality is a statement using one of the symbols $<$, $\leqslant$, $>$, $\geqslant$, which are read as follows:

Symbol	Read
$<$	is less than
$\leqslant$	is less than or equal to
$>$	is greater than
$\geqslant$	is greater than or equal to

The statement $x < 3$ would be read "x is less than 3," while the inequality $2x + y \geqslant 3y - 2$ would be read "$2x + y$ is greater than or equal to $3y - 2$."

We are sometimes interested in solving inequalities. As you might expect, a **solution** for an inequality is any particular assignment of values that will give us a true statement. The inequality $x + 3 < 8$ has many solutions. For instance, 4, $\frac{1}{3}$, $-\pi$, $\sqrt{2}$, -2.718 are five solutions. In fact, any real number smaller than 5 will be a solution for this inequality.

Frequently inequalities occur whose solutions are not obvious. When this happens we try to replace the given inequality with an equivalent inequality for which the solutions are more apparent. By **equivalent inequalities** we mean inequalities having the same variables which have exactly the same solutions. The rules for manipulating an inequality to obtain an equivalent inequality are very similar to those mentioned earlier for handling equations.

PRINCIPLES   **1. Adding or subtracting the same expression to both sides of an inequality results in an inequality which is equivalent to the original inequality.**

2. Multiplying or dividing both sides of an inequality by the same *positive* expression results in an inequality equivalent to the original one.

3. If we multiply or divide both sides of an inequality by the same *negative* expression and reverse the direction of the inequality sign, we obtain an equivalent inequality.

■ Let us solve the inequality $7x + 4 \leqslant 3x - 2$ by writing a sequence of equivalent inequalities. First we subtract $4 + 3x$ from both sides of the inequality to obtain

$$4x \leqslant -6.$$

Dividing both sides by the *positive* number 4 results in the inequality

$$x \leqslant -\tfrac{6}{4}.$$

This means that all real numbers which are less than or equal to $-\tfrac{3}{2}$ are solutions to our original inequality. □

■ Let us solve the inequality

$$\tfrac{3}{2}x - 5 < \tfrac{11}{3}x - 9.$$

To eliminate fractions, we can multiply both sides of the inequality by 6 to get

$$9x - 30 < 22x - 54.$$

Adding $30 - 22x$ to both sides gives us

$$-13x < -24.$$

Now dividing by the negative number -13, we get

$$x > \tfrac{24}{13}.$$

Note how the direction of the inequality sign was reversed when we divided by the negative number. We can conclude that the solutions to our original inequality

$$\tfrac{3}{2}x - 5 < \tfrac{11}{3}x - 9$$

are precisely all real numbers greater than $\tfrac{24}{13}$. □

Determine which of the following pairs of inequalities are equivalent. If a pair of inequalities is equivalent, explain how the second inequality can be obtained from the first using the principles given earlier.

a) $3x - 4 < 5$

$3x < 9$

b) $4x < 4$

$x < 0$

c) $5y - 2 \leqslant 3y + 6$

$2y \leqslant 8$

d) $\dfrac{t}{2} - \dfrac{1}{4} > 5$

$2t - 1 > 20$

e) $\dfrac{x}{3} - 1 < 2$

$x - 3 < 2$

f) $\dfrac{x}{3} - 1 > 2$

$-x + 3 < 2$

g) $-2z + w \leqslant 4$

$z - \dfrac{w}{2} \geqslant -2$

---

a) Yes, add 4 to both sides
b) No
c) Yes, subtract $3y$ from both sides; add 2 to both sides
d) Yes, multiply both sides by 4

e) No
f) No
g) Yes, divide both sides by -2.

As we saw in the quiz yourself exercise, our rules for manipulating inequalities can be applied to inequalities having more than one variable. For example, the inequality $8x + 5 - y \leqslant 0$ is equivalent to $8x + 5 \leqslant y$. Although we will not concern ourselves now with solving inequalities having more than one variable, this topic will be important to us in Chapter 6. At that time, we will introduce a graphical method for finding solutions for inequalities having two variables.

For the present, it will be sufficient to understand the basic rules we have introduced for handling inequalities in order to develop an acceptable procedure for apportioning seats on the Collegeville University faculty council.

EXERCISES

1. Students who desire practice working with equations should solve each of the following:

   a) $-4x - 5 = 5 + 2x$

   b) $\dfrac{4(6 + t)}{3} = 16$

   c) $\dfrac{2q + 4}{5} = q + 1$

   d) $4(3y + 3) + 5 = 1 + 6y$

   e) $\dfrac{5}{n + 2} = \dfrac{3}{n + 1}$

   f) $\dfrac{m + 3}{m - 1} = \dfrac{m - 2}{m + 4}$

2. State which of the given pairs of inequalities are equivalent.

   a) $5x - 1 \leqslant 6 - x$
      $6x \leqslant 7$

   b) $z - 1 < z + 1$
      $1 - z < -z - 1$

   c) $y - 3 \geqslant 2y - 5$
      $y \leqslant 2$

   d) $-\dfrac{w}{2} + 4 \leqslant \dfrac{w}{3}$
      $3w - 24 \leqslant -2w$

3. State which of the given pairs of inequalities are equivalent.

   a) $\dfrac{3y - 4}{-2} < 2y$
      $3y - 4 > -4y$

   b) $\dfrac{3m + 2}{-2} \geqslant \dfrac{m - 1}{-3}$
      $3(3m + 2) \leqslant 2(m - 1)$

   c) $\dfrac{3q}{4} - \dfrac{2r}{5} \geqslant 1$
      $15q - 8r \geqslant 1$

   d) $\dfrac{7z + 11 - q}{-3} \leqslant 6q - z$
      $7z + 11 - q \geqslant -18q + 3z$

In exercises 4 through 9 solve each of the given inequalities by writing a sequence of equivalent inequalities.

4. $2x + 1 \geqslant 4 - x$

5. $3w + 2 \leqslant 7 + 2w$

6. $\dfrac{3y}{2} < 5y + 1$

7. $\dfrac{q + 1}{4} \leqslant \dfrac{q}{3} - \dfrac{1}{2}$

8. $\dfrac{n - 7}{3} + \dfrac{n}{6} > \dfrac{1}{2} + n$

9. $\dfrac{3(x - 5)}{4} < \dfrac{2x - 5}{2}$

10. Solve $\dfrac{3}{x - 2} < 5$

11. Solve $\dfrac{1}{x - 2} \leqslant \dfrac{2}{x + 3}$

$x + 3 \leq 2(x - 2)$
$x + 3 \leq 2x - 4$
$3 \leq x - 4$
$7 \leq x$

AN APPORTIONMENT PRINCIPLE

We recall that an Alabama paradox can be avoided by assigning representatives one at a time. Moreover, we recall that the criterion for deciding which of two states should be given the next representative is the assignment which results in a smaller relative unfairness. The next example and the commentary which follows indicate an apportionment principle which meets this criterion.

■ Let us suppose that state A has a population of 13,680 and is entitled to exactly 5.7 representatives on a percentage basis. Also, assume that state B with a population of 6,240 deserves 2.6 representatives. If we assume that state A has already been given 5 representatives and state B has been given 2, let us determine whether state A or B is more deserving of one additional representative.

EXAMPLE

If state A received the additional representative, then it would have 6, which is slightly more than it deserves, and state B with 2 representatives would have less than it deserves. Therefore state B is more poorly represented in comparison to state A and the relative unfairness of this apportionment is

$$\frac{\left(\dfrac{\text{Population of } B}{2}\right) - \left(\dfrac{\text{Population of } A}{6}\right)}{\left(\dfrac{\text{Population of } A}{6}\right)} = \frac{3120 - 2280}{2280} = .3684.$$

Next we consider what would happen if state B received the additional representative instead of state A. Having three representatives would cause state B to be overly represented in comparison to state A. In this case, the relative unfairness of representation between states A and B is

$$\frac{\left(\dfrac{\text{Population of } A}{5}\right) - \left(\dfrac{\text{Population of } B}{3}\right)}{\left(\dfrac{\text{Population of } B}{3}\right)} = \frac{2736 - 2080}{2080} = .3154.$$

These computations tell us that the relative unfairness of the apportionment will be smaller if we give an additional representative to state B rather than to state A. Therefore, according to our criterion, state B should be assigned a third representative before state A should receive a sixth one. □

It is interesting to note that what we have done in the above example violates the Vinton method of apportionment, which would have given the extra representative to state A, since it has a larger fractional part of a representative.

Franklin D. Roosevelt signed into law the method that is presently being used to apportion the United States House of Representatives.

QUIZ YOURSELF* State *A*, with a population of 41,440, presently has 7 representatives. State *B*, with a population of 25,200, has 4 representatives. Determine the relative unfairness of apportionment given that one additional representative is allotted to state *A*. Then determine the relative unfairness of the apportionment if the extra representative is given instead to state *B*. Use your calculations to decide which state is more entitled to the additional representative.

*ANSWERS If state *A* is given one more representative, then the relative unfairness is $\frac{1120}{5180} = .2162$; Giving state *B* the extra representative results in a relative unfairness of $\frac{880}{5040} = .1746$. Thus the additional representative should be given to state *B*.

You have probably noticed that comparing the relative unfairness of the two assignments can be somewhat tedious. We will see shortly other numbers that can be used to determine more easily which of two states is more deserving of an additional representative. In order to see the form of these numbers, we will reconsider our last example in a slightly more abstract way.

Let us denote the populations of states *A* and *B* by *a* and *b*, respectively.

We saw that state B should receive its third representative in preference to state A receiving its sixth because

$$\frac{(a/5) - (b/3)}{(b/3)} < \frac{(b/2) - (a/6)}{(a/6)}. \tag{4.1}$$

Now

$$\frac{(a/5) - (b/3)}{(b/3)} = \frac{a}{5} \cdot \frac{3}{b} - 1,$$

and

$$\frac{(b/2) - (a/6)}{(a/6)} = \frac{b}{2} \cdot \frac{6}{a} - 1.$$

Therefore the inequality (4.1) can be rewritten as

$$\frac{a}{5} \cdot \frac{3}{b} - 1 < \frac{b}{2} \cdot \frac{6}{a} - 1,$$

which is equivalent to

$$\frac{a}{5} \cdot \frac{3}{b} < \frac{b}{2} \cdot \frac{6}{a}. \tag{4.2}$$

Since the letters a and b represent populations and are therefore positive, we see that $(a/6) \cdot (b/3)$ is also positive. Multiplying both sides of the inequality (4.2) by this number and simplifying gives us the equivalent inequality

$$\frac{a^2}{5 \cdot 6} < \frac{b^2}{2 \cdot 3}. \tag{4.3}$$

The point we should keep in mind is that the inequality (4.3) being true is equivalent to the relative unfairness being smaller when the additional representative is given to state B. Therefore to determine which of the two states A or B deserves the additional representative, we could compute for each state a number of the form

$$\frac{(\text{Population of the state})^2}{\left(\begin{array}{c}\text{Number of representatives} \\ \text{from the state}\end{array}\right) \cdot \left[\left(\begin{array}{c}\text{Number of representatives} \\ \text{from the state}\end{array}\right) + 1\right]},$$

and compare the two. The larger number determines which state should receive the additional representative. These observations for the states A and B motivate the following principle.

If states X and Y have already been allotted x and y representatives, respectively, then state X should be given an additional representative in preference to state Y provided that   **APPORTIONMENT PRINCIPLE**

$$\frac{(\text{population of } Y)^2}{y \cdot (y + 1)} < \frac{(\text{population of } X)^2}{x \cdot (x + 1)}.$$

Otherwise, state Y should be given the additional representative.

EXAMPLES   ■ According to the 1970 census, South Carolina had a population of approximately 2.6 million while Nebraska had a population of 1.5 million. South Carolina was given 6 representatives, while Nebraska was given 3. Let us suppose that the House of Representatives were to be increased in size so that either Nebraska or South Carolina could be given an extra representative. According to our apportionment principle, which state should receive this additional representative?

We compute the two numbers

$$\text{(South Carolina)} \qquad \frac{(2.6)^2}{6 \cdot 7} = .1610,$$

and

$$\text{(Nebraska)} \qquad \frac{(1.5)^2}{3 \cdot 4} = .1875.$$

Since $.1875 > .1610$, Nebraska is more deserving of the additional representative. □

■ Following the 1970 census, Connecticut, with a population of 3 million, received 6 representatives; Maryland, with a population of 4.1 million, received 8; Oklahoma, with a population of 2.6 million, received 6. If we assume that the House of Representatives were to be increased in size so that one of these three states could receive an additional representative, who should get it?

We compute the numbers

$$\text{(Connecticut)} \qquad \frac{3^2}{6 \cdot 7} = .2143,$$

$$\text{(Maryland)} \qquad \frac{(4.1)^2}{8 \cdot 9} = .2335,$$

$$\text{(Oklahoma)} \qquad \frac{(2.6)^2}{6 \cdot 7} = .1610.$$

Thus Maryland should be given this one extra representative, since .2335 is larger than either .1610 or .2143. □

The apportionment principle gives us a means for deciding which of two states deserves the next representative. As we saw in the last example, we can use the apportionment principle to determine which of three states should receive the next seat. In Part III of this chapter, we describe a convenient way to apply this apportionment principle to completely apportion the faculty council at Collegeville University. This procedure will therefore meet the apportionment criterion.

EXERCISES

1. According to the 1970 census, Colorado had a population of approximately 2.2 million, while Delaware had .6 million. Colorado was allotted 5 seats in the House of Representatives and Delaware was allotted 1. Suppose the House of Representatives were to increase in size so that either Colorado or Delaware would be given one more seat. Using the apportionment principle, determine which of these two states should receive the representative.

2. According to the 1970 census, Iowa had a population of approximately 2.8 million, while Utah had 1 million. Iowa was allotted 6 seats in the House and Utah was allotted 2. Suppose the House of Representatives were to increase in size so that one more seat would be given to either Iowa or Utah. Using the apportionment principle, determine which of these two states should receive that seat.

3. After the 1970 census, the populations of Alaska, New Hampshire, and Vermont were .3 million, .75 million, and .45 million, respectively. Alaska had 1 seat in the House of Representatives, New Hampshire had 2, and Vermont had 1. If the House of Representatives increased in size so that either of these states could receive one more representative, then which one should receive it according to the apportionment principle?

4. Consider the states of Alaska, New Hampshire, and Vermont, with their populations and number of representatives as given in exercise 3. If the House of Representatives increased in size so that two seats were to be distributed among these three states instead of one seat as in exercise 3, how should these two seats be given? (*Hint:* Do not assign both seats at the same time. Assign one seat and then use the apportionment principle to assign the other.)

5. After the 1970 census, Tennessee had a population of 4 million with 8 representatives; West Virginia had a population of 1.8 million with 4 representatives; and Georgia had a population of 4.6 million with 10 representatives. If two more seats were to be distributed among these three states, how should it be done according to the apportionment principle?

6. In apportioning the seats on a 10-member resident-hall council among three dorms, which we will call A, B, and C, we can use the following table of information.

Dorm	Number of students in this dorm	Percent of students	Exact number of representatives
A	235	23.5	2.35
B	333	33.3	3.33
C	432	43.2	4.32
Total	1000	100.0	10.00

Let us give two seats to A, three seats to B, and four seats to C, thus accounting for nine seats. Now determine which dorm should receive the tenth seat by (a) using the Vinton ("common-sense") method, and (b) using the apportionment principle.

7. In apportioning the seats on a 10-member resident-hall council among three dorms, which we will call E, F, and G, we can use the following table of information.

Dorm	Number of students in this dorm	Percent of students	Exact number of representatives
E	420	42	$4.\overline{2}$
F	440	44	4.4
G	140	14	1.4
Total	1000	100	10.0

Let us assign the first nine seats as follows: four to E, four to F, and one to G. Since the fractional parts are the same for dorm F and G, namely .4, it was decided to use the apportionment principle to determine which dorm would receive the tenth seat. Using the apportionment principle, which dorm would receive the tenth seat?

NEW TERMS

Absolute unfairness

Relative unfairness

Equivalent inequalities

Apportionment principle

MASTERY TEST: APPORTIONMENT AND TOOLS OF INEQUALITIES

1. Determine which pairs of inequalities are equivalent.
   a) $2x - 5 \leqslant 7$
      $x \leqslant 1$
   b) $-\frac{2}{3}y \geqslant 8$
      $y \leqslant -12$
   c) $\dfrac{3x - y}{6} \geqslant 5$
      $x \geqslant y + 10$
   d) $\dfrac{3}{2}x + \dfrac{y}{4} > 6$
      $y > 24 - 6x$

2. Solve the inequality
$$\tfrac{4}{3}x - \tfrac{1}{2} \leqslant -5.$$

3. According to the 1970 census, Idaho had a population of approximately .7 million and received 2 representatives, while Arizona with a population of 1.8 million received 4 representatives.
   a) Calculate the absolute unfairness for this apportionment.
   b) Compute the relative unfairness for this apportionment.
   c) Assume that an additional representative can be assigned to only one of these states. Use the apportionment principle to decide who should receive it.

4. According to the 1970 census, Alabama had a population of 3.5 million and received 7 representatives; Kansas had a population of 2.3 million and received 5 representatives; and New Mexico, with 1 million people, received 2 representatives. Assume that an additional representative can be assigned to one of these states. Use the apportionment principle to determine which state should receive it.

III. APPORTIONING THE COLLEGEVILLE UNIVERSITY FACULTY COUNCIL

In this part we will use the apportionment principle to assign all the representatives to the Collegeville University Faculty Council.

Recall that Collegeville University is divided into three divisions—humanities, science, and education. There are 103 faculty members in the humanities division, 63 in the science division, and 34 in the education division. As we agreed earlier, we will assign the representatives one at a time until all 20 seats on the council have been filled. It would seem unjust if any division had no representatives, so let us begin by assigning one seat to each division. This is consistent with a provision in the United States constitution which says each state must have at least one representative in the House. Now that 3 seats have been filled, we will use our apportionment principle to assign the remaining 17 seats.

Our apportionment will proceed as follows: Each time a representative is assigned to a division, we will calculate the number

$$\frac{(\text{Number of faculty in the division})^2}{\left(\begin{array}{c}\text{Number of representatives}\\ \text{division now has}\end{array}\right) \cdot \left[\left(\begin{array}{c}\text{Number of representatives}\\ \text{division now has}\end{array}\right) + 1\right]},$$

for all three divisions. The largest of these three numbers will determine which division is most deserving of the next representative. After this assignment has been made, we will recompute the above expressions for each of the three divisions and then assign another representative; we shall continue in this manner until all the representatives have been assigned.

At present each division has one representative, so we compute the numbers:

For humanities $\quad$ For science $\quad$ For education

$$\frac{(103)^2}{1 \times 2} = 5304.5 \qquad \frac{(63)^2}{1 \times 2} = 1984.5 \qquad \frac{(34)^2}{1 \times 2} = 578.$$

From these calculations we see that the humanities division should be given their second representative in preference to the science and education divisions.

We begin to summarize our apportionment in the following table.

Seat number	Goes to	Number of representatives H has	Number of representatives S has	Number representatives E has
1	H	1	0	0
2	S	1	1	0
3	E	1	1	1
4	H	2	1	1
5	?			

To determine which division should get the fifth seat, we compute the following numbers and compare them.

For humanities	For science	For education
$\dfrac{(103)^2}{2 \times 3} = 1768.2$	$\dfrac{(63)^2}{1 \times 2} = 1984.5$	$\dfrac{(34)^2}{1 \times 2} = 578.$

The above calculations show that the science division is most deserving of the fifth seat on the council. With this assignment made, we can add the following line to our table.

$$5 \quad S \quad 2 \quad 2 \quad 1$$
$$6 \quad ?$$

Now, to determine who gets the sixth seat, we need to compare again three numbers of similar form; this type of comparison will be repeated at each stage of the apportionment. In Table 4.1, we have displayed all the numbers of the desired form which will be needed to complete the apportionment.

We now calculate all the entries in Table 4.1 and give them in Table 4.2.

Since the first five seats have been assigned, let us assign the remaining seats by using Table 4.2. In determining which division should receive the sixth seat, recall that the humanities and science divisions each already have two and the education division has one. We therefore compare the circled entries of Table 4.2 to determine whether the humanities division is more deserving of its third representative than the science division is of its third or the education division is of its second. Since 1768.2 is greater than either 661.5 or 578, the humanities division should be given the next seat in preference to the other divisions.

We can thus add to the table the following line, which summarizes how the seats are being given.

Seat number	Goes to	Number of representatives H has	Number of representatives S has	Number of representatives E has
6	H	3	2	1
7	?			

In order to determine the assignment of the next representative, we again consult Table 4.2. To determine whether the humanities division is more deserving of its fourth representative than the science division its third or the education division its second, we look at row 4 under H, row 3 under S, and row 2 under E. Since 884.1 is larger than either 661.5 or 578, the humanities

Table 4.1

If given the next representative, the number a division would have is	H	S	E
2	$\dfrac{(103)^2}{1 \times 2}$	$\dfrac{(63)^2}{1 \times 2}$	$\dfrac{(34)^2}{1 \times 2}$
3	$\dfrac{(103)^2}{2 \times 3}$	$\dfrac{(63)^2}{2 \times 3}$	$\dfrac{(34)^2}{2 \times 3}$
4	$\dfrac{(103)^2}{3 \times 4}$	$\dfrac{(63)^2}{3 \times 4}$	$\dfrac{(34)^2}{3 \times 4}$
5	$\dfrac{(103)^2}{4 \times 5}$	$\dfrac{(63)^2}{4 \times 5}$	$\dfrac{(34)^2}{4 \times 5}$
6	$\dfrac{(103)^2}{5 \times 6}$	$\dfrac{(63)^2}{5 \times 6}$	$\dfrac{(34)^2}{5 \times 6}$
7	$\dfrac{(103)^2}{6 \times 7}$	$\dfrac{(63)^2}{6 \times 7}$	*
8	$\dfrac{(103)^2}{7 \times 8}$	$\dfrac{(63)^2}{7 \times 8}$	*
9	$\dfrac{(103)^2}{8 \times 9}$	$\dfrac{(63)^2}{8 \times 9}$	*
10	$\dfrac{(103)^2}{9 \times 10}$	*	*
11	$\dfrac{(103)^2}{10 \times 11}$	*	*
12	$\dfrac{(103)^2}{11 \times 12}$	*	*

Table 4.2

If given the next representative, the number a division would have is	H	S	E
2	5304.5	1984.5	578
3	1768.2	661.5	192.7
4	884.1	330.8	96.3
5	530.5	198.5	57.8
6	353.6	132.3	38.5
7	252.6	94.5	
8	189.4	70.9	
9	147.3	55.1	
10	117.9		
11	96.4		
12	80.4		

(handwritten)

```
        H   S   E
        |   |   |
        |   \   |
        |   |
        |
```

```
1 - H
2 - S
3 - E
4 - H
5 - S
6 - H
7 - H
8 - S
9 - E
10 - H
```

* We are not interested in computing these entries in the table since it is unlikely that the science division will be allotted ten or more representatives. Similarly the education division will probably not receive more than six representatives. However, if our intuition is wrong, we can easily calculate the necessary entries.

division deserves its fourth representative before the other two divisions receive another representative. Therefore the seventh seat on the council goes to the humanities division.

We can now add another line to our apportionment table.

7	H	4	2	1
8	?			

Continuing in this manner, we can assign all 20 members to the faculty council as shown in Table 4.3.

Table 4.3

Seat number	Goes to	Number of representatives H has	Number of representatives S has	Number of representatives E has
1	H	1	0	0
2	S	1	1	0
3	E	1	1	1
4	H	2	1	1
5	S	2	2	1
6	H	3	2	1
7	H	4	2	1
8	S	4	3	1
9	E	4	3	2
10	H	5	3	2
11	H	6	3	2
12	S	6	4	2
13	H	7	4	2
14	S	7	5	2
15	E	7	5	3
16	H	8	5	3
17	H	9	5	3
18	S	9	6	3
19	H	10	6	3
20	H	11	6	3
21	E	11	6	4

We have shown that in a 20-member council, the humanities division should receive 11 representatives, science division 6, and the education division 3. In a 21-member council, the education division should be given one additional representative, while the other divisions remain the same.

The procedure used in this example satisfies two important criteria. First, it prevents the Alabama paradox from occurring; and second, using Table 4.2 to determine which division is to receive the next seat, the assignment can be made so that the relative unfairness between any two divisions is minimal. Furthermore, it can be formally shown that the relative unfairness between any two states cannot be improved, even by transferring a representative from one state to another when the apportionment is done in the manner described.*

---

* E. V. Huntington, "The Apportionment of Representatives in Congress." *Trans. Am. math. Soc.* **30** (1928), pp. 85–110.

THOMAS JEFFERSON.

*"No invasions of the Constitution
are fundamentally so dangerous
as the tricks played on their own numbers."*
—THOMAS JEFFERSON.

However, there is one important criticism which can be raised concerning the described apportionment system; it *does not* necessarily guarantee that a given state will be apportioned its full quota of representatives. If a state's quota is, say 21.4 representatives, there is no guarantee that the state will be given either 21 or 22 representatives. In fact, a recent paper* gives a hypothetical example, in which Pennsylvania's quota would be exactly 24.974 representatives; yet using the method described in this chapter, it would be apportioned 26 representatives. In the same example, California's quota would be exactly 42.960 representatives, but it would be allotted 45. In spite of this defect, the method which we have explained is currently being used to apportion the United States House of Representatives.

---

*   M. L. Balinski and H. P. Young, "The Quota Method of Apportionment." *Am. math. Mon.* **82** (1975), pp. 701–730.

EXERCISES

1. Ten seats on a resident hall council are to be apportioned to dormitories A, B, and C. There are 235 students in A, 333 in B, and 432 in C. The following table was obtained in a manner similar to the one used for Table 4.2 in our Collegeville University example.

If given the next representative, the number of seats a dorm would have is	A	B	C
2	27,612.5	55,444.5	93,312
3	9,204.2	18,481.5	31,104
4	4,602.1	9,240.8	15,552
5	2,761.3	5,544.5	9,331.2
6	1,840.8	3,696.3	6,220.8
7		2,640.2	4,443.4
8			3,332.6

Use the above table to determine the apportionment of the 10 seats on the council. Begin with each dorm being allotted one representative and then assign the remaining 7 seats.

2. Refer to exercise 1. Suppose the resident-hall council was increased to 15 representatives. How should these 5 new seats be distributed? On this 15-member council, how many representatives would each dorm have?

3. Refer to exercise 1. What is the lowest number of seats we could have on the resident-hall council which would guarantee B 6 of them?

4. At Collegeville University, graduate fellowships are apportioned among the three divisions (humanities, science, and education) based upon the number of full-time graduate students in the divisions. The humanities division has 30 full-time graduate students, the science division has 40, and the education division has 10. Eight fellowships are to be given and each division is to be allotted at least one fellowship. Give a table similar to Table 4.2. Using this table, apportion the remaining five fellowships among the three divisions.

5. Redo exercise 4. However, assume now that the humanities division has 20 graduate students, the science division 16, and the education division 14.

6. Use the method described in this part to apportion 10 representatives among 3 states X, Y, and Z. Assume that X has a population of 3 million, Y 4 million, and Z 5 million. Begin with each state being given one representative.

SUGGESTED READINGS

BALINSKI, M. L., AND H. P. YOUNG, "The Quota Method of Apportionment." *Am. math. Mon.*, Aug.–Sept. 1975, pp. 701–730.

The first part of the paper contains a good historical account of the apportionment of the United States House of Representatives as well as a criticism of the current procedure. A new method is introduced and then discussed at a somewhat sophisticated mathematical level.

HUNTINGTON, E. V., "The Apportionment of Representatives in Congress." *Trans. Am. math. Soc.*, 1928, pp. 85–110.

This is an extensive mathematical treatment of apportionment. The author thoroughly discusses the procedure presently used (as presented in this chapter) and compares it to many alternate apportionment procedures.

O'ROURKE, T. B., "Reapportionment—Law, Politics, and Computers." *American Enterprise Institute for Public Research*, 1972.

An introduction to the legal and political aspects of reapportionment. The last chapter contains a discussion on the use of computer models in redistricting.

SILVA, R. C., "Reapportionment and Redistricting." *Sci. Am.*, Nov. 1965, pp. 20–27.

A general discussion on the topic with some supreme court rulings on the one-man, one-vote concept.

René Descartes, one of the most famous of the French mathematicians and philosophers, was born in La Haye, France in 1596. As a child, his health was not good so his early teachers allowed him the luxury of remaining in bed as late as he pleased. This became a lifelong habit to which he attributed his later success in philosophy and mathematics.

While still a student at the Jesuit college at La Flèche, he began to doubt whether his studies were leading him to true knowledge. His doubts prompted him to begin a search for the means by which man could learn truth. In 1637 this concern led him to publish one of his most famous works called *Discourse on Method*, in which he outlined the following basic principles.

1. Accept nothing as true which you do not fully understand.

2. Divide a complicated problem into several parts.

3. Arrange thoughts in order, beginning with the simplest and proceeding to the complex.

4. Review all calculations and arguments so thoroughly that nothing is omitted.

It is interesting to note that after 300 years these principles remain valid. Descartes had such faith in his method that he wrote,

> Those long chains of reasoning, each step simple and easy, which geometers are wont to employ in arriving even at the most difficult of

117

Rene Descartes (1596–1650)

their demonstrations, have led me to surmise that all things we human beings are competent to know are interconnected in the same manner, and that none are so remote as to be beyond our reach or so hidden that we cannot discover them . . .

In one of the appendices to his *Discourse on Method*, Descartes developed what is today called *analytical geometry*, in which objects such as points, lines, circles, and planes are represented as numbers and equations. The advantage that analytical geometry has over the traditional geometry developed by the ancient Greeks is that methods of algebra (solving equations, etc.) can be used to solve geometric problems. This new geometry became so important that philosopher John Stuart Mill wrote, ". . . it immortalized the name of Descartes and constitutes the greatest single step ever made in the progress of the exact sciences." Descartes would probably have been pleased to know his geometry is taught to virtually all students of mathematics and science.

In addition to mathematics and philosophy, René Descartes made important contributions to many other fields of intellectual endeavor. His contributions to biology were so important that some have referred to him as the "father of modern biology." He also made significant discoveries in several areas of physics, particularly in the optics field. He did research in chemistry, anatomy, embryology, medicine, astronomy, and meteorology. He is credited with being the first person to satisfactorily explain the rainbow.

Unfortunately, this genius met with an untimely end while serving as a tutor to Queen Christine of Sweden. Christine insisted on rising at 5 o'clock in the morning to study philosophy with Descartes. The early hour and the severe Swedish winter were too much for him. He caught pneumonia and died on February 11, 1650.

Descartes's belief in the power of human reasoning to obtain all knowledge had often brought him in direct conflict with the church. As a result, when his remains were eventually returned to France, the church opposed any public oration on his behalf. This led the famous mathematician Jacobi to remark, "It is often more convenient to possess the ashes of great men than to possess the men themselves during their lifetime."

I. A FISH MANAGEMENT PROBLEM: WHAT WENT WRONG AT BASS LAKE?

The State Fish and Game Commission is planning to stock the newly created Bass Lake in Fisherman's Haven State Park. The chief of the stocking operation is discussing this matter with his assistant Coswell.

"From the figures I've seen, we can expect to take about 10,000 bass out of there each year. What do you think, Coswell?"

"Yes sir. 10,000 sounds about right to me."

"You said a pair of adult bass will produce about 20 adults in a year? I thought they'd do better than that."

"Well sir. It's the cannibalism. We could expect more bass if we provide them with some other fish to eat."

"O.K. I did read somewhere that we should also stock the lake with a forage fish.* What about the golden shiner?"

"It is possible, chief. Shiners will produce small fry for the bass fingerlings to feed on and help them to grow bigger. The adult bass will also eat the small fry and therefore reduce the cannibalism. Now, I would estimate that we could expect eight additional bass for each pair of shiners."

"O.K. We'll put shiners in the lake. How many survive to adulthood?"

"Not many, I figure that for each shiner in the lake one year, there would be 3 adults the next year, if they were not preyed upon by the bass. However, it is estimated that for each 22 bass in the lake, the number of shiners surviving the next year is reduced by 1."

---

* H. S. Davis, *Culture and Diseases of Game Fishes.* Berkley: University of California Press, 1965, p. 145.

"This is very confusing, Coswell. Let's see if I can make some sense out of this. You say that we want to take 10,000 bass out of the lake each year so we have to put 1000 in at the beginning of the year. If we throw in some shiners, this will increase the number of bass that grow to adulthood, so we can take out 10,000 and still have some left to reproduce for next year. How does that sound?"

"I don't believe that would give a good balance of bass and shiners, sir. Perhaps we should look at the figures more caref . . ."

"Coswell, I don't have time for all these figures. We'll put in 1200 bass and 50 shiners. That will give us over 12,000 bass the next year."

"But chief, I think . . ."

And so in accordance with the chief's orders, Bass Lake is stocked with 1200 bass and 50 shiners. At the end of the year Coswell and the chief are discussing the output of Bass Lake.

"Well, Coswell, we had our 10,000 bass just as I said we would."

"Sir. I still think that we should look more carefully at the figures. I . . ."

"Figures! Coswell, you never learn. The proof is in the fishing."

Fishing continued to be good at Bass Lake during the second and third years; however, during the fourth year the anglers began complaining that they cannot catch any fish large enough to keep. The chief calls Coswell to his office.

"Coswell, things aren't going well at Bass Lake. I think we should look at those figures again."

"Yes sir. I agree. But I think we should continue this discussion in Part III of this chapter—after you've learned a little about solving systems of linear equations."

In analytic geometry lines and points can be represented algebraically. (Photo by Ellis Herwig, Stock Boston.)

II. TOOLS OF LINEAR EQUATIONS

LINEAR RELATIONS

We will now study an important relation that can exist between numbers. You should be familiar with mathematical relations such as "is equal to," "is greater than," and "is one less than." We must also keep in mind that each mathematical relation determines a set of pairs. For example, consider the relation "is greater than" when applied to the set {1,2,3}. Since 2 is greater than 1, we pair 2 with 1. Of course 1 is not greater than 2, so we cannot pair 1 with 2. We therefore realize that it is important to specify the order in which the numbers are paired. Thus a pair determined by a relation shall always be written as an **ordered pair**. For example, by writing (2,1) we are indicating that there are a **first component**, namely 2, and a **second component**, which is 1. We emphasize that the ordered pair (x,y) is not necessarily the same as the ordered pair (y,x).

We have seen that it is possible to start with a set of numbers and a statement on how to relate these numbers, and obtain a set of ordered pairs. The statement "is greater than" when applied to the set {1,2,3} gives us the set of ordered pairs {(2,1),(3,1),(3,2)}. Thus a relation can be considered as a set of ordered pairs.

We should recall that a set can be given by set builder notation. When a set of ordered pairs is given using this method, the open sentence will generally have two variables. It is common practice to use x as the variable name for the first component of the ordered pair and y for the second component. An example of a set of ordered pairs described in this manner is $\{(x,y): y = 2x\}$. It should be noted that we are not restricted in using x and y as the variables; for convenience, other variable names could be used.

EXAMPLE   ■ Let us write the relation which describes that the interest paid on a savings account is 7 percent of the principal.

We shall let I represent interest and P represent principal, where, of course, the principal cannot be negative. Our relation is $\{(P,I): I = 0.07P$ and $P \geq 0\}$. □

When a set is given by set builder notation, we know that an element will belong to the set when it satisfies the open sentence. This fact still holds when considering a relation. For example, consider the relation on the real numbers $\{(x,y): 2x + 3y = 2\}$. The ordered pair $(\frac{1}{2},\frac{1}{3})$ belongs to the relation whereas $(1,1)$ does not.

QUIZ YOURSELF*   For the relation $\{(x,y): 3x + 2y = 2\}$ on the real numbers, complete the following table so that the pairs belong to the relation.

x	$\frac{1}{3}$	1		2			$\frac{2}{3}$
y			1		$-\frac{1}{2}$	$-\frac{1}{3}$	

***ANSWERS**

x	$\frac{1}{3}$	1	0	2	1	$\frac{8}{9}$	$\frac{2}{3}$
y	$\frac{1}{2}$	$-\frac{1}{2}$	1	-2	$-\frac{1}{2}$	$-\frac{1}{3}$	0

When considering an equation in two variables, a solution is any ordered pair which satisfies the equation. For example, $(0,1)$ is a solution to $3x + 2y = 2$, whereas $(2,1)$ is not. The set of all solutions to an equation is called the **solution set**. Therefore we see that a relation determined by an equation is actually the solution set of the equation.

Of the many different types of relations that can exist on the real numbers, we wish to focus our attention on the linear relation. When we discuss graphing, we will see the natural property which must exist in order that two quantities be related in a linear fashion. For now we give the following definition.

DEFINITION   **An equation of the form $Ax + By = C$, where A, B, and C are constants and where A and B are not both zero, is called LINEAR. A LINEAR RELATION on the real numbers is the solution set of a linear equation.**

Some examples of linear equations are $2x + 6y = 5$, $\frac{1}{2}x - y = \frac{1}{7}$, $x = -2$, and $y = \frac{1}{2}$. The equation $y = 3 - 2x$ is also a linear equation since it is equivalent to $2x + y = 3$. However, the equation $x^2 + y = 5$ is not a linear equation because of the x^2 appearing in the equation.

When working with linear equations, there are many equivalent forms. We should keep in mind the properties of equalities: adding or subtracting the same number to both sides and multiplying or dividing both sides by the same nonzero number enable us to obtain an equivalent equation. We can also formally say that two equations are **equivalent** if they determine the same solution set. For example, $\frac{1}{2}x + \frac{1}{3}y = 1$ is equivalent to $3x + 2y = 6$, since we can multiply the first equation by 6 to obtain the second one. On the other hand, the two linear equations $x + y = 2$ and $x - y = 2$ are not equivalent since their solution sets are different. In particular, the ordered pair $(1,1)$ is a solution of the first equation but not the second.

Using our definition, it is quite easy to give an example of a linear relation; we need only make the defining property of the set a linear equation. Each of the sets $\{(x,y): 2x - 3y = 4\}$, $\{(x,y): y = 3 - 2x\}$, and $\{(x,y): y = 2\}$ is an example of a linear relation on the real numbers.

As we will see in the examples at the end of this section, we may sometimes wish to restrict the size of the numbers that can be substituted for one of the variables. For example, in the linear relation $\{(x,y): y = x + 1\}$, let us restrict x to be greater than or equal to 0. The linear relation with this restriction can easily be expressed by writing $\{(x,y): y = x + 1 \text{ and } x \geqslant 0\}$. We shall also call this set a linear relation on the real numbers. The linear relation $\{(x,y): y = -x + \frac{1}{2} \text{ and } -1 \leqslant x \leqslant 1\}$ has the restriction that x cannot be smaller than -1 nor larger than 1.

Decide whether each of the following statements is true or false. You should be prepared to give an explanation for your answer.   **QUIZ YOURSELF***

1. The two equations $2x + 3y = 10$ and $3y + 2x = 10$ are equivalent.
2. $\{(x,y): y = \frac{1}{3}x + 6\} = \{(x,y): 3y = x + 6\}$.
3. The equations $y = \frac{1}{3}x + 6$ and $3y = x + 18$ are equivalent.
4. $\{(x,y): x + y^2 = 2\}$ is not a linear relation.
5. $\{(x,y): y - 2x = 0 \text{ and } x \geqslant 1\}$ is a linear relation.
6. The equations $3x + 2y = 10$ and $2x + 3y = 10$ are equivalent.
7. The equations $2y - 5x = 6$ and $y = \frac{5}{2}x + 3$ are equivalent.

---

1. true   2. false   3. true   4. true   5. true   6. false   7. true   ***ANSWERS**

There are many situations in which we can use a linear relation as a model. However, to give the appropriate relation we must be able to write the correct linear equation, with any restrictions, which determines the relation. One thing we should keep in mind is that a mathematical statement is not just a collection of symbols; it has meaning and can be read as any English sentence. Therefore, writing a linear equation is sometimes simply a problem of translating an English sentence into a statement using mathematical symbols.

EXAMPLE   ■ Let us write the linear equation which expresses that a number is one less than twice another number.

We can represent the two numbers by x and y. One less than twice x can be written as $2x - 1$. Since y must be the same as this quantity, we can give the equation $y = 2x - 1$. □

QUIZ YOURSELF*   Express each of the following statements with a linear equation.

1. The sum of two numbers is 5.

2. Two numbers are the same.

3. A number is one more than half another number.

4. The sum of one-half of a number and one-third of another number is 10.

---

*ANSWERS   1. $x + y = 5$ 2. $x = y$ 3. $y = \frac{1}{2}x + 1$ 4. $\frac{1}{2}x + \frac{1}{3}y = 10$

EXAMPLES   ■ For some taxi companies, the fare is determined on a fixed price plus a certain amount per half mile. For example, suppose a fare is 50¢ plus 15¢ for each half mile. If we let y be the fare and x be the number of half miles traveled, then the linear equation which describes this situation is $y = 15x + 50$. □

■ A graduated income tax uses varying tax rates depending on the bracket in which the taxable income falls. For example, if the taxable income is between \$8000 and \$12,000, the tax is \$1380 plus 22% of the amount over \$8000. If the taxable income is between \$12,000 and \$16,000, then the tax is \$2260 plus 25% of the amount over \$12,000.

If we let x represent the taxable income and y the tax, we can model each income bracket with a linear relation. In particular, the two linear relations are

$$\{(x,y): y = 1380 + 0.22(x - 8000) \text{ and } 8000 \leqslant x \leqslant 12,000\};$$
$$\{(x,y): y = 2260 + 0.25(x - 12,000) \text{ and } 12,000 \leqslant x \leqslant 16,000\}. \ \square$$

■ A negative income tax is based upon the annual income being above or below a fixed amount. Suppose that a person will receive money from the government if his or her annual income is below \$7500; this is a negative tax. If, on the other hand, the annual income is above \$7500, the person will pay money to the government; this is a positive tax. Furthermore, let us suppose that the tax rate is 20% of the amount of income which is above or below the $7500 when the income is between $5000 and $10,000.

   If we let y represent the tax, negative or positive, and x the annual income, our linear relation which models this tax situation is $\{(x,y): y = 0.20(x - 7500)$ and $5000 \leqslant x \leqslant 10,000\}$. We should observe that when x is less than 7500, $x - 7500$ is a negative number and therefore y is a negative amount. However, if x is greater than 7500, then $x - 7500$ is positive and thus y is positive. $\square$

EXERCISES

1. Complete the table so that the pairs belong to the relation $\{(x,y): y = 3x - 4\}$.

x	$\frac{1}{3}$	$\frac{1}{3}$	3	-1	2	$\frac{4}{3}$
y	-3	-4	5	-7	2	0

2. Complete the following table so that the ordered pairs satisfy the linear equation $5x - 2y = 10$.

x	2	0	$\frac{4}{5}$	3	3	$\frac{2}{5}$
y	0	5	-3	$\frac{5}{2}$	$\frac{5}{2}$	

3. For each pair of linear equations, state whether or not they are equivalent. Also, be prepared to justify your answer.
   a) $x + \frac{2}{3}y = -1$ and $3x + 2y = -3$
   b) $y = \frac{2}{5}x - 1$ and $-2x + 5y = 5$
   c) $\frac{1}{3}x - \frac{1}{4}y = \frac{1}{6}$ and $4x - 3y = 2$
   d) $x - y = \frac{1}{2}$ and $-2x + 2y = -1$

4. For each of the following linear equations, give another equivalent equation which contains no fractions.
   a) $\frac{1}{2}x + y = \frac{1}{4}$ b) $\frac{1}{3}x - \frac{1}{4}y = 1$ c) $-\frac{2}{5}x + \frac{3}{4}y = \frac{1}{2}$

5. State whether or not the following pairs belong to the relation $\{(x,y): 4x + 3y = 12$ and $-2 < x \leqslant 2\}$.
   a) $(0,4)$ b) $(3,0)$
   c) $(-\frac{3}{4},5)$ d) $(-2,\frac{20}{3})$
   e) $(2,\frac{4}{3})$ f) $(1,3)$

6. Express each of the following statements with a linear equation.
   a) One number is one-third of another.
   b) The difference of two numbers is ten.
   c) One number is five less than twice another.
   d) Twice one number plus three times another is five.

7. Model each of the following statements with a linear relation, making sure to include appropriate restrictions.
   a) The interest due on a note is 8% of the principal.
   b) The new balance is the previous balance increased by 7%.
   c) On income over $1000, but not over $1500, the tax is $145 plus 16% of the excess over $1000.
   d) Fahrenheit degrees are 32 more than nine-fifths of the corresponding Celsius degrees.

8. Model each of the following statements with a linear relation.
   a) Each unit of raw potatoes contains 17 grams of carbohydrates.
   b) The sale price on each item is the list price discounted by 10%.
   c) For a certain species of snake, the total length is approximately $7\frac{1}{4}$ times the length of the tail provided that the tail length is at least 30 mm.
   d) New autos which are priced between $3000 and $5000 depreciate in value during the first year by 22% of their original price.

9. In producing a certain commodity, there is usually a **fixed cost** which is independent of the number of units produced. In addition to the fixed cost, there is a **unit cost**—the cost for each unit produced. If the fixed cost is $1000 and the unit cost is $20, then express the relation between the total cost of production and the number of units produced by a linear equation. Using this equation, determine the total production cost for 50 units; for 100 units.

10. Refer to exercise 9. Suppose that the fixed cost is $750 and the unit cost is $15 for the first 100 units and $12 for each unit thereafter. Model the total production cost using a pair of linear relations. Determine the total production cost for 50 units; for 150 units.

11. **Depreciation** is a loss in value through use. One method of obtaining the remaining value of an item is by **straight-line** depreciation. The useful life of an item is estimated and equal amounts of the original value are deducted each year of the item's useful life; the remaining value at the end of the useful life is 0.
    Suppose an item originally costs $350 and has a useful life of 5 years. Using the straight-line depreciation method, express the remaining value and the number of years of useful life with a linear equation. Determine the remaining value at the end of 1 year; at the end of $3\frac{1}{2}$ years; and at the end of 4 years.

12. For taxable income between $16,000 and $20,000, the tax is $3260 plus 28% of the excess over $16,000. Model this tax bracket with a linear relation.

GRAPH OF A LINEAR RELATION

Before discussing the graph of a linear relation, let us first review how to display a set of ordered pairs as points in a plane.

First, we draw two perpendicular lines in a plane and identify the point of intersection, which is called the **origin**, with the ordered pair (0,0). Next, we choose a unit length for each line; usually, the same unit length is used for both lines. Now to locate a unique point in this plane which corresponds to a particular ordered pair, we recall that the *first* component of the pair gives the direction and distance to move *horizontally* from the origin, and the *second* component indicates the direction and distance to move *vertically*. Furthermore, the *positive horizontal* direction is to the *right* while the *positive vertical* direction is *up*. The plane with these two lines is called a **Cartesian plane**.

The Cartesian plane is one of man's most important and useful mathematical inventions. (Photo by Fredrik D. Bodin, Stock Boston.)

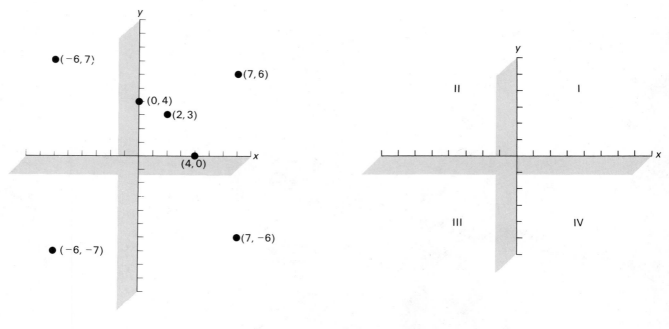

Figure 5.1   **Figure 5.2**

In Fig. 5.1, we graphed and labeled a sample of ordered pairs.

A Cartesian plane is a good way to display ordered pairs because there is a **one-to-one correspondence** between ordered pairs and points; that is, for each ordered pair of real numbers, there is a unique point in the Cartesian plane and only one ordered pair corresponds with each point. Since this correspondence exists, we can use the terms ordered pair and point interchangeably. For example, there is no confusion if we call (2,3) a point.

The pair of numbers associated with a specified point in a Cartesian plane is called the **coordinates** of the point. Also, a convenient way to indicate the coordinates (a,b) of a point P is to write $P(a,b)$.

It is common to call the horizontal number line in a Cartesian plane the *x*-**axis** and the vertical number line the *y*-**axis**. The two number lines, the x and y axes, divide the plane into four regions called **quadrants**. These quadrants are labeled I, II, III, and IV, starting with the upper right-hand region and going counterclockwise as shown in Fig. 5.2.

The following observations should be made: The coordinates of any point in quadrant I are both positive. The coordinates of a point in the third quadrant are both negative. A point with a negative first coordinate and a positive second coordinate lies in quadrant II, whereas a point with a positive first coordinate and a negative second coordinate is in the fourth quadrant.

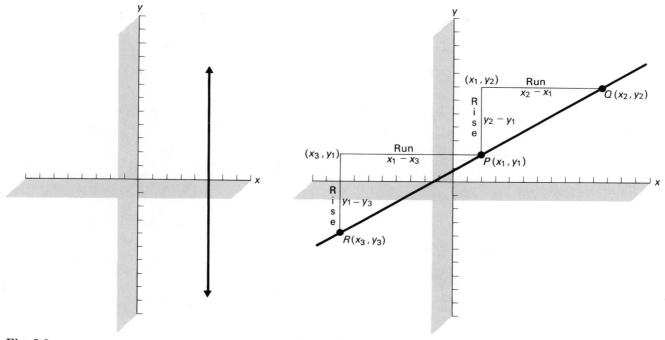

Fig. 5.3
Graph of $\{(x,y): x = 5\}$

Figure 5.4

We should also observe that for any point on the x-axis the second coordinate is zero. Furthermore, the first coordinate of each point on the y-axis is zero.

We shall begin our discussion of graphing a linear equation with those determined by an equation of the form $x = M$, where M is a constant. For example, $x = 5$ is an equation of this type. A point is in the graph of the relation $\{(x,y): x = 5\}$ if the first coordinate is 5. We recognize that the points in the plane for which x is 5 are on a line parallel to the y-axis. Thus the graph of this relation is the vertical line 5 units to the right of the y-axis. (The graph is given in Fig. 5.3.) In general, we realize that the graph of a linear relation $\{(x,y): x = M\}$, where M is a constant, is a line parallel to the y-axis—that is, a vertical line.

Before demonstrating that the graph of a linear relation $\{(x,y): Ax + By = C\}$, where $B \neq 0$ is also a line, let us first consider the concept of slope. Suppose $P(x_1, y_1)$, $Q(x_2, y_2)$, and $R(x_3, y_3)$ are three points on a nonvertical line. We now draw two right triangles as shown in Fig. 5.4. Using a fact about similar triangles, we know that the ratio of the lengths of the legs is the same for each triangle. In particular,

$$\frac{y_1 - y_3}{x_1 - x_3} = \frac{y_2 - y_1}{x_2 - x_1}.$$

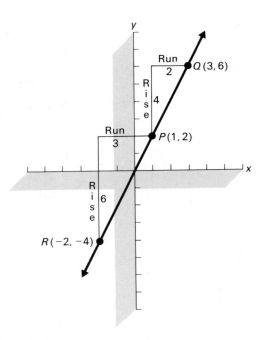

Figure 5.5

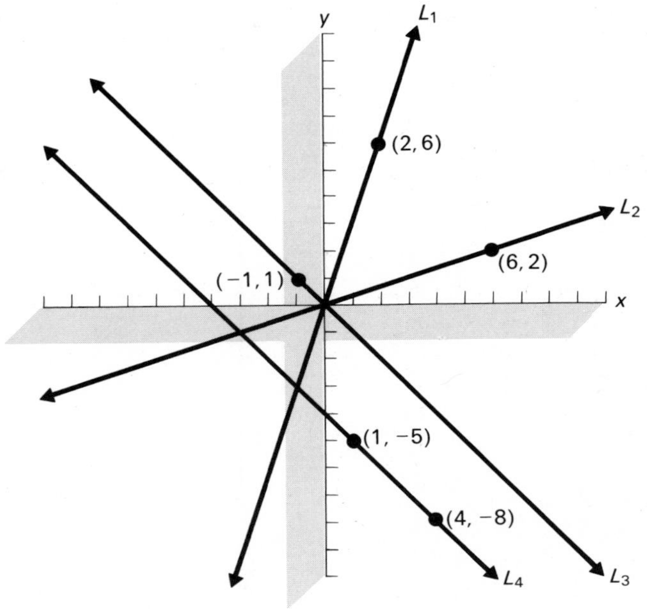

Figure 5.6

We can also express this fact in the following way. When moving from point R to point P, the **ratio** of the **rise** over the **run** is the same as when going from point P to point Q.

It is easy to verify with an example that our discussion is correct. The three points $P(1,2)$, $Q(3,6)$, and $R(-2,-4)$ are on the same line, as shown in Fig. 5.5. Using the coordinates of points P and R, we find that the ratio of the rise over the run is $\frac{6}{3}$. The ratio of the rise over the run when moving from point P to point Q is $\frac{4}{2}$. We see that each of these ratios is equal to 2, and therefore they are the same.

From the above discussion we realize that no matter which two points we use from the same nonvertical line, we will always obtain the same ratio of the rise over the run. We call this ratio the slope of the line.

DEFINITION

Let $P(x_1, y_1)$ and $Q(x_2, y_2)$ be any two points on a nonvertical line. The SLOPE of the line is the ratio of the rise over the run obtained when going from one of the points to the other. In particular, the slope of the line is the number

$$\frac{y_2 - y_1}{x_2 - x_1}.$$

Generally speaking, the slope of a line tells us how "steep" the line is. (Photo by Frank Siteman, Stock Boston.)

We see that if point P and point Q lie on the same vertical line, then $x_1 = x_2$. In this case, the denominator of

$$\frac{y_2 - y_1}{x_2 - x_1}$$

is zero and this quotient does not exist. We therefore do not define a slope for a vertical line.

When obtaining the slope of a line, we should keep several things in mind.

1) The slope is *not defined* for a vertical line.
2) For a nonvertical line, it does not matter which two points we use to compute the slope.
3) The easiest formula to remember and use for slope is RISE/RUN

Determine the slope of each line given in Fig. 5.6.

L_1: 3, L_2: 1/3, L_3: -1, L_4: -1

*ANSWERS

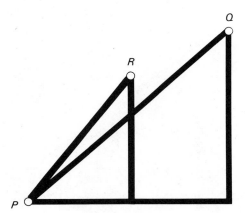

Figure 5.7

Thus far we have demonstrated that if three points P, Q, and R are on the same nonvertical line, then the slope determined from point P to point Q is the same as that determined from point P to point R. We would now like to assure ourselves that the opposite is also true; that is, if these slopes are equal, then the three points are on the same line.

Consider the drawing shown in Fig. 5.7. We see that if point R is not on the line joining points P and Q then the right triangle with vertices at points P and Q cannot be similar to the right triangle with vertices at points P and R. Therefore for the two triangles the ratio of the lengths of the legs cannot be equal; that is, the slope determined from points P and Q cannot be equal to the slope determined from points P and R. Therefore, from what we said, if the slopes are equal we must conclude that the three points are on the same line.

In summary, our discussion leads us to the following principle.

PRINCIPLE   **That three points P, Q, and R lie on the same nonvertical line is equivalen. to saying that the slope of the line through points P and Q is equal to the slope of the line through points P and R.**

We shall now use this principle to demonstrate that the graph of a linear relation $\{(x,y): Ax + By = C\}$, where $B \neq 0$, is a line. At this point we will find it more convenient to speak of the graph of the equation instead of the graph of the relation determined by the equation. Also, our demonstration shall be given in terms of the specific example, $3x + 2y = 6$.

First, we recognize that the points $(0,3)$ and $(2,0)$ are solutions to the equation $3x + 2y = 6$, and therefore they belong to the graph. Now consider the line through these two points; its slope is $-\frac{3}{2}$. We shall show that any other solution to the equation is a point on this line. Since an equivalent form of the equation is $y = -\frac{3}{2}x + 3$, any solution will be an ordered pair of the form $(x, -\frac{3}{2}x + 3)$. Since the slope determined from this point and the point (0,3) is

$$\frac{(-\frac{3}{2}x + 3) - 3}{x - 0} = -\frac{3}{2},$$

we have by our principle that $(x, -\frac{3}{2}x + 3)$ is on the line. We can therefore conclude that the graph of $3x + 2y = 6$ is a line with slope $-\frac{3}{2}$.

The above type of argument can be used with any linear equation $Ax + By = C$, where $B \neq 0$. Thus the following fact should be easy to accept.

The graph of a linear equation $Ax + By = C$ is a line in the Cartesian plane. **PRINCIPLE**
If $B \neq 0$, the line is nonvertical and the slope is equal to $-A/B$. If $B = 0$, the graph is a vertical line and the slope is not defined.

Two points are sufficient to determine the graph of a linear equation. (Photo by Cary Wolinsky, Stock Boston.)

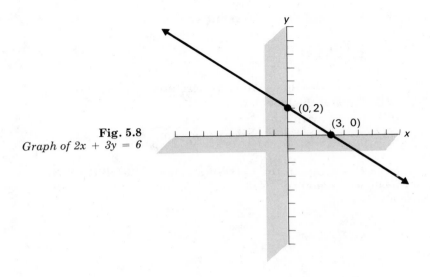

Fig. 5.8
Graph of 2x + 3y = 6

Since we know that the graph of a linear equation is a line, there is a simple procedure for obtaining it. We need only to keep in mind that *a line is determined by two points.* Therefore the method we can use is

1) Obtain *any* two points which are solutions to the equation.
2) Plot the two points in a Cartesian plane.
3) Draw the line that goes through these two points.

This will be the graph of the equation.

EXAMPLE ■ We see that the two points (3,0) and (0,2) satisfy the equation $2x + 3y = 6$. The graph of the equation is the line containing these two points (see Fig. 5.8). □

In graphing $2x + 3y = 6$, we could use any pair of numbers which would satisfy the equation; however, we used two special points called the **intercepts**. The **x-intercept** is the point where the graph crosses the x-axis and the **y-intercept** is the point where the graph crosses the y-axis. The x-intercept of $2x + 3y = 6$ is (3,0) and the y-intercept is (0,2). The x- and y-intercepts are sometimes convenient points because to obtain each we let one of the variables be zero. However, it is not necessary to use the intercepts; we can use *any* two solutions to graph the line.

Also, when we have an equation $Ax + By = C$ with $B \neq 0$, it can be rewritten as

$$y = -\frac{A}{B}x + \frac{C}{B}.$$

When *written in this form*, we immediately see that the slope $-A/B$ is the number multiplying x. For example, $-2x + 5y = 3$ is equivalent to $y = \frac{2}{5}x + \frac{3}{5}$, and we realize that the slope is $\frac{2}{5}$.

Graph each of the following linear equations and give the slope of each line.

QUIZ YOURSELF*

a) $x - y = 2$

b) $4x + 3y = 3$

c) $y = 3x - 4$

d) $y = -3$

***ANSWERS**

(a)

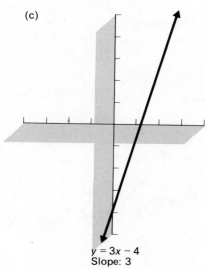

$x - y = 2$
Slope: 1

(b)

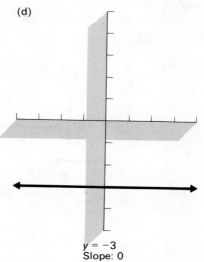

$4x + 3y = 3$
Slope: $-4/3$

(c)

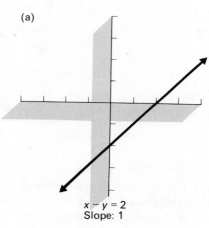

$y = 3x - 4$
Slope: 3

(d)

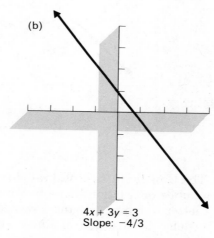

$y = -3$
Slope: 0

We have said that a linear equation in which the size of one of the variables is restricted still determines a linear relation. The graph of such a relation is a portion of a line.

EXAMPLES   ■ Consider the relation $\{(x,y): y = x - 1 \text{ and } 2 \leqslant x < 6\}$. The graph of this relation is given in Fig. 5.9. We should observe that since the pair $(6,5)$ does not belong to the relation, the corresponding point is not included in the graph. □

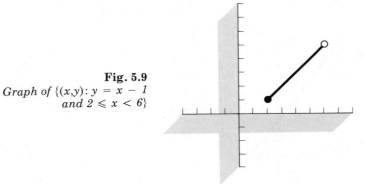

Fig. 5.9
Graph of $\{(x,y): y = x - 1$
and $2 \leqslant x < 6\}$

■ The United States postal rate for First Class Mail at one time was 10¢ for each ounce or fraction thereof. For example, the amount of postage needed on a letter weighing $1\frac{1}{2}$ ounces was 20¢. The postage on First Class Mail can be modeled by a union of linear relations. If we consider weights up to 3 ounces, our model is $\{(x,y): y = 10 \text{ and } 0 < x \leqslant 1\} \cup \{(x,y): y = 20 \text{ and } 1 < x \leqslant 2\} \cup \{(x,y): y = 30 \text{ and } 2 < x \leqslant 3\}$. The graph of this set is given in Fig. 5.10. □

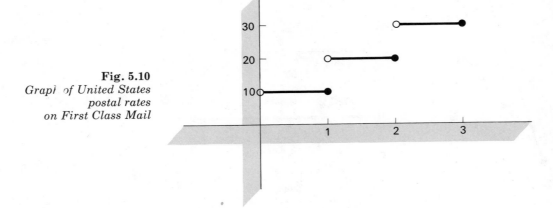

Fig. 5.10
Graph of United States
postal rates
on First Class Mail

We close this section by taking another look at the meaning of slope. In the case where the slope is positive, if we start at a point on the line and move to the right, then we must go up to get back to the line. More specifically, if the slope of the line is $M > 0$, then each increase of one unit in x will cause an increase of M units in y. For example, the equation $y = 2x + 5$ has solutions $(0,5)$ and $(1,7)$. By increasing x from 0 to 1, y increases by 2 units from 5 to 7. If we increase x by 1 more unit from 1 to 2, then y will increase from 7 to 9.

In the case where the slope M is negative, each increase of one unit in x will cause a decrease of M units in y.

Lastly, when the slope of a line is zero, the line is horizontal. Thus an increase in x must be accompanied by no change in y.

Now considering this understanding of slope, we see that the relation between two quantities can be modeled by a linear equation whenever each unit change in one quantity causes a constant change in the other quantity. The following examples illustrate this point.

EXAMPLES

■ A book publisher sells each copy of a particular textbook at the same price. Suppose this price is \$8. Total sales grow at the constant rate of \$8 for each additional copy sold; thus we see that the total sales y and number of copies sold x are related in a linear fashion. In particular, $y = 8x$. □

■ Part of a Purple Martin's diet is mosquitoes. Suppose that the mosquito population in a small area is 100,000 and a single Purple Martin eats mosquitoes at the constant rate of 10 per minute. Let P represent the mosquito population after a Martin is introduced in the area and t represent the time in minutes. By the assumed constant rate of decline in the population, we are actually assuming that the two quantities P and t are related by a linear equation. Thus the population is decreasing by 10 per minute and it is easy to see that the equation is $P = 100,000 - 10t$. □

In the publishing example, it is reasonable to say that the total dollar sales from a particular textbook and the number of copies sold are related by a linear equation. The justification is that generally each copy sells for the same amount. On the other hand, in the Purple Martin example, expressing the relation between the population and time as linear is most likely unrealistic. It is, of course, the assumption that a Purple Martin eats mosquitoes at a constant rate, which is questionable. Thus the following point must be kept in mind. To say that two quantities are related by a linear equation is equivalent to saying that there is a *constant* rate of change; that is, each unit change in one quantity causes the same amount of change (not necessarily one unit) in the other quantity.

EXERCISES

1. Describe where the point (x,y) is located in a Cartesian plane in each of the following:
   a) $x > 0$
   b) $x < 0$ and $y > 0$
   c) $y = 0$
   d) $x = 0$ and $y < 0$
   e) $x > 0$ and $y \leqslant 0$
   f) $x \geqslant 0$ and $y \geqslant 0$

2. For each of the following, determine whether or not the three points lie on the same line.
   a) $(0,0)$, $(2,3)$, and $(6,4)$
   b) $(-2,0)$, $(0,6)$, and $(1,3)$
   c) $(-1,3)$, $(3,-1)$, and $(-7,9)$
   d) $(4,3)$, $(-2,-2)$, and $(10,8)$

3. Graph each of the following linear equations and give its slope.
   a) $4x - y = 6$
   b) $x - \frac{1}{4}y = \frac{3}{2}$
   c) $y = -2$
   d) $y = x + 1$

4. Graph each of the following linear equations and give its slope.
   a) $5x - 2y = 10$
   b) $y = \frac{1}{3}x - 6$
   c) $x = 3$
   d) $\frac{1}{2}x + \frac{1}{3}y = 1$

5. Graph each pair of linear equations in the same Cartesian plane.
   a) $3x + y = 5$ and $y = -3x + 2$
   b) $x - y = 3$ and $x + y = 3$
   c) $x + \frac{1}{2}y = 3$ and $\frac{1}{3}x + \frac{1}{6}y = 1$

6. Graph each of the following linear relations.
   a) $\{(x,y): y = 5x + 7 \text{ and } x \geqslant 0\}$
   b) $\{(x,y): y = 4 \text{ and } -4 \leqslant x \leqslant 4\}$
   c) $\{(x,y): 5x - 2y = 10 \text{ and } 2 < x \leqslant 10\}$

7. Graph each of the following sets.
   a) $\{(x,y): y = 5 \text{ and } 0 \leqslant x < 2\} \cup \{(x,y): y = 10 \text{ and } 2 \leqslant x < 4\}$
   b) $\{(x,y): 3x + y = 14\} \cap \{(x,y): y = 3x + 2\}$
   c) $\{(x,y): y = 2x \text{ and } 0 \leqslant x < 3\} \cup \{(x,y): 2y - x = 9 \text{ and } 3 \leqslant x \leqslant 6\}$

8. During 1976 the United States rate for First Class Mail was 13¢ for the first ounce or fraction thereof. For each additional ounce or fraction thereof, the postage was increased by 11¢. Graph the cost of the postage on First Class letters weighing up to 4 ounces.

9. Suppose that the telephone rate between two points is 40¢ for the first three minutes and 15¢ for each additional minute (or fraction thereof). Graph the charge on telephone conversations of up to six minutes in length between these two points.

10. Below is a portion of a United States income tax table. Graph in a Cartesian plane the set of ordered pairs (I,T) obtained from this table.

If income I is:		Tax T is:	
	not over $500	14% of income	
Over	But not over		Of excess over
$500	$1000	$70 + 15%	$500
$1000	$1500	$145 + 16%	$1000

11. For each pair of quantities, state whether or not it is reasonable for them to be related by a linear equation. Be prepared to justify your answer.
    a) The number of ounces of fresh potatoes and the number of calories.
    b) The population of the United States and the year where 1776 is taken to be year zero.

c) The balance in an account paying 2% per quarter after only one deposit is made and the number of quarters.
d) The average resale value of a particular model of an American car and its age.
e) The height of mercury in a particular thermometer and the temperature.
f) Total United States consumer demand for beef in pounds and the average price per pound.

SOLVING A SYSTEM OF LINEAR EQUATIONS

In this section, we will obtain a procedure for finding a point which satisfies two linear equations. Since the graph of each equation is a line, the problem is the same as determining the point of intersection of the two lines.

We shall refer to a pair of linear equations as a **system of linear equations**. An ordered pair which satisfies both equations will be called a **solution to the system**.

When graphing a system of linear equations, one of the following must occur. (See Fig. 5.11.)

1) The two graphs are parallel lines and therefore the system has no solutions.

2) The two graphs happen to be the same line; in this case the two equations are equivalent.

3) The two graphs are lines which intersect at a single point and therefore the system has one solution, namely, the coordinates of this point.

In order for the lines to be parallel or to coincide, the slopes for each of the equations must be the same. However, if the slopes are *not* equal, then

Fig. 5.11
Graph of a system of equations

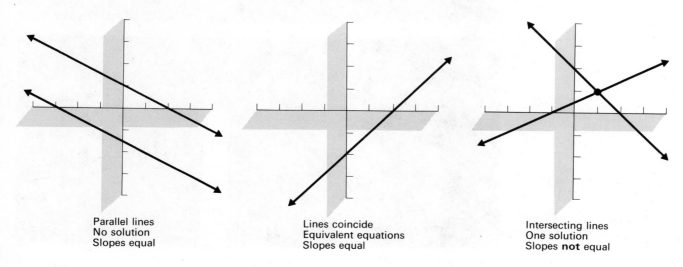

Parallel lines
No solution
Slopes equal

Lines coincide
Equivalent equations
Slopes equal

Intersecting lines
One solution
Slopes **not** equal

The graph of a system of linear equations will be two parallel lines, two intersecting lines, or a single line. (Photo by Frank Siteman, Stock Boston.)

the lines will intersect at a single point and the system of equations will have a single solution.

EXAMPLES

■ The two lines $2x + y = 5$ and $x - y = 6$ have different slopes. In particular, the slope for the first line is -2 and for the second, 1. There is therefore one point which satisfies both equations. □

■ Consider the following system of equations:

$$3x + 2y = 5,$$
$$6x + 4y = 7.$$

The slope for $3x + 2y = 5$ is $-\frac{3}{2}$, which is the same as $-\frac{6}{4}$, the slope of the second equation. Since the pair $(1,1)$ satisfies the first equation but not the second, we see that the equations are *not* equivalent. Therefore the graph of the system is two parallel lines and the system has no solution. □

For each of the following, state whether the system has a single solution, no solution, or that the two equations are equivalent.

QUIZ YOURSELF*

a) $3x + 2y = 16$
$\quad -x + y = 3$

b) $3x + 2y = 16$
$\quad 3x + 2y = 10$

c) $x - 2y = 5$
$\quad -2x + 4y = 6$

d) $4x - 3y = 5$
$\quad y = \frac{1}{2}x + 4$

e) $2y = -x + 1$
$\quad -6y = 3x - 3$

f) $x - y = 2$
$\quad 2x + 2y = 4$

a) One solution   b) No solution   c) No solution   d) One solution   e) Equivalent
f) One solution

*ANSWERS

Let us turn our attention to those systems in which we know there is a single solution. We recognize this case when the slopes of the two lines are different.

■ Solve the system

EXAMPLE

$$3x + 2y = 16, \tag{1}$$
$$y = x + 3. \tag{2}$$

Since the slopes are different, this system has a single solution. From Eq. (2), we see that the y-coordinate of this solution must be $x + 3$. Placing this value for y into the first equation, we obtain

$$3x + 2(x + 3) = 16,$$
$$3x + 2x + 6 = 16,$$
$$5x = 10,$$
$$x = 2.$$

Therefore the first coordinate of the solution must be 2. Now since $y = x + 3$, we see that the second coordinate of the point must be 5. The solution of the system is the point (2,5). (See Fig. 5.12). $\square$

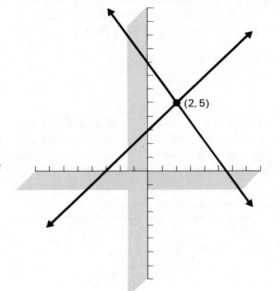

Fig. 5.12
Graph of the system
$3x + 2y = 16, y = x + 3$

(2, 5)

The general procedure we shall apply is to use the system of two equations to obtain one equation in one variable. This was accomplished in the previous example by substitution.

EXAMPLES   ■ Solve

$$4x + 5y = 3 \tag{1}$$
$$x + 3y = 1. \tag{2}$$

First, the equations have a single solution since the slopes are different. Next, from Eq. (2),

$$x = 1 - 3y.$$

Putting this into Eq. (1), we have

$$4(1 - 3y) + 5y = 3,$$
$$-12y + 4 + 5y = 3,$$
$$-7y = -1,$$
$$y = \tfrac{1}{7},$$

and

$$x = 1 - 3(\tfrac{1}{7}) = \tfrac{4}{7}.$$

The solution to the system is $(\tfrac{4}{7}, \tfrac{1}{7})$. $\square$

■ Solve the system

$$2x + 9y = 3, \tag{1}$$
$$5x + 7y = -8. \tag{2}$$

For this system, let us multiply Eq. (1) by 5 to obtain the equivalent equation, $10x + 45y = 15$. Next, by multiplying Eq. (2) by 2 we get the equivalent equation, $10x + 14y = -16$. Let us solve the equivalent system

$$10x + 45y = 15, \tag{1'}$$
$$10x + 14y = -16. \tag{2'}$$

We can obtain one equation in the variable y by subtracting Eq. (2') from Eq. (1'), which gives us

$$31y = 31,$$
$$y = 1.$$

Using this value for y in Eq. (1'), we have

$$10x + 45(1) = 15,$$
$$10x = -30,$$
$$x = -3.$$

The solution of the system is $(-3, 1)$. □

■ Solve the system

$$\tfrac{1}{2}x + \tfrac{1}{3}y = -1, \tag{1}$$
$$5x - 2y = 7. \tag{2}$$

By multiplying the first equation by 6, we obtain the equivalent system,

$$3x + 2y = -6, \tag{1'}$$
$$5x - 2y = 7. \tag{2'}$$

To obtain one equation in one variable, add the two equations, which gives us $8x = 1$ and $x = \tfrac{1}{8}$.

Using this value for x in Eq. (1'), we have

$$3(\tfrac{1}{8}) + 2y = -6,$$
$$2y = -6\tfrac{3}{8},$$
$$y = -\tfrac{51}{16}.$$

Solution: $(\tfrac{1}{8}, -\tfrac{51}{16})$. □

EXERCISES

By comparing slopes, first determine whether each of the following systems has no solution, the two equations are equivalent, or the system has a single solution. If the system has a single solution, then solve for it.

1) $3x + y = 14$
$\quad y = 3x + 2$

2) $x = y - 2$
$\quad 2x + y = 5$

3) $x - y = \frac{1}{2}$
$\quad x - y = 1$

4) $x - y = 4$
$\quad 3x + 4y = 12$

5) $4x - y = 6$
$\quad x - \frac{1}{4}y = \frac{3}{2}$

6) $x = 3$
$\quad 2x + y = 1$

7) $x + 2y = 13$
$\quad 4x - y = 4$

8) $\frac{2}{3}x + \frac{1}{2}y = -1$
$\quad 4x + 3y = -6$

9) $x + \frac{1}{2}y = 9$
$\quad \frac{1}{3}x + \frac{1}{6}y = -3$

10) $x + 3y = 11$
$\quad x - y = 17$

11) $\frac{2}{5}x + \frac{1}{7}y = 0$
$\quad \frac{1}{3}x - \frac{3}{5}y = 0$

12) $\frac{1}{3}x - \frac{1}{4}y = -1$
$\quad \frac{1}{4}x - \frac{1}{3}y = 1$

NEW TERMS

Ordered pairs

First component, Second component

Solution set

Linear equation

Equivalent equations

Cartesian plane

Origin

Straight-line depreciation

Fixed cost, Unit cost

X-axis, Y-axis

Quadrants

Coordinates

Slope

X-intercept, Y-intercept

System of linear equations

Solution of the system

MASTERY TEST: TOOLS OF LINEAR EQUATIONS

1. State whether each of the following statements is true or false.
   a) $y = \frac{5}{3}x + 6$ is a linear equation.
   b) The equations $6x + 2y = 3$ and $y = -3x + \frac{3}{2}$ are equivalent.
   c) $(2,1) \in \{(x,y): y - x = 1\}$.
   d) $\{(x,y): \frac{1}{3}x + \frac{1}{2}y = 1\} = \{(x,y): 2x + 3y = 6\}$.

2. Complete the table so that the pairs belong to the relation $\{(x,y): 3x + 2y = 1\}$.

x	0			3		$\frac{5}{3}$	
y		0	-1		2		$-\frac{1}{4}$

3. For each of the linear equations, give an equation equivalent to it being written in the form $y = mx + b$.
   a) $5y = 2x + 1$ b) $\frac{1}{3}y = x - 2$ c) $2x + y = -1$ d) $x - y = 3$
   e) $3x - 2y = 4$

4. For each of the following, write a linear equation to describe the relation between the two quantities.
   a) The weight of an object in pounds and its weight in ounces.
   b) Distance measured in kilometers and measured in miles.
   c) A state income tax and the income where the tax rate is 2.3% .
   d) The temperature measured in Celsius (centigrade) degrees and measured in Fahrenheit degrees.

5. Model each of the following by a linear relation.
   a) Using straight line depreciation, the remaining value of a sewing machine originally costing $200 and having a useful life of 4 years.
   b) The real estate tax where the rate is $81 per $1000 of assessed value (81 mills).
   c) The total production cost where the fixed cost is $1500 and the unit cost is $12 per item.
   d) The income tax on income between $16,000 and $20,000, where the tax is $3260 plus 28% of the excess over $16,000.

6. Give the coordinates of each of the points appearing in the Cartesian plane of Fig. 5.13.

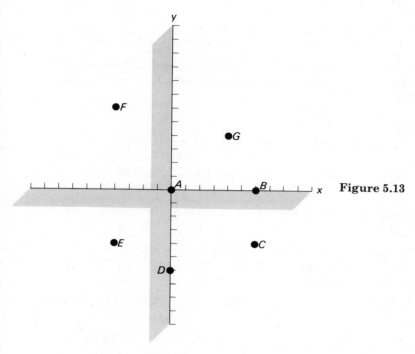

Figure 5.13

7. For each of the following pairs of points, give the slope of the line which contains these points.
   a) (0,0) and (−2,4)
   b) (0,0) and (4,2)
   c) (1,1) and (6,8)
   d) (−2,1) and (−8,−10)

8. Graph each of the following linear equations.
   a) $x = 3$
   b) $y = -2x + 5$
   c) $3x - 4y = 12$
   d) $\frac{1}{2}x + \frac{1}{5}y = 1$

9. Solve each of the following system of equations.
   a) $y = \frac{1}{2}x + \frac{3}{4}$
      $5x + 4y = -4$
   b) $2x - 7y = 18$
      $2x + y = 2$
   c) $3x + 6y = 12$
      $x - 2y = -\frac{8}{3}$
   d) $4x - 3y = -10$
      $3x + 2y = 1$

III. A MODEL FOR THE FISH-MANAGEMENT PROBLEM

"O.K., Coswell, now that I know how to solve a system of linear equations, I can't for the life of me see what it has to do with the fish in Bass Lake."

"You will, sir. First, it should be clear that the number of bass in the lake at the end of the year depends on the number of both bass and shiners at the beginning of the year."

"I am not sure I follow you, Coswell."

"Well, we assumed that the rate at which the bass population grows is constant. In particular, if there are no shiners in the lake, then for each pair of bass at the beginning of the year there will be 20 bass at the end of the year."

"I see. The bass population at the end of the year will be 10 times what it was at the beginning."

"There's more to it than that, Chief. We also agreed that introducing shiners into the lake will produce small fry for the fingerlings to feed on and also reduce cannibalism by the adults."

"But Coswell, how can we incorporate this into our estimate of the growth of the bass population?"

"Well, by adding this food source, I estimate that there will be eight additional bass at the end of the year for each pair of shiners. Now . . ."

"I have it! We can increase our estimate of the bass population at the end of the year by four times the number of shiners. Can't we summarize what we have said thus far?"

"Yes, sir. Let us represent the number of bass at the beginning of the year by B and the number of shiners by S. Then . . ."

"Our estimate of the growth of the bass population is $10B + 4S$. Of course, there will be 10,000 bass taken out of the lake each year, which leaves $10B + 4S - 10,000$ at the end of the year. . . Now, ah . . . this will be the number of bass at the beginning of the next year. . . . But it seems possible that even with this estimate the bass population could grow out of hand."

"Yes, chief, but that's the point. We want to start with an initial bass and shiner population so that the bass population will be stable from year to year. Therefore. . ."

"I'll take it, Coswell. Since we want the bass population to be the same from year to year, we must stock enough bass and shiners so that the number of bass at the end of the year is the same as what we started with, that is,

$$10B + 4S - 10,000 = B.$$

But . . . look . . . if we don't put any shiners in the lake we could still accomplish our goal by stocking the lake with about 1111 bass."

"Chief, you're getting better at mathematics but you are forgetting something. You said yourself that the lake needed a forage fish and the shiner would serve that purpose."

"O.K. I did say we needed some shiners. But how many? I guess if we start with too many the bass may not be able to control their numbers . . . if we start with too few, then they may become extinct from the lake. Well, . . . I suppose we must look at the growth of the shiner population in relation to the bass population."

"Right! We agreed that for each shiner in the lake one year there will be three the next. But the bass controls the growth of this population. In particular, every 22 bass in the lake will reduce the growth of the shiner population by 1. Now . . ."

"Coswell, I'm with you again. So, the shiner population at the end of the year will be $3S - \frac{1}{22}B$. Since this population should also be the same from year to year, we want $3S - \frac{1}{22}B = S$."

"Right! Therefore to maintain stable populations from year to year, the number of bass and shiners at the beginning of the year must satisfy

$$10B + 4S - 10,000 = B,$$
$$3S - \tfrac{1}{22}B = S.$$

Now, if we solve . . ."

"Coswell, you talk too slowly. The solution is $B = 1100$ and $S = 25$ . . . But the numbers of bass and shiners with which we stocked the lake were not too far from these. What went wrong?"

"Well, lets start with your numbers, 1200 bass and 50 shiners. At the end of the first year, using our growth rates, the number of bass was $10 \cdot 1200 + 4 \cdot 50 - 10,000 = 2200$, and the number of shiners was $3 \cdot 50 - \frac{1}{22} \cdot 1200 = 95$. Using these results, we can continue computing the fish population for the succeeding years. In particular, what happened, Chief, was that during the third year the shiners became extinct in Bass Lake and the lake became overstocked with bass."

Year	Beginning of year Bass	Shiners	End of year Bass	Shiners
1	1,200	50	2,200	95
2	2,200	95	12,380	185
3	12,380	185	114,540	*

* Here the number becomes negative so we assume the shiners disappeared.

"Coswell, now I see what was wrong with my figures. But isn't it possible to start with some other bass and shiner populations which would remain the same from year to year?"

"Not if the growth rates are always what we said. Any such pair of populations must be a solution to our system of equations, and this system has only the one solution $S = 25$ and $B = 1100$."

"O.K. But what if we started with populations of different sizes, couldn't they eventually stabilize at 25 shiners and 1100 bass?"

"Chief, I'll let you think about that until we get to the exercises."

A system of linear equations can be used as a model for other types of problems. Several more examples are given below.

EXAMPLE   ■ Consider that the American economy is divided into three sectors. Two of these sectors, agriculture and industry, are engaged in production, while the third sector is made up of consumers. The consumer sector does not add to the economy by producing goods; it only consumes the goods produced by industry and agriculture. Industry and agriculture also use each other's goods in their production processes. For example, machinery is used in the production of farm goods. Thus not only must the production sectors provide enough goods to satisfy consumer demand but at the same time they must produce goods for their own use.

Suppose that for each $1 of agricultural goods produced, agriculture uses $.25 worth of their own products and $.15 worth of industrial products. Also, when producing $1 of industrial goods, industry uses $.05 worth of agricultural goods and $.40 worth of their own goods. This information is displayed in the following table.

	Supplies to	
	Agriculture	Industry
Agriculture	$.25	$.05
Industry	$.15	$.40

Let x and y represent the following quantities:

x = dollar value of *all* goods produced by agriculture (gross agricultural production),

y = dollar value of *all* goods produced by industry (gross industrial production).

Using these numbers, the combined value of agricultural goods used in production by both agriculture and industry is

$$.25x + .05y.$$

Also, the combined value of industrial goods used in production is

$$.15x + .40y.$$

The total value of agricultural goods produced, when reduced by the amount of goods used in production, gives the amount left for consumers. Thus the value of agricultural goods left for consumers is

$$x - (.25x + .05y).$$

Also, the amount left from the gross industrial production for consumers is

$$y - (.15x + .40y).$$

Suppose the consumer sector requires $15 billion worth of agricultural goods and $100 billion worth of industrial goods. What must be the gross agricultural and industrial production to meet these demands? To find the answer to this question, we must solve the system of equations,

$$x - (.25x + .05y) = 15,$$
$$y - (.15x + .40y) = 100.$$

Using the "tools" of Part II, we arrive at the answer, rounded to the nearest tenth, $x = 31.6$ and $y = 174.6$. Therefore the agricultural sector of the economy must produce approximately $31.6 billion worth of goods and the industrial sector must produce approximately $174.6 billion worth of goods in order to supply the production processes and to meet the consumer demand. □

The following is another example of a problem in which the solution is obtained by solving a system of equations. We will see a variety of these "mixture-" type problems in the exercises.

■ A recommended daily dietary requirement for an active male weighing   **EXAMPLE**
about 160 pounds includes 3000 calories and 70 grams of protein. Suppose a male who fits into the above category plans to eat only a combination of rice and almonds while on a backpacking trip. It is known that 100 grams of rice supplies 360 calories and 7.5 grams of protein while 100 grams of almonds supplies 598 calories and 18.6 grams of protein. How many grams of each food should he eat daily to satisfy exactly his requirement in calories and protein?
    Let x and y represent the following quantities,

x = grams of rice in the daily diet,
y = grams of almonds in the daily diet.

Since each gram of rice supplies $\frac{360}{100}$ calories and each gram of almonds supplies $\frac{598}{100}$ calories, the number of calories in this diet of x grams of rice and y grams of almonds is

$$\frac{360}{100}x + \frac{598}{100}y.$$

Since the number of calories in this diet must be equal to 3000, we have

$$\frac{360}{100}x + \frac{598}{100}y = 3000.$$

The grams of protein in this diet are expressed by the number

$$\frac{7.5}{100}x + \frac{18.6}{100}y.$$

Since the diet must supply 70 grams of protein, we can write

$$\frac{7.5}{100}x + \frac{18.6}{100}y = 70.$$

Now, to find the solution to our problem, we solve the following system of equations,

$$\frac{360}{100}x + \frac{598}{100}y = 3000,$$

$$\frac{7.5}{100}x + \frac{18.6}{100}y = 70.$$

By applying the methods for solving a system of equations, we obtain the solution, rounded to the nearest tenth,

$$x = 630.5 \quad \text{and} \quad y = 122.1.$$

Thus by eating 630.5 grams (22.1 ounces) of rice and 122.1 grams (4.3 ounces) of almonds daily, the backpacker will satisfy the stated requirement in calories and protein.

We may question the need to have a diet which meets the requirements exactly. It seems more natural to expect the diet to contain no less than the required number of calories and protein. In Chapter 6, Linear Programming, we will consider this question. □

In the retail sale of consumer products, usually the demand for the product increases as the price decreases. Also, the amount a supplier is willing to make available to the consumer will increase as the price increases. A problem of interest is to determine the price in which the demand equals the supply. This price is called the **equilibrium price**. As an illustration of this concept, consider the next example.

*Many real-life situations can be modeled by linear equations. However, for two
quantities to be related by a linear equation there must be a constant rate of change.*
(Photo by Patricia Hollander Gross, Stock Boston.)

■ Suppose a refinery is willing to produce and supply to consumers 1000   **EXAMPLE**
gallons of gasoline per day for each penny per gallon in price, provided that
this price is at least 60¢ per gallon. It is known that the average daily demand
is 112,500 gallons when priced at 60¢ per gallon. It is estimated that the demand would be 100,000 gallons when priced at 65¢ per gallon.

Suppose we have only two pieces of information about demand, that is,
the points (60; 112,500) and (65; 100,000). Due to a lack of any other information, we wish to characterize the relation between demand and price by the
line going through these two points. Thus let us assume that the relation
between demand and price is linear and find the equilibrium price.

Let x and y represent the following quantities,

$$x = \text{the price per gallon,}$$
$$y = \text{the number of gallons of gasoline.}$$

It is not difficult to see that the relation between the number of gallons the refinery is willing to supply and the price per gallon is

$$y = 1000x, \qquad \text{when } x \geqslant 60. \tag{s}$$

Let us now write the linear equation which relates demand and price. The equation can be written in the form $y = mx + b$, where m is the slope of the line going through the points (60; 112,500) and (65; 100,000). Computing this slope, we have that

$$m = \frac{112{,}500 - 100{,}000}{60 - 65} = -2500.$$

So far we have as our demand equation

$$y = -2500x + b.$$

Since the point (60; 112,500) must satisfy this equation, then

$$112{,}500 = -2500 \cdot 60 + b,$$

and $b = 112{,}500 + 150{,}000 = 262{,}500$. Therefore our demand equation is

$$y = -2500x + 262{,}500. \tag{d}$$

In order to find the equilibrium price, we want the x and y in the supply equation (s) and the demand equation (d) to be the same. We see then to find the equilibrium price, we must solve the following system of equations,

$$y = 1000x, \qquad x \geqslant 60 \tag{s}$$
$$y = -2500x + 262{,}500. \tag{d}$$

Solving the above system of equations, we arrive at $x = 75$ and $y = 75{,}000$. The interpretation of this solution is that if the refinery supplies 75,000 gallons of gasoline daily at 75¢ per gallon, they should sell the entire supply, assuming, of course, the demand relation is accurate. $\square$

EXERCISES

1. A craftsperson working in leather can produce three pocketbooks and two wallets from a hide costing $40. From a smaller hide which costs $25, she can fashion two pocket-books and three wallets. Suppose she needs to fill orders for 9 pocketbooks and 11 wallets. If she is operating her business on a limited budget and wants to buy only enough leather to fill these orders, then what is the least amount of money she can spend?

2. An appliance dealer plans to have a one-day promotional sale. One item he will reduce in price is a particular model of a portable television which normally sells for $200. He is willing to sell only a limited number of these. In fact, for each $100 of the sale price, he is willing to sell 6 sets. The dealer knows that on the average he sells 2 of these sets daily when priced at $200. Also, he estimates that he could sell 11 sets if they were priced at $100 apiece. Assume that the relation between demand and price is linear and find the equilibrium price.

3. Suppose a rancher and a meat packing house are negotiating a beef sale. The rancher states that he would sell 10 head of cattle for each nickel per pound. On the other hand, the packing house offers to buy 200 head at 30¢ per pound, but only 75 head at 60¢ per pound. Assuming that the demand and supply relations are linear, what price should they agree on which would be satisfactory to both?

4. Dauber is an outfitter who operates wilderness canoe trips. For each trip he outfits and guides, the costs to him are $400 in fixed cost plus $100 for each person going on the trip. In order to be competitive, the fee Dauber plans to charge is $180 per person. How many people should he have on a trip in order to break-even—that is, so that the cost to him is the same as the amount he receives back in fees? The answer to this type of problem is called the **break-even point.**

5. Resha, a small private business which manufactures and sells dog equipment, is planning to sell sled-dog harnesses. If they make and sell up to 20 harnesses, their total cost is $116 plus $2 per harness. If they sell more than 20, the total cost is $156, the cost for the first 20, plus $1.50 for each additional harness over 20. Suppose they plan to sell a harness for $6. How many harnesses must they make and sell before they realize a profit? What is their profit on each harness beyond the break-even point?

6. An individual who wishes to invest $10,000 has decided to divide the money between two types of investments, one that returns 13% annually and the other, a lower risk investment, that yields 8% annually. The investor wishes to diversify the investment by not putting all the money into just one investment. This individual would like to realize an annual income of at least $1000 on the $10,000 invested. What is the maximum amount that can be invested at 8% to achieve this goal?

7. Suppose the investor mentioned in exercise 7 would like to realize an annual income of at least $1100 on the initial $10,000. What is now the maximum amount that can be invested at 8% to achieve this new investment goal?

8. Suppose in our diet problem example, the backpacker decides that his complete food supply will consist of oatmeal and dates. Each 100 grams of oatmeal supply 390 calories and 14.2 grams of protein, while each 100 grams of dates supply 274 calories and 2.2 grams of protein. How many grams of each food should he eat daily to satisfy the requirement in calories and protein?

9. At a copper smelter, water drawn from a river is used in certain stages of the production process. As it passes through the smelter it becomes 3% acid. An environmental agency requires that the acid concentration of any water returned to the river be no more than 0.1% acid. Therefore the water coming from the smelter is put into a holding pond and mixed with fresh water before being released into the river. The capacity of the holding pond is 150,000 gallons. How much fresh water and acidic water must be mixed together in this pond so that when the water is released into the river it is no more than 0.1% acid?

10. Collegeville University wishes to maintain a student-faculty ratio of 1 faculty member for every 15 undergraduates and 1 faculty member for every 5 graduate students. It is also estimated that it costs the University $200 each year for each undergraduate, and $400 each year for each graduate student over and above the money received in tuition. The additional funds for paying the added costs come from sources such as endowments, gifts, and state appropriations. If the University will have 350 faculty members and $1,000,000 in additional funds available, then how many undergraduate and graduate students can the University maintain?

11. Consider the following table of the values of goods needed to produce $1 of agricultural goods and $1 of industrial goods.

	Supplies to	
	Agriculture	Industry
Agriculture	.25	.15
Industry	.10	.45

Also, suppose that the consumer sector requires $15 billion worth of agricultural goods and $100 billion worth of industrial goods. Find the gross amount of agricultural and industrial production needed to meet these requirements.

12. Using the information given in exercise 11, suppose that in addition to the demand by American consumers, there is foreign demand for our agricultural and industrial products. If foreign demand is for $2 billion worth of grain and $5 billion worth of steel and machinery, what is the gross production needed to meet these added demands?

13. Suppose there are two countries, "Us" and "Them," engaged in trade with each other. Assume that each country's wealth is measured by the amount of gold it holds. Therefore after each trade period, a country's wealth is measured by the amount of gold it retains plus the amount it receives from the other country for trade goods. For a trade period, "Us" retains $\frac{3}{4}$ of it's gold and spends the other quarter on imports while "Them" retains $\frac{2}{5}$ of it's gold and gives the other $\frac{3}{5}$ to "Us" for imports. What is the ratio of gold held by "Us" to the amount held by "Them" so that after a trade period each country has the same amount it started with?

14. Considering "Us" and "Them" in exercise 13, suppose that during a trade period each country also adds to its wealth gold mined within its own country. Suppose "Us" mines 1700 bars of gold and "Them" mines 400 bars. How many bars of gold must each country hold at the beginning of a period so that it is twice as wealthy at the end of the trading period?

15. The population is divided into two sectors, urban and nonurban. Let us assume that over a single time period, three-fifths of the urban population moves into nonurban areas, while only one-seventh of the nonurban population moves into urban areas. Assume that changes in the population distribution are due only to moving. What must be the ratio of urban population to nonurban population so that the number of people living in each sector remains the same?

16. Answer the Chief's question. Is it possible to start with fish populations other than 25 shiners and 1100 bass that would eventually stabilize at these populations?

SUGGESTED READINGS

BATSCHELET, E., *Introduction to Mathematics for Life Sciences* (2nd ed.). New York: Springer-Verlag, 1976.

In Chapter 3, several examples are given which can be modeled by a linear relation.

BOYER, C. B., "Invention of Analytic Geometry." *Sci. Am.*, January 1949, pp. 40–45.

An interesting history of analytic geometry from the early Greeks to Descartes and Fermat.

GALE, D., *The Theory of Linear Economic Models.* New York: McGraw-Hill, 1960.

The linear models presented in Chapters 8 and 9 are interesting economics applications. However, an understanding of the mathematics is dependent on a familiarity with matrix notation and computation.

HUANG, D. S., *Introduction to the Use of Mathematics in Economic Analysis.* New York: John Wiley, 1964.

Contains an easy to read mathematical introduction to input-output analysis.

LEONTIEF, W. W., "Input-Output Economics." *Sci. Am.*, October 1951, pp. 15–22.

A nonmathematical introduction to early work done by economists in constructing input-output tables [matrices] to model the United States economy.

SEARLE, S. R., *Matrix Algebra for the Biological Sciences.* New York: John Wiley, 1966.

Chapter 7 introduces a linear model for population dynamics with consideration given to stability. The discussion is in a setting of matrices and latent roots.

6 Linear programming

Linear programming, as referred to in Fig. 6.1, is a branch of applied mathematics. More specifically, linear programming is concerned with the study of interrelated components of a complex system so that limited resources can be used as efficiently as possible in reaching a desired goal. An example of a situation in which linear programming might be used is in determining how a manufacturer could best use its supply of raw materials to obtain the largest possible profit.

Although pioneering work was done by Kantorovich in 1939, the origins of the subject can be traced as far back as 1758, when the economist Francois Quesnay wrote a paper, entitled "Tableau Economique." In this paper, Quesnay attempted to describe the economic relationship between landlord, peasant, and artisan, using what would today be considered a crude example of a linear programming model.

It was not until World War II that the theory of linear programming began to be developed extensively. In order to maintain the huge war effort, it was crucial that troops be recruited, trained, and deployed to battle on schedule and that vital equipment be produced as needed; all of this, of course, required extensive planning. An Air Force research group, later known as Project SCOOP (Scientific Computation Of Optimum Programs), developed elaborate planning techniques which, shortly after the war, resulted in the theory of linear programming. George Dantzig, one of the leaders in this research group, is credited with being the "father" of linear programming.

RUSSIAN ASSAILS SOVIET ECONOMICS

Berates Colleagues for Lack of Achievement Through Marxist Approach

By HARRY SCHWARTZ

The Soviet Union's leading mathematical economist, whose theories had been ignored for two decades, has made a scathing attack on that country's traditional Marxist economists, charging that they have contributed virtually nothing of value since the Bolshevik Revolution.

The scholar, Prof. L. V. Kantorovich, said in a debate that Soviet economists had been inspired by a fear of mathematics that left the Soviet Union far behind the United States in applications of mathematics to economic problems. It could have been a decade ahead, he contended.

Professor Kantorovich, in the late Nineteen Thirties, invented one of the most powerful tools of mathematical economics, linear programming, which has been widely applied here in recent years but whose study and use have been only begun in the last year or so in the Soviet Union.

Figure 6.1

In a peacetime economy, an effort was made by mathematicians and economists to apply linear programming techniques to nonmilitary uses. It soon became clear that the sophisticated organizational procedures developed by Project SCOOP had wide application in industry. Significant uses of linear programming were made in operating petroleum refineries, exploring and producing oil, planning efficient transportation routes of manufactured goods, producing steel, awarding contracts, etc.

In fact, linear programming is so widely used today in commerce and industry that it would be impossible to list all its many applications. If the value of a mathematical theory is measured by its ability to solve problems, then surely linear programming must be considered one of the most important areas of modern day applied mathematics.

Analagous to the way in which an architect uses a physical model to envision the final appearance of a construction project, a city planner may use a linear programming model to predict nonphysical aspects such as cost, tax revenues to be generated, etc. (Photo by Mike Mazzaschi, Stock Boston.)

I. AN URBAN REDEVELOPMENT PROBLEM

A city, which we will call Urbanopolis, has decided to use $90 million of federal funds for a center-city redevelopment project. The city council has decided that at least 15 blocks of Urbanopolis will be developed into new apartment and commercial buildings. The actual number of blocks of each type to be built is limited by the cost of each block and the total funds available. Furthermore, to encourage people to reside within the city, the city council has decided that at least one-third of the blocks must be developed into apartments.

(Photo by Laurence Lowry.)

(Photo by Laurence Lowry.)

Of course, the city council must also concern itself with additional city services (water, police-protection, sanitation, schools, etc.) that must be supplied; these services are supported by tax revenue. However, the city council does not wish to increase the existing real-estate tax rates that are in use throughout the city. Therefore to ensure sufficient income from the project, the city council may ask, "Considering all the limitations, how many blocks of each type building should the project contain so that Urbanopolis will receive the largest possible tax revenue from this project?"

Although this is an hypothetical problem and somewhat simplified, it was introduced to point out that land-development problems are being modeled mathematically.* In the case of private land development, the goal is generally to achieve the largest possible return on the investment.

Of course, a way to achieve this goal would be to construct a high concentration of the most profitable type of dwelling. However, in addition to the funds and acreage that are available, the developer must take the following items into consideration: local zoning laws; acreage to be set aside for roads and community use; stores for the residents' convenience; requirements of people at different income levels; etc.

All of the above items place restrictions on the number and type of buildings that can be constructed. The restrictions imposed by the local environment are also a land-development concern. This did not seem to be a concern in the 1950s when the New Jersey coastal lands were being developed. However, pollution created by the large concentration of people moving onto the lands adjoining the bays had a detrimental effect on the shell-fishing industry.

Our Urbanopolis Project is introduced to exemplify the use of the "tools" of linear programming. We shall return to our problem in Part III of this chapter.

---

* See, for example, "Linear Programming and Financial Analysis of the New Community Development Process," R. L. Heroux and W. A. Wallace, *Management Science* **19** (1973), pp. 857–72.

Provided the assumptions made in forming a model are valid and are properly converted into mathematical language, we can expect that it will give accurate predictions. However, if these considerations are treated carelessly, the model will not reflect reality and will be essentially useless. (Photo by Laurence Lowry.)

II. TOOLS OF LINEAR PROGRAMMING

LINEAR INEQUALITIES IN THE PLANE

In Chapter 5 we learned how to graph linear equations in a Cartesian plane. We will now learn how to graph **linear inequalities** in a plane.

A LINEAR INEQUALITY in two variables is a statement which can be written in one of the following forms: **DEFINITION**

$$Ax + By \geqslant C,$$
$$Ax + By > C,$$
$$Ax + By \leqslant C,$$
$$Ax + By < C,$$

where A and B cannot both be zero.

Each of the following statements is an example of a linear inequality.

$$2x + 3y \geqslant 6,$$
$$2x + 3y < 6,$$
$$x \geqslant 0,$$
$$y < 5,$$
$$4x - 5y - 20 \leqslant 0.$$

■ Let us begin our discussion of graphing with the linear inequality   **EXAMPLE**
$x + y \geqslant 5$. We wish to graph in a Cartesian plane the set $\{(x,y): x + y \geqslant 5\}$. The definition of $\geqslant$ tells us that an ordered pair (x,y) belongs to this set if either the number $x + y$ is the same as 5 or is greater than 5. The points for which the first condition is true are those on the line $x + y = 5$.

In order to locate the points for which the number $x + y$ is larger than 5, it is useful to first start with a point on the line $x + y = 5$. For example, (2,3) is one of the many points on this line (see Fig. 6.2). As we move *up* from (2,3), we obtain points whose x-coordinate is still 2, but the y-coordinate is larger than 3 and therefore $x + y$ is larger at these points than at the point (2,3). For example, (2,4) is above (2,3) and $2 + 4 = 6$ is larger than $2 + 3 = 5$. Thus the points above (2,3) are in the graph of $x + y \geqslant 5$.

We can use this same type of reasoning with other points on the line $x + y = 5$. By starting at a point on the line, we realize as we move above this point (keeping x the same and making y larger) that the number $x + y$ becomes larger. Therefore since the sum of the x- and y-coordinates is 5 at a point on the line, we have that the sum $x + y$ is larger than 5 at the points above the line. Thus the graph of the linear inequality $x + y \geqslant 5$ is the set

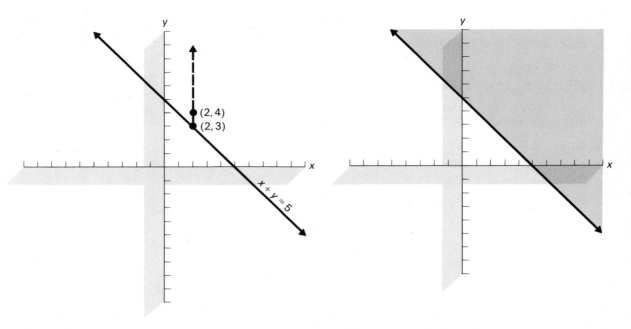

Figure 6.2   **Fig. 6.3** *Graph of* $x + y \geqslant 5$

of points on the line $x + y = 5$ and those points above it. We indicate the
graph by shading this region as in Fig. 6.3. □

In the above example, the line $x + y = 5$ is a **boundary line** for our
graph and is included in the graph.

EXAMPLE   ■ In this example we shall graph the linear inequality $5x - 2y > 3$. As in
the previous example, we start with a point on the boundary line $5x - 2y = 3$,
even though points on this line do not belong to the graph. For example, a
point on the line is (3,6). As we move above this point, the x-coordinates
remain equal to 3, but the y-coordinates become larger, that is, (3,7) is above
(3,6) but now $5 \cdot 3 - 2 \cdot 7 = 1$ is *smaller* than $5 \cdot 3 - 2 \cdot 6 = 3$.

What we should realize is that as y becomes larger, $-2y$ becomes smaller
and therefore $5x - 2y$ becomes smaller. Thus in order to make the number
$5x - 2y$ larger than 3, we must move *below* the line. For example, (3,5) is
below (3,6) and $5 \cdot 3 - 2 \cdot 5 = 5$, which is larger than $5 \cdot 3 - 2 \cdot 6 = 3$.

The graph of $5x - 2y > 3$ is the region below the line $5x - 2y = 3$. Since
the points on the boundary line do not belong to the graph, we draw a dotted
line and shade the region below it, as shown in Fig. 6.4. □

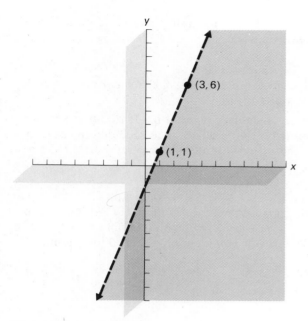

Fig. 6.4 *Graph of 5x − 2y > 3*

1. a) What is the value of the number $5x + 3y$ at the point (2,1)?   **QUIZ YOURSELF***
   b) What is the value of the number $5x + 3y$ at the point (2,3)?
   c) The point (2,0) is on the line $5x + 3y = 10$. How does the number $5x + 3y$ compare to 10 at points above (2,0)?
   d) On which side of the line $5x + 3y = 10$ does the graph of $5x + 3y > 10$ lie?

2. a) What is the value of the number $3x + 2y$ at the point (4,1)?
   b) What is the value of the number $3x + 2y$ at the point (4,0)?
   c) The point (4,2) is on the line $3x + 2y = 16$. How does the number $3x + 2y$ compare to 16 at points below (4,2)?
   d) On which side of the line $3x + 2y = 16$ does the graph of $3x + 2y < 16$ lie?

1. a) 13	b) 19	c) larger than 10	d) above
2. a) 14	b) 12	c) smaller than 16	d) below

***ANSWERS**

The discussion thus far should convince you that the graph of a linear inequality $Ax + By \geqslant C$ or $Ax + By \leqslant C$ is the region on one of the two sides of the line $Ax + By = C$ and includes the line. For the linear inequalities $Ax + By > C$ and $Ax + By < C$, the graph does not include the boundary line $Ax + By = C$. The graph of a linear inequality can be obtained rather easily by using the following procedure.

A procedure for graphing a linear inequality:
1. Graph the boundary line. Draw a solid line if the inequality is either $\geqslant$ or $\leqslant$, otherwise, draw a dotted line.
2. Pick one point on *either* side of the boundary line. If this point satisfies the inequality, then all points on that side are also included in the graph. Otherwise, the points on the other side of the boundary line belong to the graph.

EXAMPLE   ■ Let us graph the inequality $3x - 2y \geqslant 4$. First, we graph $3x - 2y = 4$ as a solid line. Next, we test the inequality at the point $(0,0)$, which is above the boundary line. We see that $3 \cdot 0 - 2 \cdot 0 = 0$, which is *not* greater than 4. Therefore the region we shade is below the boundary line. The graph of the inequality is given in Fig. 6.5. □

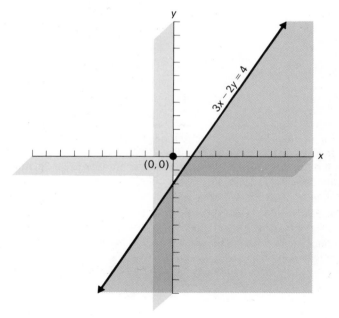

Fig. 6.5 *Graph of* $3x - 2y \geqslant 4$

In order to solve the problems of the next section, we will need to graph a system of linear inequalities. A system contains two or more linear inequalities; the graph is the set of points in the plane which satisfy each of these inequalities. Since the graph of each linear inequality in the system is a region to one side of a line, the graph of the system is the *intersection* of these regions.

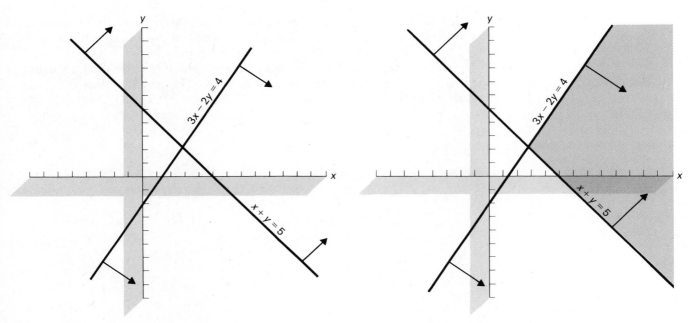

Figure 6.6

Fig. 6.7 *Graph of the system*
$x + y \geqslant 5,\ 3x - 2y \geqslant 4$

■ Let us graph the system,

$$x + y \geqslant 5,$$
$$3x - 2y \geqslant 4.$$

EXAMPLES

First, we graph the region determined by the inequality $x + y \geqslant 5$. Instead of shading the region as we did in Fig. 6.3, we draw two arrows to indicate the side of the boundary line on which the region lies. Also, we graph on the same plane the region corresponding to $3x - 2y \geqslant 4$. We obtained this graph earlier (in Fig. 6.5), but again we draw only arrows in place of the shading (see Fig. 6.6).

Finally, we look for the intersection of these two regions. This intersection is below the line $3x - 2y = 4$ *and* above the line $x + y = 5$. We complete our graph by shading this area as shown in Fig. 6.7. The boundaries of this shaded area are included in the graph. □

■ Graph the system of inequalities

$$2x + 3y \leqslant 24,$$
$$2x + y \leqslant 16;$$
$$x \geqslant 0,$$
$$y \geqslant 0.$$

Since $x \geqslant 0$ and $y \geqslant 0$, we recognize that the graph is in the first quadrant. We now graph in this quadrant the other boundary lines and indicate with arrows to which side of the boundary lines the region lies.

Upon doing this, we shade the desired area, which is in the first quadrant below the lines $2x + 3y = 24$ and $2x + y = 16$. The graph of the system is given in Fig. 6.8. $\square$

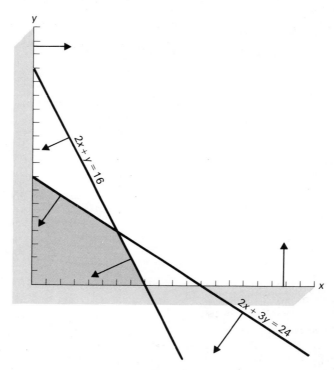

Fig. 6.8 *Graph of the system* $2x + 3y \leqslant 24$, $2x + y \leqslant 16$; $x \geqslant 0, y \geqslant 0$

EXERCISES

1. a) What is the value of the number $4x - y$ at the point $(3,1)$?
    b) What is the value of the number $4x - y$ at the point $(3,2)$?
    c) The point $(3,0)$ is on the line $4x - y = 12$. How does the number $4x - y$ compare to 12 at points above $(3,0)$?
    d) The region above the line $4x - y = 12$ is the graph of which inequality, $4x - y > 12$ or $4x - y < 12$?

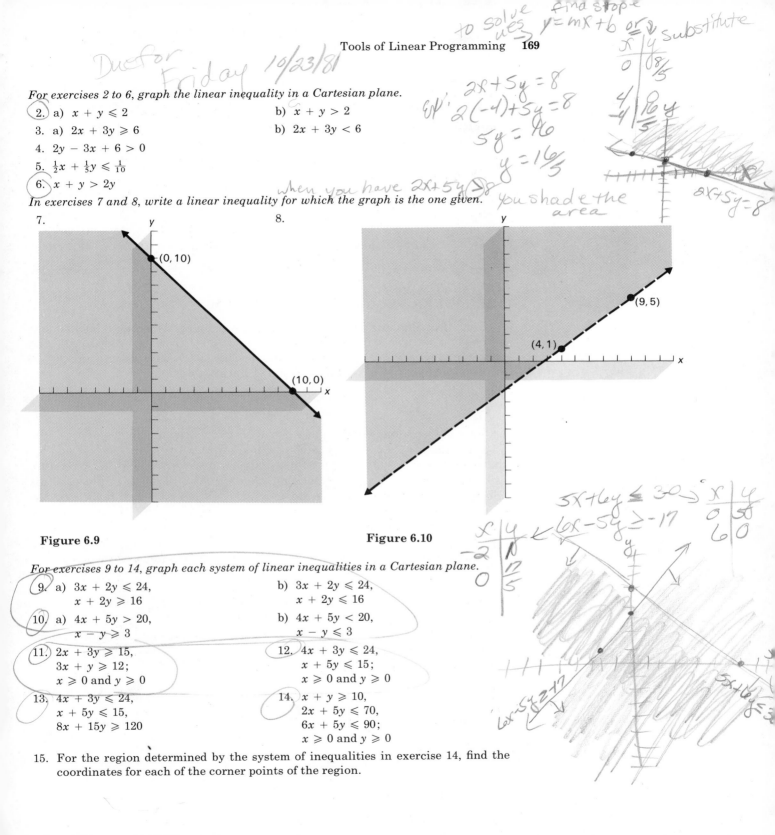

For exercises 2 to 6, graph the linear inequality in a Cartesian plane.

2. a) $x + y \leqslant 2$ b) $x + y > 2$

3. a) $2x + 3y \geqslant 6$ b) $2x + 3y < 6$

4. $2y - 3x + 6 > 0$

5. $\frac{1}{2}x + \frac{1}{5}y \leqslant \frac{1}{10}$

6. $x + y > 2y$

In exercises 7 and 8, write a linear inequality for which the graph is the one given.

7.

8.

Figure 6.9

Figure 6.10

For exercises 9 to 14, graph each system of linear inequalities in a Cartesian plane.

9. a) $3x + 2y \leqslant 24,$
$\quad x + 2y \geqslant 16$

   b) $3x + 2y \leqslant 24,$
$\quad x + 2y \leqslant 16$

10. a) $4x + 5y > 20,$
$\quad x - y \geqslant 3$

   b) $4x + 5y < 20,$
$\quad x - y \leqslant 3$

11. $2x + 3y \geqslant 15,$
$\quad 3x + y \geqslant 12;$
$\quad x \geqslant 0$ and $y \geqslant 0$

12. $4x + 3y \leqslant 24,$
$\quad x + 5y \leqslant 15;$
$\quad x \geqslant 0$ and $y \geqslant 0$

13. $4x + 3y \leqslant 24,$
$\quad x + 5y \leqslant 15,$
$\quad 8x + 15y \geqslant 120$

14. $x + y \geqslant 10,$
$\quad 2x + 5y \leqslant 70,$
$\quad 6x + 5y \leqslant 90;$
$\quad x \geqslant 0$ and $y \geqslant 0$

15. For the region determined by the system of inequalities in exercise 14, find the coordinates for each of the corner points of the region.

16. For the region determined by the system of inequalities in exercise 12, find the coordinates for each of the corner points of the region.

17. Find all ordered pairs (x,y) where both x and y are *integers* and satisfy the system of inequalities given in exercise 12.

18. Find all ordered pairs of integers which satisfy the system of inequalities given in exercise 14.

19. Consider the line $3x + 2y = 15$.
    a) What is the value of the number $3x + 2y$ at a point on the line?
    b) What is the value of the number $3x + 2y$ at points 1 unit vertically above the line?
    c) The points for which $3x + 2y = 21$ lie how many units vertically above the line?

THE LINEAR PROGRAMMING PROBLEM

In Part III of Chapter 5 we considered a diet problem in which a backpacker planned to eat a mixture of rice and almonds on a trip. In order to find the proper combination of foods to obtain *exactly* 3000 calories and *exactly* 70 grams of protein, we solved the system of linear equations

$$3.6x + 5.98y = 3000,$$
$$0.075x + 0.186y = 70,$$

where x is the grams of rice and y is the grams of almonds in the diet.

The number $3.6x + 5.98y$ gave the amount of calories in the diet and we wanted to pick x and y so that this number was *exactly* 3000. We therefore expressed this requirement by writing $3.6x + 5.98y = 3000$. If in place of this condition we require that the diet contain *at least* 3000 calories, then the mathematical sentence we would write to express this is the linear inequality

$$3.6x + 5.98y \geqslant 3000.$$

Also, if the protein requirement was *at least* 70 grams, we would write

$$0.075x + 0.186y \geqslant 70.$$

Since x and y are amounts of food, we have also the natural conditions that $x \geqslant 0$ and $y \geqslant 0$. Thus, to obtain diets which would contain no less than the minimum requirement of calories and protein, we would look for solutions to the system of linear inequalities,

$$3.6x + 5.98y \geqslant 3000,$$
$$0.075x + 0.186y \geqslant 70;$$
$$x \geqslant 0,$$
$$y \geqslant 0.$$

The graph of this system is given in Fig. 6.11.

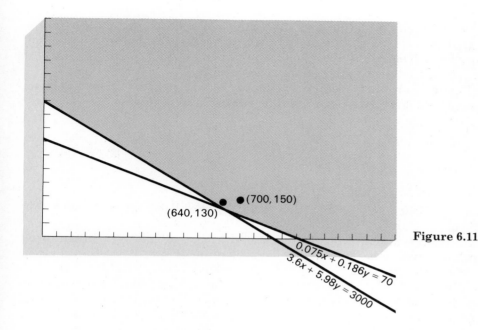

Figure 6.11

Any point in the shaded region of Fig. 6.11 represents a diet that supplies at least 3000 calories and 70 grams of protein. For example, the point (640,130) is in the region. Therefore 640 grams of rice and 130 grams of almonds would satisfy the requirements. In particular,

$$3.6 \cdot (640) + 5.98 \cdot (130) = 3081.4,$$

and

$$0.075 \cdot (640) + 0.186 \cdot (130) = 72.18.$$

Another solution to the system of linear inequalities is (700,150). This combination of rice and almonds would also give more than 3000 calories and more than 70 grams of protein.

When the requirement was for the diet to contain *exactly* 3000 calories and 70 grams of protein, we saw that there was only one solution to the system of equations. In replacing the equalities with inequalities, we see that there are many solutions. We may now ask which diet is best. If no further demand is made on the choice of the mixture other than satisfying the minimum nutritional requirements, then any one of the points in the region determined by the system of inequalities represents a feasible diet.

Let us now consider a further requirement on the choice of the mixture. Weight is a major concern to a backpacker. Therefore we now wish to obtain a solution to the system of linear inequalities that also represents the smallest total weight. We see that with this added condition on the problem (640,130) gives a better solution than (700,150).

For a diet of x grams of rice and y grams of almonds, the total weight is $x + y$. Thus we can model the problem of picking a mixture of rice and almonds which meets the nutritional requirements and represents the smallest weight by the following mathematical problem.

Choose a pair of numbers (x,y) to make the quantity $W = x + y$ minimum but subject to the restrictions that the pair satisfy the system

$$3.6x + 5.98y \geqslant 3000,$$
$$0.075x + 0.186y \geqslant 70;$$
$$x \geqslant 0,$$
$$y \geqslant 0.$$

In the exercises at the end of this section you will be asked to solve this problem.

The mathematical problem just stated is an example of a **linear programming problem**. The number of problems arising in the real world that can be modeled by a linear programming problem is very extensive. We will return in Part III of this chapter to a sample of problems which can be solved using this model. For this introduction to linear programming, we restrict our attention to problems which can be stated for two variables, x and y.

DEFINITION   **A LINEAR PROGRAMMING PROBLEM is to determine from among the solutions to a system of linear inequalities a solution for which a number of the form $Ax + By$ is smallest (largest). The set of solutions to the system of inequalities is called the set of FEASIBLE SOLUTIONS. A feasible solution for which $Ax + By$ is smallest (largest) is called an OPTIMAL SOLUTION.**

In defining a linear programming problem we did not say find *the* optimal solution since there could be more than one, as we will see in the third example which follows.

For the remainder of this section we shall develop a procedure for solving a linear programming problem.

EXAMPLE   ■ Let us solve the linear programming problem: Minimize the number
$$C = x + y,$$
where x and y are subject to the restrictions
$$3x + y \geqslant 27,$$
$$x + 2y \geqslant 24;$$
$$x \geqslant 0 \quad \text{and} \quad y \geqslant 0.$$

We start by graphing the set of feasible solutions as determined by the system of inequalities (see Fig. 6.12).

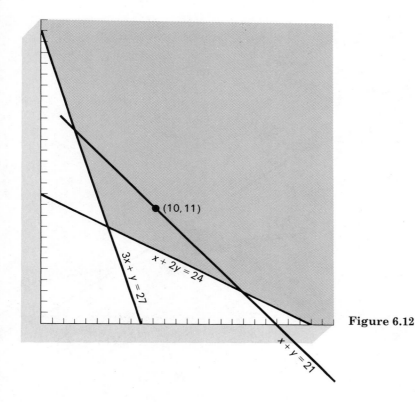

$3x + y = 27$

$x + 2y = 24$

$(10, 11)$

$x + y = 21$

Figure 6.12

Next we choose any one of the many points in the set of feasible solutions and compute the value of C for this point. For example, $(10,11)$ is a point in the region and the value of C at this point is $10 + 11 = 21$. If we now graph the line $x + y = 21$, we will find all points for which the number $x + y$ is 21.

We must keep in mind that the problem is to locate a point or points in the region at which $x + y$ is smallest. Our knowledge of graphing inequalities tells us that the points for which $x + y < 21$ are below the line $x + y = 21$. We see from Fig. 6.12 that there are many points in the set of feasible solutions for which $x + y$ is smaller than 21. Therefore the number $C = x + y$ can be made smaller than 21 by using one of these points. We now ask how much smaller can we make $x + y$?

If we start at a point on the line $x + y = 21$ and move to a point 1 unit below it (keeping x the same and decreasing y by 1 unit), the number $x + y$ will be 20. For example, $(10,10)$ is 1 unit below $(10,11)$ and $10 + 10 = 20$. If we drop 2 units below the line $x + y = 21$, the value of C at these points will be 19. Therefore, since we want $C = x + y$ to be as small as possible, we ask what points in the region of feasible solutions lie farthest below the line $x + y = 21$?

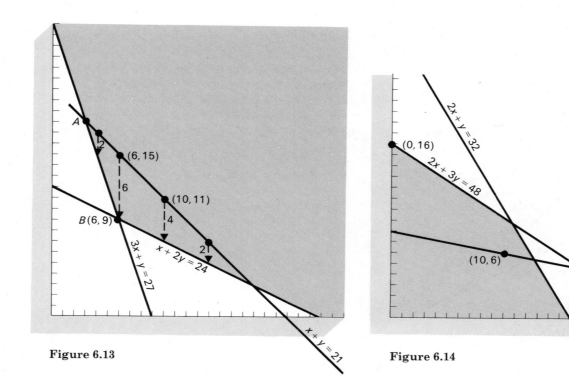

Figure 6.13

Figure 6.14

Let us observe in Fig. 6.13 that as we move away from the point A along the boundary line $3x + y = 27$ we are dropping farther below the line $x + y = 21$, until we reach the point B. Now moving from B along the boundary line $x + 2y = 24$, we are moving back towards the line $x + y = 21$. Therefore $B(6,9)$, is the point *in the region* which lies farthest below the line $x + y = 21$. The solution to our first linear programming problem is $x = 6$ and $y = 9$ and the minimum value of C is $6 + 9 = 15$. $\square$

Let us keep in mind as we continue with our development that the optimal solution for the above problem is a corner of the region of feasible solutions.

EXAMPLES   ■ Let us solve the following linear programming problem: Maximize the quantity

$$P = x + 5y,$$

where x and y are subject to the restrictions

$$2x + 3y \leqslant 48,$$
$$2x + y \leqslant 32;$$
$$x \geqslant 0 \quad \text{and} \quad y \geqslant 0.$$

First we graph the region of feasible solutions as shown in Fig. 6.14. Within the region we pick a point—for example, (10,6). When we obtain the value of P at this point, we have $10 + 5 \cdot 6 = 40$. Next we graph the line $x + 5y = 40$.

Our objective for this problem is to find, if possible, a point in the region of feasible solutions for which the number $P = x + 5y$ is largest. If we start at any point on the line $x + 5y = 40$ and move up, keeping x the same and increasing y, the number $x + 5y$ will become larger. Therefore we seek a point in the region which is farthest above the line. Observing how the boundary lines $2x + y = 32$ and $2x + 3y = 48$ move away from the line $x + 5y = 40$, we see in Fig. 6.14 that the point (0,16) is the one we seek. We also observe that (0,16) is a corner point of the region of feasible solutions.

The optimal solution for this problem is $x = 0$ and $y = 16$ and the maximum value of P is 80. □

■ We shall solve the following linear programming problem: Maximize

$$P = 2x + y,$$

subject to the restrictions

$$2x + 3y \leqslant 48,$$
$$2x + y \leqslant 32;$$
$$x \geqslant 0 \quad \text{and} \quad y \geqslant 0.$$

Observe that the region of feasible solutions is the same as in the previous example, but the number to be maximized is changed. Let us again select the point (10,6), one of the feasible solutions. The value of P at this point is $P = 2 \cdot 10 + 6 = 26$.

Now consider the line $2x + y = 26$ as shown in Fig. 6.15. We will vary our approach slightly in solving this problem. We realize that we can make the value of $2x + y$ larger by keeping y the same and increasing x. Therefore by moving to points to the right of the line $2x + y = 26$, the number $2x + y$ will be larger than 26. Since our objective is to maximize $P = 2x + y$, we look for a point *in the region* that is farthest to the right of the line $2x + y = 26$. Considering the graph in Fig. 6.15, we see that there are many such points, namely, all the points on the line $2x + y = 32$ which are between (16,0) and (12,8). Thus our linear programming problem does not have just one solution but many solutions. For example, the points (16,0), (14,4), and (12,8) are optimal solutions. For each of these points the value of P is 32. □

In each of the previous linear programming problems we saw that the optimal solution was one of the corners of the region of feasible solutions or, as in the last example, all the points on one of the edges of the region. Keeping in mind the way in which we obtained the solutions, it should not be difficult to accept this as always being the case. Therefore we state now a procedure that we will use for solving a linear programming problem.

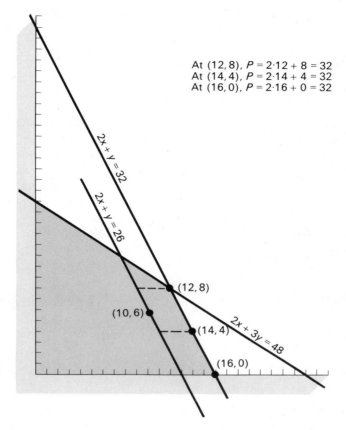

At $(12, 8)$, $P = 2 \cdot 12 + 8 = 32$
At $(14, 4)$, $P = 2 \cdot 14 + 4 = 32$
At $(16, 0)$, $P = 2 \cdot 16 + 0 = 32$

Figure 6.15

The graphical method of solution:

1. Graph the region of feasible solutions and determine the corner points of the region.

2. Each of these corners is the point of intersection of two boundary lines. Find the coordinates of each corner point by solving a system of linear equations.

3. Determine the value of the number which is to be maximized (or minimized) at each of these corner points. The point at which the value is largest (or smallest) is the optimal solution. If two corner points are solutions to the problem, then all points on the edge connecting these two corners are also optimal solutions.

EXAMPLES   ■ Solve the linear programming problem: Maximize

$$P = 5x + 3y,$$

subject to the restrictions

$$2x + 3y \leqslant 15,$$
$$3x + y \leqslant 12;$$
$$x \geqslant 0 \quad \text{and} \quad y \geqslant 0.$$

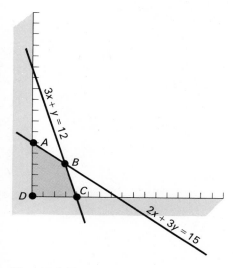

Figure 6.16

Using the graphical method to solve this problem, we first graph the region of feasible solutions and determine the corner points, as shown in Fig. 6.16.

The corner A is on the boundary line $x = 0$ and the line $2x + 3y = 15$. It is easy to see that the coordinates are $(0,5)$.

In order to find the coordinates of the point B, we must solve the system of equations

$$2x + 3y = 15,$$
$$3x + y = 12.$$

We find upon solving this system that the coordinates of B are $(3,3)$.

The point C is on the line $3x + y = 12$ and the x-axis and therefore its coordinates are $(4,0)$.

Since the objective is to maximize $P = 5x + 3y$, it is clear that D is not optimal. Now let us evaluate the number P at each of the other three corners. We have

at $(0,5)$, $P = 5 \cdot 0 + 3 \cdot 5 = 15$,
at $(3,3)$, $P = 5 \cdot 3 + 3 \cdot 3 = 24$,
at $(4,0)$, $P = 5 \cdot 4 + 3 \cdot 0 = 20$.

Thus the optimal solution is $x = 3$ and $y = 3$ and the maximum value of P is 24. $\square$

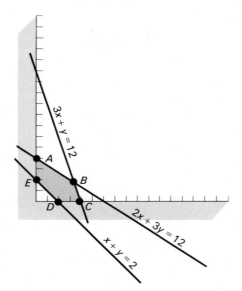

Figure 6.17

■ Minimize

$$W = 2x + 5y,$$

subject to the restrictions

$$x + y \geqslant 2,$$
$$2x + 3y \leqslant 12,$$
$$3x + y \leqslant 12;$$
$$x \geqslant 0 \quad \text{and} \quad y \geqslant 0.$$

The region of feasible solutions with the corners indicated is given in Fig. 6.17. A is the y-intercept of the boundary line $2x + 3y = 12$ and therefore its coordinates are $(0,4)$. Since the corners C, D, and E are also x- or y-intercepts, it is easy to determine their coordinates. In particular, we have $C(4,0)$, $D(2,0)$, and $E(0,2)$.

The coordinates of the corner B are the solution to the system of equations

$$3x + y = 12,$$
$$2x + 3y = 12.$$

The solution to this system is $x = \frac{24}{7}$ and $y = \frac{12}{7}$.

We now determine the value of W at each of these five corners.

At $(0,4)$, $W = 2 \cdot 0 + 5 \cdot 4 = 20$.
At $(\frac{24}{7}, \frac{12}{7})$, $W = 2 \cdot \frac{24}{7} + 5 \cdot \frac{12}{7} = \frac{108}{7}$.
At $(4,0)$, $W = 2 \cdot 4 + 5 \cdot 0 = 8$.
At $(2,0)$, $W = 2 \cdot 2 + 5 \cdot 0 = 4$.
At $(0,2)$, $W = 2 \cdot 0 + 5 \cdot 2 = 10$.

We see that the smallest value of W occurs at the corner $(2,0)$ with a value of 4.

If our linear programming problem were to maximize $W = 2x + 5y$ over the same region, then the optimal solution is $x = 0$ and $y = 4$ with the maximum value being 20. □

EXERCISES

1. Consider the line $2x - y = 10$.
   a) What is the value of the number $2x - y$ at a point on this line?
   b) What is the value of the number $2x - y$ at points 2 units vertically below the line?
   c) The points for which $2x - y = 15$ lie how many units vertically below the line?
2. Consider the graph in Fig. 6.18.
   a) Which point in the shaded region lies farthest above the line $3x + 2y = 31$ and what is the value of the number $3x + 2y$ at this point?
   b) At what point in the shaded region does $3x + 2y$ have the largest value?

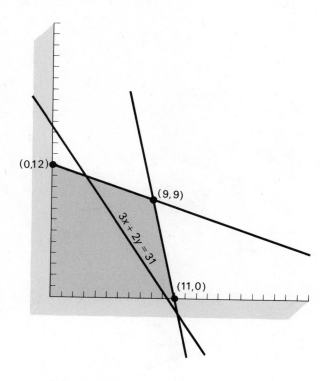

Figure 6.18

3. Consider the graph in Fig. 6.19.
   a) Which point in the shaded region is farthest below the line $x + 2y = 20$ and what is the value of $x + 2y$ at this point?
   b) At what point in the shaded region does $x + 2y$ have the smallest value?

4. Consider the region determined by the system of inequalities

$$3x + 2y \leqslant 24,$$
$$x + 2y \leqslant 16;$$
$$x \geqslant 0 \quad \text{and} \quad y \geqslant 0.$$

Determine the point (or points) in this region at which the following has its largest value.
   a) $P = 2x + y$ b) $P = x + y$ c) $P = x + 2y$

5. Consider the region determined by the system of inequalities

$$2x + y \geqslant 32,$$
$$2x + 3y \geqslant 48;$$
$$x \geqslant 0 \quad \text{and} \quad y \geqslant 0.$$

Determine the point (or points) in this region at which the following has its smallest value.
   a) $C = 3x + 5y$ b) $C = 2x + 3y$ c) $C = 3x + 2y$

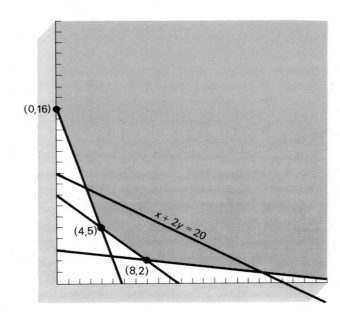

Figure 6.19

For exercises 6 to 12, solve the linear programming problem using the graphical method.

6. Minimize
$$W = x + y,$$
subject to the restrictions
$$3.6x + 5.98y \geqslant 3000,$$
$$0.075x + 0.186y \geqslant 70;$$
$$x \geqslant 0 \quad \text{and} \quad y \geqslant 0.$$

Note: The graph of feasible solutions is given in Fig. 6.11.

7. Maximize
$$P = 3x + 4y,$$
subject to the restrictions
$$x + 2y \leqslant 16,$$
$$x + y \leqslant 10,$$
$$3x + 2y \leqslant 26;$$
$$x \geqslant 0 \quad \text{and} \quad y \geqslant 0.$$

8. Maximize
$$P = 5x + 4y,$$
subject to the restrictions given in exercise 7.

9. Minimize
$$C = x + y,$$
subject to the restrictions
$$x + 2y \geqslant 16,$$
$$x + y \geqslant 10,$$
$$3x + 2y \geqslant 26;$$
$$x \geqslant 0 \quad \text{and} \quad y \geqslant 0.$$

10. Maximize and minimize
$$D = 5x + 6y,$$
subject to the restrictions
$$5x + 7y \geqslant 105,$$
$$7x + 5y \geqslant 105,$$
$$3x + 5y \leqslant 105;$$
$$x \geqslant 0 \quad \text{and} \quad y \geqslant 0.$$

11. Minimize and maximize
$$D = 4x + 5y,$$
subject to the restrictions
$$x + y \geqslant 10,$$
$$2x + 5y \leqslant 70,$$
$$6x + 5y \leqslant 90;$$
$$x \geqslant 0 \quad \text{and} \quad y \geqslant 0.$$

12. Minimize and maximize
$$D = 6x + 5y,$$
subject to the restrictions given in exercise 11.

MASTERY TEST: TOOLS OF LINEAR PROGRAMMING

1. Graph the inequality $4x + 5y \geqslant 20$.
2. Graph the inequality $3x + 2y \leqslant 15$.
3. Graph the system of inequalities,

$$3x + 4y \leqslant 36,$$
$$5x + 2y \leqslant 30;$$
$$x \geqslant 0 \quad \text{and} \quad y \geqslant 0.$$

4. Graph the system of inequalities,

$$x + 2y \geqslant 18,$$
$$x + y \geqslant 14,$$
$$2x + y \geqslant 19;$$
$$x \geqslant 0 \quad \text{and} \quad y \geqslant 0.$$

5. Solve the linear programming problem: Maximize

$$P = 4x + 3y,$$

subject to the restrictions

$$3x + 2y \leqslant 30,$$
$$7x + 3y \leqslant 65,$$
$$x + y \leqslant 13;$$
$$x \geqslant 0 \quad \text{and} \quad y \geqslant 0.$$

6. Solve the linear programming problem: Minimize

$$C = 3x + 4y,$$

subject to the restrictions

$$7x + 3y \geqslant 65,$$
$$x + y \geqslant 13;$$
$$x \geqslant 0 \quad \text{and} \quad y \geqslant 0.$$

NEW TERMS

Linear inequality

Boundary line

System of linear inequalities

Linear programming problem

Feasible solution

Optimal solution

Graphical method

III. A SOLUTION FOR THE URBAN REDEVELOPMENT PROBLEM

As you recall from Part I of this chapter, Urbanopolis is planning a redevelopment project which consists of new apartment and commercial buildings. Given certain conditions on available funds and land utilization, the city wishes to plan the project so as to realize the largest tax revenue. The necessary information is given below.

Cost

The total amount to be spent is limited by the available funds ($90 million), with apartment buildings costing $6 million per block and commercial buildings costing $2 million per block.

Land utilization

a) The total number of blocks to be redeveloped must be at least 15.

b) The number of blocks allocated to apartment buildings must be at least one-third of the total number redeveloped.

Tax revenue

A block of apartment buildings yields $30,000 per year in taxes, while a block of commercial buildings yields $20,000 per year in taxes.

To give the mathematical description of this redevelopment problem, let us start by representing the number of blocks of apartments and blocks of commercial buildings to be built by x and y, respectively. That is, let

x = the number of blocks of apartment buildings to be built, and

y = the number of blocks of commercial buildings to be built.

The objective of the project is to maximize the tax revenue that will be realized. Using x and y, we can express the tax revenue as

$$T = 30{,}000x + 20{,}000y.$$

Of course, the tax revenue will increase as we increase the values for x and y. However, the choice for (x,y) must be feasible in that it must satisfy the cost and land utilization conditions. In particular, the combined cost of constructing blocks of apartments and commercial buildings is $6x + 2y$. Since this cannot exceed the available funds, (x,y) must satisfy the linear inequality

$$6x + 2y \leqslant 90.$$

The total number of blocks that will be redeveloped is $x + y$, and this must be at least 15. Therefore we have

$$x + y \geqslant 15.$$

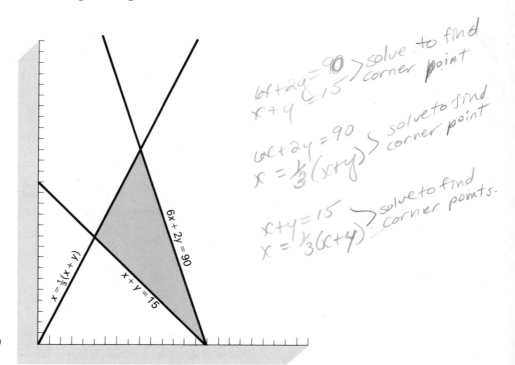

Figure 6.20

The handwritten annotations read:

$6x + 2y = 90$ ⎫ solve to find
$x + y \le 15$ ⎭ corner point

$6x + 2y = 90$ ⎫ solve to find
$x = \frac{1}{3}(x+y)$ ⎭ corner point

$x + y = 15$ ⎫ solve to find
$x = \frac{1}{3}(x+y)$ ⎭ corner points.

The number of blocks of apartments, x, must be greater than or equal to one-third of the total. Thus

$$x \geqslant \tfrac{1}{3}(x + y).$$

Finally, we have also the natural restrictions that both x and y cannot be negative. In summary, we see that the mathematical problem we can solve to obtain a solution to this redevelopment problem is the linear programming problem: Maximize

$$T = 30{,}000x + 20{,}000y,$$

subject to the restrictions

$$6x + 2y \leqslant 90,$$
$$x + y \geqslant 15,$$
$$x \geqslant \tfrac{1}{3}(x + y);$$
$$x \geqslant 0 \quad \text{and} \quad y \geqslant 0.$$

The graph of the set of feasible solutions is given in Fig. 6.20.

We see from Fig. 6.20 that the corners of the region occur at the following locations.

1) The intersection of the boundary lines $x = \frac{1}{3}(x + y)$ and $x + y = 15$. The coordinates of the point are (5,10).

2) The intersection of the boundary lines $x = \frac{1}{3}(x + y)$ and $6x + 2y = 90$. The coordinates of this point are (9,18).

3) The intersection of the boundary lines $x + y = 15$ and $6x + 2y = 90$. The coordinates of this point are (15,0).

Since the solution of the linear programming problem will be at one or more of the corner points, we now obtain the amount of tax revenue for each of these possible solutions.

At (5,10), $T = 30{,}000 \cdot 5 + 20{,}000 \cdot 10 = 350{,}000.$ → minimise

At (9,18), $T = 30{,}000 \cdot 9 + 20{,}000 \cdot 18 = 630{,}000.$ → maximise

At (15,0), $T = 30{,}000 \cdot 15 + 20{,}000 \cdot 0 = 450{,}000.$

Therefore the optimal plan for Urbanopolis is to build 9 blocks of apartments and 18 blocks of commercial buildings; the city council would thus realize the largest possible tax revenue of $630,000 per year.

We should note that the solution to our Urbanopolis problem is a pair of integers. Even though it is not required that the optimal solution be integers, there are situations in which this would be a natural requirement. For example, this requirement is imposed when finding the number of cars of each of two types to be built.

In cases where integer solutions are needed, we could first ignore the requirement and solve the problem as an ordinary linear programming problem. If the optimal solution to the linear programming problem is a pair of integers, then this is the solution we seek. (Why?) However, if we do not obtain integers, then we must resort to the techniques of *integer programming*, which are different from what we have done thus far. We shall not consider this more difficult question.

■ Regardless of the amount of land planted, a farmer has certain fixed **EXAMPLE** costs (taxes, mortgage payments, payments on equipment, etc.) and therefore needs a minimum amount of farm income per year. Suppose the farmer needs at least $6400 per year, which can be obtained by producing corn at $160 per acre or buckwheat at $100 per acre. At most 50 acres of the farm are available for planting a combination of these two crops.

The farmer is concerned about chemical fertilizers entering the streams through runoff and the pollution they create; but to achieve the desired yield, the farmer would use 400 pounds of chemical fertilizer for each acre of corn and 100 pounds of the fertilizer for each acre of buckwheat. How many acres of corn and acres of buckwheat should the farmer plant to minimize the amount of fertilizer used?

Similar to many areas of applied mathematics, linear programming received great initial impetus because of its military applications. However, after World War II, it was realized that linear programming also had great potential for large industries such as the automobile industry. (Photo by Owen Franken, Stock Boston.)

In order to obtain a mathematical statement of this problem, let

x = the number of acres of corn to be planted, and

y = the number of acres of buckwheat to be planted.

The objective of the problem is to minimize the amount of fertilizer that will be needed. This quantity can be written as

$$F = 400x + 100y.$$

Of course, F will have the value 0 if the farmer does not plant any corn or buckwheat. However, taking $x = 0$ and $y = 0$ is not feasible due to the minimum income that is needed. Therefore we must consider the restrictions imposed on the choice of (x,y).

Since there are at most 50 acres available and since the total number of acres to be planted is $x + y$, we can write

$$x + y \leqslant 50.$$

Each acre of corn yields \$160 and each acre of buckwheat yields \$100; we can therefore represent the income by $160x + 100y$. Since the farmer needs at least \$6400 in income, we also have the following condition on the problem:

$$160x + 100y \geqslant 6400.$$

Finally, we have the natural conditions that

$$x \geqslant 0 \quad \text{and} \quad y \geqslant 0.$$

In summary, we can solve this farm problem by solving the linear programming problem: Minimize

$$F = 400x + 100y,$$

subject to the restrictions

$$x + \quad y \leqslant 50,$$
$$160x + 100y \geqslant 6400;$$
$$x \geqslant 0 \quad \text{and} \quad y \geqslant 0.$$

The region of feasible solutions determined by the restrictions is graphed in Fig. 6.21.

We see from Fig. 6.21 that the corners of the region are located in the following places:

1) the x-intercept of the boundary line $160x + 100y = 6400$, which is $(40,0)$;

2) the x-intercept of the boundary $x + y = 50$, which is $(50,0)$;

3) the intersection of the two boundary lines $160x + 100y = 6400$ and

4) $x + y = 50$. Solving this system of equations, we find that this point of intersection is $(23\frac{1}{3}, 26\frac{2}{3})$.

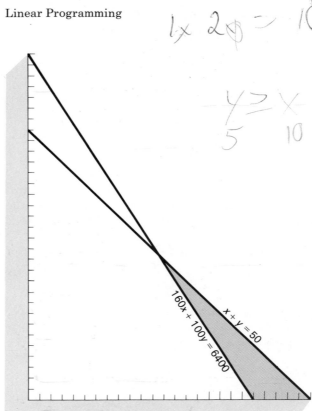

Figure 6.21

To determine the value of the number F at each of these corners, we have

at $(40,0)$, $F = 400 \cdot 40 + 200 \cdot 0 = 16,000$;
at $(50,0)$, $F = 400 \cdot 50 + 100 \cdot 0 = 20,000$;
at $(23\frac{1}{3}, 26\frac{2}{3})$, $F = 400 \cdot (23\frac{1}{3}) + 100 \cdot (26\frac{2}{3}) = 12,000$.

Thus we see that in order for the farmer to use a minimum amount of fertilizer, 12,000 pounds, $23\frac{1}{3}$ acres should be planted in corn and $26\frac{2}{3}$ acres in buckwheat. □

EXERCISES

1. Consider the Urbanopolis problem and suppose the restrictions on land utilization and costs remain the same. But if each block of apartments now yields $60,000 per year, while each block of commercial buildings yields $20,000 per year in

taxes, then how many blocks of each type of building should be built to obtain the largest possible tax revenue?

2. A land developer is planning to subdivide a 30-acre tract into one-quarter-acre and one-half-acre building lots. Each one-quarter-acre lot costs the developer $1000 for improvements, while the cost for each one-half-acre lot is $2000. The developer has $120,000 available for these improvements. If the developer prices the lots so as to make a profit of $3000 on each one-quarter-acre lot and $5000 on each one-half-acre lot, how many of each should be in the development in order to achieve the largest profit?

3. A builder wishes to construct a high-rise apartment building in a certain town. The builder's interest is in making a profit, whereas the town fathers' concern lies in the increase in the number of children who would be attending an already crowded school system.

   The plans are for the building to consist of two and three-bedroom apartments. Each two-bedroom apartment will contain 600 square feet of floor space and each three-bedroom apartment will contain 700 square feet. The builder feels that there should be at least 26,000 square feet of floor area devoted to apartments in order for the building to be profitable. In addition, it is thought that there should be at least as many three-bedroom apartments as two-bedroom apartments.

   The town fathers predict that, on the average, there would be one school-age child for every two-bedroom apartment and two school-age children for every three-bedroom apartment. In order for the predicted number of school-age children moving into the building to be minimal, how many apartments of each type should there be?

4. The daily diet for a pregnant woman requires at least 15 milligrams of iron, 100 grams of protein, and 2000 milligrams of calcium. Suppose a mixture of two types of food are used in obtaining a diet that meets these requirements. One unit of Type B contains, on the average, 4 milligrams of iron, 20 grams of protein, and 20 milligrams of calcium, whereas one unit of Type D contains, on the average, 1 milligram of iron, 10 grams of protein, and 770 milligrams of calcium. If Type-B food costs $1.50 per unit and Type-$D$ food costs $.60 per unit, how many units of each type should be used in the daily diet so that the cost is minimal?

5. In planning a diet, two types of food are used, which we will call food A and food B. The dietary requirements are no less than 2000 calories and no more than 3000 calories. The diet must also contain between 60 and 80 grams of protein. One unit of food A supplies 300 calories and 5 grams of protein, whereas one unit of food B supplies 200 calories and 10 grams of protein. Recent studies indicate that nondigestible fiber in a diet helps to reduce the likelihood of cancer of the colon.

   If 30% of food A contains nondigestible fiber and 40% of food B contains nondigestible fiber, how many units of each food must be used in the diet to maximize the amount of nondigestible fiber?

6. Refer to exercise 5. In addition to the units of calories and protein the foods contain, if each unit of food A contains 3 milligrams of vitamin C and each unit of food B contains 4 milligrams of vitamin C, in place of maximizing the amount of fiber, how many units of each food must be used to maximize the amount of vitamin C in the diet?

7. A farmer has 150 acres available for planting in corn or wheat or a combination of both. The planting costs (seed, fertilizer, etc.) are $100 per acre for corn and $30 per acre for wheat. The farmer has at most $9000 available for planting costs. If the anticipated income is $375 per acre for corn and $200 per acre for wheat, how many acres of each should the farmer plant to maximize the income?

8. If all information given in exercise 7 remains the same except that in addition the farmer receives $250 in federal subsidies for each acre *not* planted, then how many acres of corn and wheat should the farmer plant to maximize the income?

9. A clothing manufacturer makes both men's and women's suits. Each suit goes through three phases of production: cutting, sewing, and finishing. A man's suit requires $\frac{1}{2}$ hour for cutting, $1\frac{1}{2}$ hours of sewing, and 1 hour for finishing. A woman's suit requires $\frac{1}{2}$ hour for cutting, 1 hour of sewing, and $1\frac{1}{2}$ hours for finishing. The number of hours a day available for these processes are 15 hours of cutting time, 38 hours of sewing time, and 40 hours of finishing time. If the profit on a man's suit is $12 and on a woman's suit is $16, then how many of each type of suit should be made per day to achieve the largest possible profit?

10. A manufacturer makes two items, which we will call A and B. Natural gas is used in the production of each item. In order to produce one unit of Item A, 4 cubic feet of gas are required whereas 5 cubic feet of gas are required for one unit of Item B.

During the winter, the manufacturer's allotment of natural gas is restricted to 1000 cubic feet per month. In order to keep the plant operating, at least $18,900 worth of goods must be produced per month. A unit of Item A sells for $81 and a unit of Item B sells for $90. The amount of labor needed to produce one unit of Item A is 2 hours and for Item B, it is 3 hours. How many units of each item can the manufacturer produce to maximize the number of hours spent producing them? (This is the same as minimizing unemployment at the plant.)

SUGGESTED READINGS

COOKE, W. P., "Two-dimensional Graphical Solutions of Higher Dimensional Linear Programming Problems." *Mathl Mag.*, March 1973, pp. 70–76.

Contains examples of linear programming problems in more than two dimensions in which solutions are obtained through two-dimensional graphical techniques.

DANTZIG, G. B., *Linear Programming and Extensions*, Princeton, N.J.: Princeton University Press, 1963.

The first three chapters give a very readable exposition on the concept, origin and formulation of the linear programming problem.

KEMENY, J. G., J. L. SNELL, AND G. L. THOMPSON, *Introduction to Finite Mathematics* (3rd ed.), Englewood Cliffs, N.J.: Prentice-Hall, 1974.

Contains an interesting linear programming problem called "Optimal Harvesting of Deer." Also gives an easy-to-read proof that a solution to a two-dimensional linear programming problem occurs at a corner point.

KOHN, R. E., "A Mathematical Programming Model for Air Pollution Control." *School Sci. and Math.*, June 1969, pp. 487–499.

Presents a two dimensional problem involving a minimization of cost on controls for a hypothetical air shed.

LEVINSON, H. C., AND A. A. BROWN, "Operations Research." *Sci. Am.*, March 1951, pp. 15–17.

An excellent history of the beginnings of operations research with particular military applications. The article contains three examples.

Things to know for test!!
Find Slope
Write equations
Solve 2 equations w/2 unknowns.
word problems - cost, Break even, Depreciation
Graph inequality — Know How to do.
Graph system of inequalities
Linear Program problems.

slope
$6x + 2y = 3$ $y = mx + b$
$y = -3/2x + 1/2$
$2y = -6x + 3$
$y = -3x + \frac{3}{2}$
$6x + 2(0) = 3$
$6x = 3$
$x = 1/2$

1 type writing equations
A(4, 3) B(−2, 7)
$m = \frac{y_2 - y_1}{x_2 - x_1} = -\frac{2}{3}$
$y = mx + b$
$y = -2/3x + b$
$3 = -2/3(4) + b$
$3 = -8/3(4) + b$
$3 = -8/3 + b$
$3 + 2\frac{2}{3} = b$
$b = 5\frac{2}{3}$ or $\frac{17}{3}$
$y = -2/3x + \frac{17}{3}$
$3y = -2x + 17$

Solving 2 equations
① $2x - 9y = -48$
$8x - y = 8$ — intersect
$2(\frac{12}{7}) - 9y = -48$
$\frac{24}{7} - 9y = -48$
$-9y = 48 - \frac{24}{7}$
$y = \frac{40}{7}$
Substitution
$8x - 8 = y$
$2x - 9(8x - 8) = -48$
$2x - 72x + 72 = -48$
$-70x = -120$
$x = \frac{12}{7}$

② $2x - y = 5$
$2x - y = -4$ — parallel - because coefficients are the same.

③ $3x - 2y = -7$
$2y - 3x = 7$ — on same line - because they coincide

to solve
1. add
2. subtract or
3. substitute.

depreciation
8936.00
$V = 936 - \frac{936}{12}A$

cost
fixed cost = \$2480
cost of item = \$2.50
Sell each item = \$5.60
$2480 = 5.60A - 2.50A$
$P.C. = 2480 + 2.50A$
$R = 5.60A$
$5.60A = 2480 + 2.50A$
$3.10A = 2480$
$A = 800$

7 Counting

As you might suspect when you see an entire chapter of a book devoted to the subject of counting, counting means something entirely different to a mathematician than it does to you. A mathematician usually does not count the number of objects in a collection one at a time, but rather seeks to determine patterns and relationships among the objects which allow them to be counted in an indirect way. Counting in this manner occurs in many parts of mathematics and often involves quite sophisticated methods.

Some of the earliest counting formulas can be traced back to the Hindu mathematician Brahmagupta in the seventh century. Little was done to develop this theory of counting until the sixteenth century, when mathematicians began to analyze certain games of chance. In attempting to answer questions about throwing dice and drawing cards, some of the leading European mathematicians of that time started to organize their results into a formal theory of counting. One of the chief figures in this development was the French mathematician Blaise Pascal, who wrote a paper in 1654 dealing with the theory of combinations.

Pascal was a child prodigy who did important work in several areas of mathematics. At the age of 12 he became interested in geometry and quickly mastered *Euclid's Elements*; in the process of doing so, he independently rediscovered and proved several important theorems. Within four years he was doing original research and wrote a paper of such quality that some of the   **193**

blaise pascal
1623~1662

Invented the first mechanical adding
machine at age 19.

Blaise Pascal (1623–1662)

leading mathematicians of the time refused to believe it had been written by a 16-year-old boy.

Unfortunately, shortly after writing this paper, Pascal became afflicted with dyspepsia, a digestive disorder, which was to cause him great pain and suffering throughout the rest of his life. In spite of his illness, he continued his work, not only in mathematics, but also in physics. Later in his life he decided to abandon mathematics and science and devote himself completely to philosophy and religion. In 1658, however, while unable to sleep because of a toothache, he decided to think about geometry to take his mind off the pain and, surprisingly, the pain stopped. Pascal took this as a sign from heaven that he should return to mathematics. For a short while he again devoted himself to research, but soon after he became seriously ill. Pascal spent the remaining years of his life in excruciating pain, doing little work until his death at the age of 39 in 1662.

The work done by Pascal and others has now developed into a branch of mathematics called combinatorial analysis. We will study two elementary aspects of this subject, permutations and combinations, which have applications in the area of probability theory.

I. WHAT IS THE NUMBER OF SHORTEST ROUTES?

The postmaster of a certain city wishes to improve the mail service within his district. One area that he is considering is the amount of time necessary to transfer the morning mail from a substation to the main post office. Figure 7.1 is a partial map of the city which indicates the streets and avenues between the substation, point A (corner of 1st and Maple), and the main post office, point B (corner of 4th and Hemlock).

Due to varying traffic-flow on the different streets, the time needed to travel one route from point A to point B may not be the same as the time needed to travel another route. The postmaster is aware of this and wishes to determine the best of all possible routes. Therefore he could have the driver follow a different route each day and record the time required. If each possible route is tested once, then how many days are needed for the postmaster to complete his study?

Very often the mathematician must determine the number of objects in a large collection. The counting principles you will encounter provide some of the useful tools for doing this. (Photo by Laurence Lowry.)

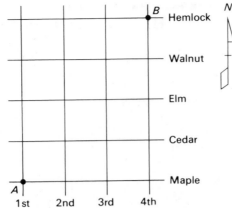

Figure 7.1

Let us consider only those routes on which travel is to the north and east. Since there is only a small number of streets and avenues between points *A* and *B*, you may want to attempt an answer at this time by finding all possible routes—however, be forewarned that there are more than 30 of them.

Our study of counting techniques in the following parts of this chapter will enable us to answer the above question along with many others.

If we can recognize patterns, it often enables us to count the objects in a collection quickly in an indirect way. (Photo by Fredrik D. Bodin, Stock Boston.)

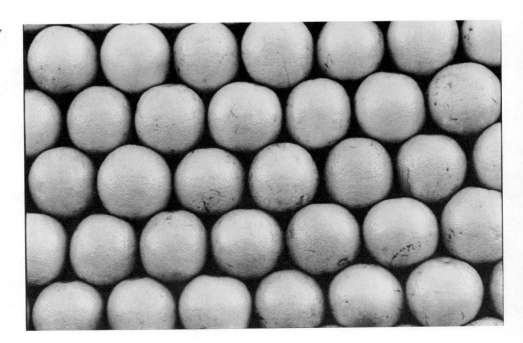

II. THE TOOLS OF COUNTING

COUNTING PRINCIPLES AND PERMUTATIONS

The task of counting the number of objects in a finite set needs no lengthy introduction. A simple example would be counting the number of people in a room. Even though the objects we encounter in this chapter may not be as familiar as a roomful of people, the counting techniques we use are the same. In order to identify these techniques, let us look at the problem of counting the number of students in your class.

Assuming that the seats in the room are arranged in rows, the class is divided into disjoint subsets, each subset being a row of students. Therefore we may count the number of students in each subset and add these quantities together.

■ How many three-digit numbers can be formed by using the digits 1, 2, and 3 with each of these digits occurring in the number only once?   **EXAMPLE**

Let S be the set of objects to be counted. We can subdivide S into three disjoint subsets. One subset consists of the numbers with 1 in the hundreds position, which we will call E_1; E_2 will be the subset of numbers with 2 in the hundreds position; and, lastly, E_3 will be the subset of those numbers with 3 in the hundreds position. Since each pair of subsets is disjoint and since the union of the three is S, we should recognize that $n(S) = n(E_1) + n(E_2) + n(E_3)$.*

To determine the number of elements in E_1, it is a matter of counting the number of ways we can arrange the 2 and the 3 in the tens and units positions. It should be clear that there are two such ways, in particular, E_1 contains the two elements, 123 and 132. Since the number of elements in E_2 and E_3 is also the same as the number of ways to arrange the remaining two digits in the tens and units positions, we have that $n(E_2)$ and $n(E_3)$ are also equal to 2. Therefore the number of elements in S is $n(S) = 2 + 2 + 2 = 3 \cdot 2 = 6$. □

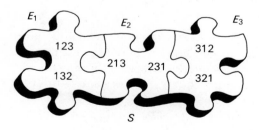

E_1 E_2 E_3

123 312
213 231
132 321

S

---

* Recall from Chapter 2 that $n(S)$ denotes the number of elements in the set S.

In determining the number of elements in a set, we should look for patterns and thereby avoid having to list all the elements of the set. In particular, for the sets E_2 and E_3 of the above example, we recognize that the number of elements in each of these sets is the same as the number of ways to arrange two distinct objects in a row. We also see that the set of objects to be counted is partitioned into subsets with each pair being disjoint; since each of these is the same size, we have added the same number several times and we could therefore express this sum as a multiplication. In fact, since this situation will occur often, we should keep in mind that *repeated addition can always be expressed as a multiplication.*

EXAMPLE   ■ A scientist wants to test the effects of a particular drug on animals by comparing the dosage used to the size of the animal. He chooses to give three different sized doses to each of five types of animals of different weights. How many different experiments will he conduct?

Each type of animal will be tested with three doses of the drug. Since there are five types of animals, our answer is three added five times, which is the same as $5 \cdot 3$. □

We have illustrated thus far a very important counting principle.

A COUNTING PRINCIPLE   **Let S be a finite set of elements, where each element can be thought of as the outcome of performing a first task followed by performing a second task. Suppose that the number of ways to do the first task is N. If for each way the first task can be performed there are K ways to perform the second task, then the number of elements in S is $N \cdot K$—that is, $n(S) = N \cdot K$.**

In the example on drug testing, the first task can be thought of as picking a certain type of animal, which can be done in five ways. The second task of selecting the dosage has three possible outcomes and therefore our answer for that example is $5 \cdot 3$.

We have stated our counting principle for two tasks; however, it can be extended to more than two, as we do in the next example.

EXAMPLE   ■ A college registrar is constructing schedules for students who must elect one history, one English, and one mathematics course. There are four history, five English, and three mathematics courses available where there is no conflict of time between any two departments' courses. How many different possible schedules are there?

There are three tasks to be performed in succession: first, pick a history course, then an English course, and finally a mathematics course. Since for *each* choice of a history course there are five choices for the English course followed by three choices for the mathematics course, our answer is $4 \cdot 5 \cdot 3 = 60$. □

1. How many four-digit numbers can be formed by using 2, 4, 6, and 8, allowing any one of these digits to be used more than once in the number?

2. In the drug-testing example, how many experiments would be performed if four different dosages are used on each of the five types of animals?

3. How many choices of a meal are there if you can pick from four appetizers, five main dishes, and three desserts?

4. How many numbers with three or less digits can be formed using the digits 1, 2, 3, and 4, not allowing any digit to be used twice in a number?

1. 256   2. 20   3. 60   4. 40

The type of counting question we will now consider concerns the arrangement of objects. For example, how many ways can three filters be arranged in a water purification column? We should realize that the question of arrangement might not be explicitly stated in the problem. "The number of three-digit numbers formed by using 1, 2, and 3" is the same as "the number of arrangements in a line of the three objects 1, 2, and 3."

We call an arrangement of n objects in a row a PERMUTATION of the n objects. **DEFINITION**

It is easy to see that the number of permutations of two distinct objects is $2 \cdot 1$. For example, there are two arrangements of the letters A and B in a line, namely, AB and BA.

Now consider three distinct objects, which we will label A, B, and C. To determine the number of permutations of these three objects, we use our counting principle. The first task is to fill the first position in the arrangement. This can be done in three ways:

$$A - -   B - -   C - -$$

For each of these three outcomes, there are two ways to arrange the remaining two objects in the next two positions. We see therefore that there are $3 \cdot (2 \cdot 1) = 6$ ways to arrange three distinct objects in a row.

$$ABC \quad BAC \quad CAB$$
$$ACB \quad BCA \quad CBA$$

Without too much effort we realize that the number of permutations of four distinct objects can also be obtained from our counting principle. As before, the first task is to fill the first position in the arrangement, which can be done in four ways. The second task is to arrange the other three objects in the remaining three positions. We have already seen that the second task can be done in $3 \cdot 2 \cdot 1$ ways; thus by our counting principle, the number of permutations of four distinct objects is $4 \cdot (3 \cdot 2 \cdot 1) = 24$.

A permutation of the objects in a set is an arrangement of its members in a straight line. (Photo by Cliff Garboden, Stock Boston.)

Using the same reasoning, we observe that the number of ways to arrange five distinct objects in a row is $5 \cdot (4 \cdot 3 \cdot 2 \cdot 1) = 120$. In general, the number of permutations of n distinct objects is $n \cdot (n - 1) \cdot (n - 2) \cdots 2 \cdot 1$. For convenience, we introduce a special symbol for this number.

For any positive integer n, we can write $n! = n \cdot (n - 1) \cdot (n - 2) \cdots (2) \cdot (1)$, where $n!$ is read "n factorial".   DEFINITION

The number of permutations of n distinct objects is $n!$.   PRINCIPLE

When we ask, "How many ways can the objects of a set be arranged in order?", we are asking for the number of permutations of the elements. (Photo by William R. Devine.)

From our discussion, we should realize that n factorial is the same as n times $(n - 1)$ factorial; that is, $n! = n \cdot (n - 1)!$.

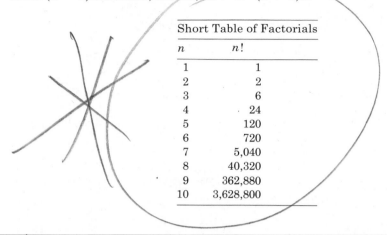

Short Table of Factorials	
n	$n!$
1	1
2	2
3	6
4	24
5	120
6	720
7	5,040
8	40,320
9	362,880
10	3,628,800

EXAMPLES

■ How many ways can three different filters be arranged in a water purification column?

The answer to this question is the number of permutations of three distinct objects, which we know to be $3! = 6$. □

■ How many batting orders are there for a particular baseball team of nine players?

It is easily seen that the answer is the number of permutations of nine distinct objects, which we know to be $9! = 362,880$. □

■ Suppose for a baseball team of nine players, the batting order must be such that the pitcher bats last and the catcher bats fourth. How many possible batting orders are there for this team?

Since two of the objects, the pitcher and the catcher, must be placed in specific positions in the batting line-up, there are only seven players to be arranged among the remaining seven batting positions. Therefore, the answer to this question is $7! = 5040$. □

■ There are four persons who are eligible for the four offices of a club. How many executive boards of officers are possible?

Since we are asked to arrange the four persons among the four positions, President, Vice President, Secretary, and Treasurer, the answer to this question is 4! = 24. □

■ In planning a housing development, a contractor has five different models which he will build along each street. How many streets can there be in this development so that no two streets will have the same arrangement of the five models?

There are 5! = 120 different permutations of the five models. Therefore if there are no more than 120 streets in the development, then no arrangement of the five models has to be repeated. □

Another arrangement question which we will look at is that of counting the number of permutations of a subset of objects chosen from a given set of distinct objects. A question of this type is given in the next examples.

■ How many two-digit numbers can be formed from the digits 1, 2, 3, and 4, where no digit is used more than once in the number?

EXAMPLES

Let S be the set of objects we wish to count. The elements of S can be thought of as the result of performing two tasks in succession. The first task is to select a digit for the tens position and the second task is to select a digit for the units position from among the remaining three digits. The number of ways to select a digit for the tens position is four. Because for *each* way we can accomplish the first task there are three ways to do the second task, we have that the number of elements in S is $4 \cdot 3$. □

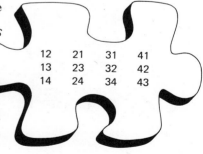

S

12	21	31	41
13	23	32	42
14	24	34	43

■ A consumer survey is made on five brands of coffee. Each person surveyed is asked to rate three of the brands, giving a first, second, and third choice. How many different ratings are possible?

Let S be the set of possible ratings. Each element of S can be thought of as the outcome of performing three tasks. The first task is to select the most preferred brand; the second task is to pick the second most preferred brand from the remaining four; and the third task is to give the third choice from the remaining three brands. Now, there are five ways to satisfy the first task. Since for each of these there are four ways to accomplish the second task and three ways to accomplish the last task, we have $n(S) = 5 \cdot 4 \cdot 3 = 60$. □

With these two examples in mind, we see from our counting principle that the number of permutations of four objects selected from six distinct objects is $6 \cdot 5 \cdot 4 \cdot 3$. We should now recognize that the number of ways to select k objects from a group of n distinct objects and to arrange these objects in a row is the product of k numbers; the first number in this product is n, the second number is $n - 1$, etc. We shall denote this product by $P[n,k]$.

DEFINITION $P[n, k]$ **is the number of permutations of k objects taken from a set of n distinct objects. We read $P[n,k]$ as the NUMBER OF PERMUTATIONS OF n OBJECTS TAKEN k AT A TIME.**

PRINCIPLE   **The number of permutations of n objects taken k at a time is a product of k numbers starting with n, the next number being $(n - 1)$ and so forth.**

From the above principle,

$$P[n,k] = \underbrace{n(n - 1) \cdots (n - k + 1)}_{k \text{ factors}}.$$

Therefore we can also write this number as

$$P[n,k] = \frac{n!}{(n - k)!}$$

QUIZ YOURSELF*   1. $P[5,2] = ?$

2. What is the number of permutations of five things taken two at a time?

3. What is the number of permutations of six things taken three at a time?

4. $P[7,2] = ?$

5. $P[5,5] = ?$

---

*ANSWERS   1. $5 \cdot 4 = 20$ 2. $P[5,2] = 20$ 3. $P[6,3] = 6 \cdot 5 \cdot 4 = 120$ 4. $7 \cdot 6 = 42$
5. $5 \cdot 4 \cdot 3 \cdot 2 \cdot 1 = 120$

EXAMPLES   ■ How many different choices for a basketball team are there if the team is selected from among 11 players, where any one of the eleven can play any of the 5 positions?

For this question, we are asked to find the number of permutations of 11 objects taken 5 at a time. Therefore the answer is $P[11,5] = 11 \cdot 10 \cdot 9 \cdot 8 \cdot 7 = 55{,}440.$ □

■ A psychologist presents a subject with the names of ten well-known persons, among them are military leaders, politicians, and humanitarians. The subject is to respond by ranking in order the four he could most easily identify with. How many different responses are possible?

The subject is to select four out of the ten names and arrange them in some order. Therefore the number of possible responses is $P[10,4] = 10 \cdot 9 \cdot 8 \cdot 7 = 5040.$ □

(Photo by Jeff Albertson, Stock Boston.)

■ An anagram is a rearrangement of the letters of a word. How many two- or three-letter anagrams are there if the letters are chosen from the word **love**?

The number of ways to select two letters from the four and arrange them is $P[4,2] = 12$. The number of three-letter anagrams chosen from **love** is $P[4,3] = 24$. Therefore there are 36 two- or three-letter anagrams chosen from the word **love**. □

■ In an art class, a student is shown six paintings and five sculptures. The student is asked to list in order of preference three of the paintings and two of the sculptures. How many different preferential lists are possible?

The number of possible ways to rank three of the six paintings is $P[6,3] = 120$. For *each* of these, there are $P[5,2] = 20$ ways to select and rank two of the five sculptures. Therefore, by our counting principle, there are $120 \cdot 20 = 2400$ possible lists. □

As you do the following exercises, you may find that in some cases it is not immediately apparent which formula or procedure to use in obtaining an answer. Keep in mind that in this case it is a good idea to list some of the elements from the set to be counted. If you can recognize how the elements could be organized within a list, then you should acquire a feeling for some method of counting that could be used.

EXERCISES

1. How many four-digit numbers can be formed by using the digits 1, 2, 3, and 4 if no digit can be used twice in the same number?

2. How many four-digit numbers can be formed by using the digits 1, 2, 3, and 4 if there is no restriction on the number of times a digit is used in forming a number?

3. How many four-digit numbers can be formed by using the digits 1, 2, 3, and 4 if a digit can be used more than once in the number and if we insist that no two consecutive digits are allowed to be the same (for example, 1413 is allowed but 1443 is not)?

4. At a restaurant a full-course dinner consists of appetizer, entree, and dessert. If a diner has a choice of three appetizers, six entrees, and five desserts, how many different full-course dinners are available?

5. A car manufacturer wants to give a customer the following options on one of its car models:
   a) He may choose any one of 14 colors;
   b) He may have standard or radial tires;
   c) He may have no radio, an AM radio, or an AM-FM radio;
   d) He may have an air bag, seat belts, or both.
   How many different types of cars are available to any customer?

6. In how many ways can seven people be lined up for a group picture?

7. In how many ways can four girls and three boys be lined up for a group picture if there must be a boy between each pair of girls?

8. A night watchman at an office building must punch his time clock at each of six stations as he makes his rounds. Because of the type of work the watchman is doing, he does not always want to check the stations in the same sequence. How many different routes are available to the watchman?

9. In one question on a history test, the student is asked to match ten dates with ten events; each date can only be matched with one event. How many different ways can this question be answered?

10. How many three-digit numbers can be formed using the digits 1, 2, 3, 4, and 5 with no digit being used twice in a number?

11. In a sled-dog race, prizes are awarded to the first-, second-, and third-place teams. In how many ways can the awards be made if there are 15 teams competing in the race?

12. When using mathematics to create music, a composer might be interested in determining the length of a musical composition in which no two phrases are the same. (A phrase is an arrangement of notes.) The composer decides that each phrase must consist of 5 notes chosen from the 12 possible notes. This composition could consist of how many phrases?

13. On a preferential ballot, a person is asked to rank three of seven candidates for the office of chairperson, giving first, second, and third choices. What is the minimum number of ballots that must be cast in order to guarantee that at least two ballots are the same?

14. A certain IBM data card is divided into five fields. One field is used for a student's number, another contains the student's name. The remaining three fields are

for address, credits earned, and QPA. In how many ways can these fields be arranged on the card? In how many ways can the information be arranged on the card, if the field for the address must immediately follow the field for the student's name?

15. A college wishes to use the same number of digits for each student's ID number. If the college has 9539 students, then what is the smallest number of spaces that should be set aside on an IBM card for the ID number?

16. How many numbers greater than 2000 can be formed using the digits 1, 2, 3, 4, 5, and 6 with no digit being used twice in a number?

17. How many numbers are there between 1 and 999 which can be divided evenly by 5?

COMBINATIONS

In the previous section, we obtained a formula for the number of permutations of n objects taken k at a time. Sometimes we are only interested in counting the number of ways to select the k objects from the n distinct objects and are not interested in the arrangements of the k objects. For example, if we want to select a committee of 5 people from a group of 11 people, then the order in which the people are selected is not important. Counting the number of different possible committees is the same as counting the number of different subsets of 5 people that can be selected from the set of 11 people. Each of these subsets is called a **combination**.

The **NUMBER OF COMBINATIONS OF n THINGS TAKEN k AT A TIME is the number of subsets of size k contained in the set of n objects. This number is denoted by the symbol $C[n,k]$.**   DEFINITION

In our first example we will obtain the value for the number $C[4,2]$.

■ Let us count the number of combinations of two objects chosen from the   EXAMPLE
set $\{A,B,C,D\}$. One such combination is the subset $\{A,B\}$. Corresponding to this combination there are 2! permutations, namely AB and BA. In fact, for each such combination there will be 2! permutations.

Combinations	Permutations	
$\{A,B\}$	AB	BA
$\{A,C\}$	AC	CA
$\{A,D\}$	AD	DA
$\{B,C\}$	BC	CB
$\{B,D\}$	BD	DB
$\{C,D\}$	CD	DC

We see that 2! times the number of combinations of four things taken two at a time is the same as the number of permutations of four things taken two at a time. We can rewrite this symbolically as $2! \cdot C[4,2] = P[4,2]$. Therefore dividing by $2!$, we obtain

$$C[4,2] = \frac{P[4,2]}{2!} = \frac{4 \cdot 3}{2} = 6. \ \square$$

We can use the same type of procedure to determine, for example, the value of the number $C[5,3]$. $C[5,3]$ is the number of combinations of three objects selected from a set of five objects. Corresponding to *each* of these combinations there are $3!$ permutations of the three objects. By our counting principle, the number $P[5,3]$ is the same as $3!$ times $C[5,3]$. Therefore, dividing by $3!$, we obtain

$$C[5,3] = \frac{P[5,3]}{3!} = \frac{5 \cdot 4 \cdot 3}{6} = 10.$$

Our discussion points out that $P[n,k]$ is the same as $k!$ times $C[n,k]$. Therefore to compute $C[n,k]$ we need only divide $P[n,k]$ by $k!$

PRINCIPLE
$$C[n,k] = \frac{P[n,k]}{k!}.$$

For example, the number of combinations of six things taken three at a time is

$$C[6,3] = \frac{P[6,3]}{3!} = \frac{6 \cdot 5 \cdot 4}{6} = 20.$$

Since a combination is a subset and since the empty set is a subset of every set, it is not difficult for us to accept that for any n, we have $C[n,0] = 1$.

EXAMPLES   ■ A committee of four persons is to be selected from a group of 12 people. How many different possible committees are there?

This is a combinations problem with the answer being

$$C[12,4] = \frac{12 \cdot 11 \cdot 10 \cdot 9}{4!} = 495. \ \square$$

■ A committee of six people is to be selected from the following group of 13 people: 6 lawyers, 2 teachers, and 5 businessmen. If the committee must consist of three lawyers, one teacher, and two businessmen, then how many such committees are there?

In order to select the committee, we may first choose the lawyers, then the teacher, and finally the businessmen.

The number of ways to fill three seats on the committee with lawyers is $C[6,3]$. For *each* choice of the lawyers, the number of ways to select one teacher and two businessmen for the remainder of the committee is $C[2,1] \cdot C[5,2]$. Therefore the answer to this question is

$$C[6,3] \cdot C[2,1] \cdot C[5,2] = 20 \cdot 2 \cdot 10 = 400. \quad \square$$

■ An English instructor gives an assignment in which a student must read and review any five novels chosen from a list of seven novels. If there are 25 students in the class, is it possible that no two students will select the same five novels?

The number of ways that a student can select his five novels from the seven available is $C[7,5]$. To simplify the computation, we should realize that this number is the same as $C[7,2]$, for when the student selects the five novels he wishes to review, he is at the same time picking two novels that he will not review. Now, $C[7,2] = 21$.

Since there are only 21 ways in which to pick 5 novels, there are at least two students in this class of 25 who must choose the same 5. $\quad \square$

In the above example, we stated that the number $C[7,5]$ is the same as $C[7,2]$. In fact, it is always true that $C[n,k] = C[n,n - k]$.

Another important consideration to keep in mind is the difference between $C[n,k]$ and $P[n,k]$. In a counting question, whenever the *order* in which the objects are selected is *not important*, then we want to find the number of *combinations*, $C[n,k]$. In our example above, the order in which the five novels were selected was not important and that is why we found the number of combinations and not the number of permutations.

If we choose objects from a collection without regard to the order in which they are picked, then we are considering combinations of the objects rather than permutations. (Photo by Frank Siteman, Stock Boston.)

EXERCISES

1. In Morse code, a letter of the alphabet is designated by an arrangement of dots and dashes; for example, *A* is · –. How many letters could be coded using three dots and one dash?

2. A college committee must submit to the board of trustees the names of three persons for the position of college president. The committee has obtained seven names, any three of which they feel would be a good choice. If the committee decides to pick three at random, how many possible choices are available to them?

3. A pollster has 50 people available from which he selects 5 to survey for an opinion on a particular issue. How many polls can be done before the pollster has to survey the same group of 5 people?

4. An urn contains 20 different chips of various colors. A sample of three is selected from the urn. In how many possible ways can the sample of three be drawn? If five of the chips are red, six are blue, and nine are green, how many ways can the sample be drawn in which each chip is of a different color? In how many ways can the sample of three be drawn if two of the chips are to be red?

5. A deck of playing cards contains 52 distinct cards divided equally into four suits—hearts, diamonds, clubs, and spades. How many different possible five-card combinations are contained in the deck, where all five cards are hearts? How many possible five-card combinations are there where all five cards are of the same suit?

6. The thirteen cards of a particular suit have values 2, 3, 4, . . . , 10 (number cards) and J, Q, K, and A (face cards). How many different possible five-card combinations are contained in the deck, where four of the cards are jacks (J)? How many possible five-card combinations are there, where three of the cards are face cards and two are number cards?

7. A committee of six people is to be chosen to represent a school. It is agreed that one administrator, two science teachers, and three humanities teachers are to be on the committee. How many different committees are there to choose from if there are three administrators, seven science teachers, and 11 humanities teachers at the school?

8. In how many ways can a jury of 12 with 2 alternates be selected from a pool of 25 prospective jurors?

9. A mountain climbing team of twelve people divides into three groups. Five of the twelve remain in base camp, three take the easier route to the summit, and four make the more difficult ascent. In how many different ways can this division be made? In how many different ways can this division be made if a particular exhausted climber must remain in base camp?

10. How many committees of six people are there available to choose from if the members come from a group of seven men and five women and if the committee must have at least three women?

11. There are four weights available, 2, 3, 4, and 8 ounces. Using a balance and these weights, how many items of various weights can be measured?

due for wed. Dec. 2nd

MASTERY TEST: TOOLS OF COUNTING

& pg. 230

1-5

1. Find the value for each of the following numbers.
   - a) $P[5,3]$
   - b) $C[5,3]$
   - c) $C[5,2]$
   - d) $P[5,2]$
   - e) $C[10,4]$
   - f) $P[10,4]$
   - g) $C[6,6]$
   - h) $C[6,0]$

2. What number times $C[5,3]$ would give $P[5,3]$? What number times $C[10,4]$ would give $P[10,4]$?

3. A certain state plans to change its license plates so as to consist of three letters followed by three digits. If only the digits 1, 2, 3, . . . , 9 can be used in the number portion and if any of these digits and any of the 26 letters can be used more than once in a particular plate, then how many autos can be registered?

NEW TERMS

Permutation

Combination

***n* Factorial (*n* !)**

Number of permutations of *n* things taken *k* at a time ($P[n,k]$)

Number of combinations of *n* things taken *k* at a time ($C[n,k]$)

What patterns do you see that would help you to determine the number of bricks in this plaza? (Photo by Jean-Claude Lejeune, Stock Boston.)

4. A committee is to cast a preferential ballot for its chairman; there are four nominees. A preferential ballot is one in which the voter indicates his first choice, second choice, etc. How many different ballots could be cast?

5. Suppose in exercise 4, there are seven nominees and a voter must rank four of the seven. How many different ballots could be cast?

6. An electronic computer represents a number by a sequence of "on" or "off" switches. The number 0 is represented when all the switches are off. The number 1 is represented when the first switch is on and the remaining switches are off. How many numbers can be represented with 4 switches? How many switches are needed to represent all the numbers from 0 to 63 inclusively?

7. In a psychological test, a subject is shown the names of 15 famous people and asked to select three with whom he can identify. How many different responses are possible?

8. In a small town, which is divided by railroad tracks, there are 100 families living on the north side of the tracks and 150 families living on the other side of the tracks. A pollster samples 25 of these families. How many samples are there to choose from if 10 of the families are selected from those living on the north side of the tracks?

9. In order to avoid boredom when working on an assembly line, a production worker's assignment is changed from week to week. Each day of the week he is assigned to one of five work stations and, during any given week, he will work at each of the five stations. How many weeks can the production worker spend at the plant before a week's assignment would be exactly the same as one of the previous weeks' assignments?

10. A telephone number consists of 7 digits, where the first digit is 2, 3, 4, 5, 6, 7, 8, or 9. In order to have direct distance dialing, area codes were assigned. Would it be possible to have direct distance dialing without assigning area codes? Why?

III. A MODEL FOR COUNTING THE SHORTEST ROUTES

We now have at our disposal enough counting "tools" to answer the question posed in Part I of this chapter.

You will remember that a postmaster of a certain city wished to determine the best route from a substation to the main post office. He also wanted the driver to follow a different route each day and record the time required. How many days are needed for the postmaster to complete his study of all possible routes?

Figure 7.2 gives a partial street map of the city which includes the substation at point A and the main post office at point B.

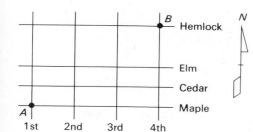

Figure 7.2

We realize that for the truck to travel from point A to B as directly as possible it must travel three blocks east and four blocks north. When using the street map in Fig. 7.2, we recognize that the lengths of the blocks are not important in determining the number of routes. We can specify a possible route by giving a sequence of seven directions, of which three are to the east and four are to the north. For example, one route would be to travel east, north, east, north, east, north, north. To give any other route, we must choose from the seven directions to be traveled, three of them to the east. Thus we see that we can use the tools of combinations to solve our counting question. In particular, the number of possible routes to be tested is the same as the number of combinations of seven items taken three at a time.

Since $C[7,3] = 35$, we now know that it will take the postmaster 35 days to complete his study.

Let us consider some interesting variations of the above question.

■ Using the map given in Fig. 7.2, find the number of routes from point A to point B which include Walnut Street between 3rd and 4th Avenues and in which travel is always east and north.   **EXAMPLES**

Obviously, if the truck is to travel over the desired street, it must first reach the intersection of 3rd Avenue and Walnut Street. To specify one of these routes from point A to 3rd Avenue and Walnut Street, we must give five directions, of which two are to the east. Therefore there are $C[5,2]$ = 10 such routes.

For *each* of the 10 routes which take us to 3rd Avenue and Walnut Street, there is only one way to point B that goes over Walnut Street. Therefore there are $10 \cdot 1 = 10$ routes from point A to point B which utilize Walnut Street between 3rd and 4th Avenues. □

■ Refer to the map given in Fig. 7.2. If Walnut Street between 3rd and 4th Avenues is under construction and cannot be used, how many shortest routes are now possible from point A to point B?

For the moment let us assume that Walnut Street between 3rd and 4th Avenues could be used. We know from our earlier discussion that there are 35 shortest routes from point A to point B.

Now, because of the construction, some of these 35 routes cannot be used; from our previous example we saw that there are 10 such routes. Therefore the number of routes from point A to point B that do not utilize Walnut Street between 3rd and 4th Avenues is $35 - 10 = 25$. □

Another complication, which could arise, is illustrated in the next example.

EXAMPLE   ■ If Walnut Street is made a one-way street going west, what is the number of shortest routes from point A to point B?

To obtain a shortest route travel must always be to the east and north. Since the flow of traffic on Walnut is to the west, it cannot be considered as part of any route. Therefore for our purposes we can eliminate Walnut Street from the map.

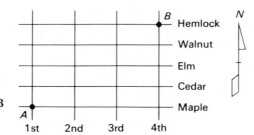

Figure 7.3

Using the modified map in Fig. 7.3, we see that a possible route from point A to point B can be given by a sequence of six directions, of which three are to the east. Therefore the solution to this problem is $C[6,3] = 20$. □

The way in which we obtained the answer to the shortest-routes question can also be applied to other counting problems.

■ A coin is flipped five times. In how many ways can exactly three heads appear?

An example of one of the possibilities in which three heads appear is HTTHH. To give another example we realize that we would have to choose three of the flips to show heads. Therefore we see that the total possible ways in which three heads can occur on the five flips is the number of combinations of five items taken three at a time. Thus the answer to our question is $C[5,3] = 10$. □

■ On a 20-question true-false test, a passing score is at least 16 correct answers. In how many ways can the questions be answered to pass the test?

The number of ways in which exactly 16 questions are answered correctly is $C[20,16]$. Also, the test can be passed by obtaining exactly 17 correct answers, which can occur in $C[20,17]$ ways. Continuing, we see that the total number of ways in which 16 or more questions can be answered correctly is $C[20,16] + C[20,17] + C[20,18] + C[20,19] + C[20,20]$.

Recall that $C[20,16] = C[20,4]$, $C[20,17] = C[20,3]$, etc. Therefore to simplify the computation, we can rewrite the above as $C[20,4] + C[20,3] + C[20,2] + C[20,1] + C[20,0] = 4845 + 1140 + 190 + 20 + 1 = 6196$. □

EXERCISES

Exercises 1 to 3 refer to the street map in Fig. 7.4.

1. How many routes are there from the postal substation located at point A to the main post office at point B if all streets can be used and travel is only to the north and east?

2. How many shortest routes are there from point A to point B if the street joining points X and Y cannot be used?

3. How many shortest routes are there from A to B if any street can be used except the one between points W and U?

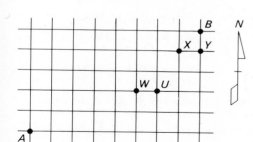

Figure 7.4

Exercises 4 and 5 refer to the street map in Fig. 7.5, in which all the streets are one-way with the direction indicated by the arrows.

4. What is the number of shortest routes from point A to point B?

5. What is the number of shortest routes from point A to point C?

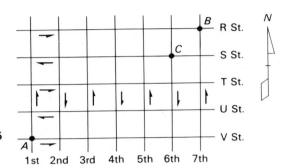

Figure 7.5

6. Consider a ten-question test, where each question can be answered either right or wrong. How many different ways are there to answer the questions so that exactly eight are right and two are wrong? *Note*: One way is for the first eight questions to be right and the last two wrong. How many ways are there to answer the questions so that at least 8 are right?

7. A drug company wants to test a new vaccine. They select ten animals and inject each with the vaccine. How many ways can the experiment show at least seven positive results?

8. A person, who claims to be a wine connoisseur, is given five glasses of wine labeled a, b, c, d, and e. For each glass, he is asked to determine whether it is from New York State or California. If he is correct at least four times, then he passes the test and is accepted as a wine connoisseur. In how many ways can he pass the test?

SUGGESTED READINGS

GARDNER, M., "Mathematical Games: A Handful of Combinatorial Problems Based on Dominoes." *Sci. Am.*, Dec 1969, p. 122.
On counting the number of ways to arrange a set of dominoes using graph tracings.

GARDNER, M., "Mathematical Games: The Combinatorial Richness of Folding a Piece of Paper." *Sci. Am.*, May 1971, pp. 110–116.
Presents the question of how many ways to fold a rectangular map; an upper bound is given as n!.

KEMENY, J. G., J. L. SNELL, AND G. L. THOMPSON, *Introduction to Finite Mathematics*, 3rd ed. Englewood Cliffs, N.J.: Prentice-Hall, 1974.
Several examples and a theorem are given on counting the number of shortest routes in a rectangular grid.

RUEFF, M., AND M. JEGER, *Sets and Boolean Algebra*, New York: American Elsevier, 1970.
Contains a development of the standard counting formulas by way of mappings on a set.

Probability

The origin of probability theory can be traced back to prehistoric times. Archaeologists have found numerous small animal bones, called astrogali, which are believed to have been used by early men in playing various games of chance. Similar bones were later ground into the shape of cubes, decorated in various ways, and eventually evolved into modern dice. Games of chance were played for thousands of years, but it was not until the sixteenth century that a serious attempt was made to study them using mathematics.

One of the first to consider these questions was the Italian Girolamo Cardano, one of the most colorful men in the history of mathematics. In addition to being a scientist, Cardano was also an astrologer, which caused him frequent misfortune. He once cast the horoscope of 15-year-old King Edward VI of England, in which he made specific predictions of events that would occur on the seventeenth day of the third month of the fifty-fifth year of Edward's life. Unfortunately, Cardano was quickly discredited when Edward died the following year at the age of 16. He then cast the horoscope of Jesus and for this blasphemy he was sent to prison. Shortly thereafter, he predicted the exact date of his own death and, when the day arrived, Cardano made his prediction come true by taking his own life.

The work begun by Cardano was continued by others in the seventeenth century. A nobleman of the time, Antoine Grombauld, the Chevalier de Méré, asked Blaise Pascal (1623–1662) several questions concerning gambling.

"How many times should a single die be thrown before we could reasonably expect two sixes?"

217

Girolamo Cardano (1501–1576)

"How should the prize money in a contest be fairly divided in the case that the contest, for some reason, cannot be completed?"

In order to reply, Pascal and his friend Pierre Fermat (1601–1665) began a systematic study of the games of chance, which is now generally recognized as the beginning of modern probability theory.

Following Pascal and Fermat, some of the most distinguished mathematicians of the seventeenth and eighteenth centuries made further contributions in this area. In 1812, Pierre-Simon de Laplace (1749–1827) wrote *Théorie Analytique des Probabilités*, in which he presented what is now recognized as classical probability theory. Convinced of the applicability of his work, Laplace boldly affirmed that all knowledge could be obtained by using the principles he set forth.

During the nineteenth and twentieth centuries, many important applications of probability theory have been found. Physicists use it in studying various gas and heat laws and in the theory of radiation and atomic physics, while biologists apply it in genetics, in studying natural selection, and in mathematical-learning theory. In addition, probability theory provides a theoretical basis for statistics, a subject used in social, scientific, and industrial research.

Since the time of Cardano, Pascal, Fermat, and Laplace, a great deal of work has been done in probability theory, making it one of the most active and applied areas of contemporary mathematics.

I. SHOULD THE CLAIM BE ACCEPTED?

Suppose that during the cold season 50 percent of the population contracts the sniffles at least once; or in other words, the chance of any one person catching a cold is $\frac{1}{2}$. The Cure-All Pharmaceutical Company claims they have discovered a vaccine to help protect a person from catching a cold. While the company is not claiming that they have "cured" the common cold, they can at least reduce the chances of a person's catching a cold.

To protect the consumer from fraud, we wish to design an experiment to test the validity of this claim. Obviously, the vaccine cannot be tested on the entire United States population; however, when testing it on a small group we must be careful in interpreting our results. It would not, for example, be reasonable to say that the chances of anyone catching a cold is zero if we vaccinate ten people and none of them catch a cold.

It would be difficult to design an experiment to determine the chances of a person's catching a cold when using the vaccine; however, we can design an experiment so that, if we accept the company's claim, the likelihood of accepting a worthless vaccine is small. In this chapter, we will develop the "tools" of probability that will enable us to test the claim of the Cure-All Company.

$$Pr(E) = \frac{2}{3}$$

$$Pr(E') = \frac{1}{3}$$

1,1	1,2	1,3	1,4	1,5	1,6
2,1	2,2	2,3	2,4	2,5	2,6
3,1	3,2	3,3	3,4	3,5	3,6
4,1	4,2	4,3	4,4	4,5	4,6
5,1	5,2	5,3	5,4	5,5	5,6
6,1	6,2	6,3	6,4	6,5	6,6

II. TOOLS OF PROBABILITY

PROBABILISTIC MODELS

The following statement should be familiar to you. "The chance of obtaining a head on the flip of a coin is $\frac{1}{2}$." The rolling of a die is another familiar process in which the likelihood of an outcome is considered.

Flipping a coin and rolling a die are two examples of **experiments** which are considered in a study of probability.

An **EXPERIMENT** is a chance process, which can be performed many times and which has a finite number of possible **OUTCOMES**. The set of possible outcomes is called the **POSSIBILITY SET** of the experiment.   **DEFINITION**

Probability theory is used by oil companies to decide not only what land is likely to produce oil but also to determine how much to bid on a tract and still ensure a reasonable return on the investment. (Photo by Owen Franken, Stock Boston.)

In any experiment we can say, "We know what *could* happen, but we do not know what *will* happen prior to performing the experiment."

■ We give the following as examples of experiments:

a) rolling a die and noting the number of dots on the upturned face; → *getting a 2 Probability would be $\frac{1}{6}$*

b) tossing a coin and observing the upturned face;

c) selecting an item from a production line and testing to see if it is defective or not;

d) observing the path a mouse runs in a T-shaped maze.

For each experiment, there is a finite set of possible outcomes. The possibility sets for the above examples of experiments are as follows:

a) $\{1,2,3,4,5,6\}$;

b) $\{T,H\}$;

c) {defective, nondefective};

d) {right, left}. □

Give the possibility set for each of the following experiments.

1. A card is drawn from an ordinary deck of 52 cards and the suit of the card is observed.

2. Money is put into a vending machine which gives either the item or the money back.

3. A pair of dice is thrown and the sum of the appearing dots is noted.

---

1. {heart, diamond, club, spade}   2. {item, money}   3. $\{2,3,4,\ldots,12\}$

In the flipping-a-coin experiment, it is generally accepted that the chance of a head appearing is $\frac{1}{2}$. Let us look at what is meant by the number $\frac{1}{2}$. Suppose a coin is flipped and a tail appears. Since the chance of a head appearing is $\frac{1}{2}$, can we say with assurance that on the next flip a head will appear? Obviously, we do not expect that on any two trials of flipping a coin one flip will yield a head and the other a tail, even though we would like to consider $\frac{1}{2}$ as the ratio of the number of heads appearing to the number of flips.

Try the following: Flip a coin 50 or more times and tally the number of heads that appear. The ratio of "the number of times a head appears" to "the number of times the coin is flipped" is called the **relative frequency** of a head appearing. In flipping the coin, say 50 times, you should find that the relative frequency is almost $\frac{1}{2}$. In fact, as the coin is flipped more and more times, the relative frequencies will remain close to $\frac{1}{2}$. We would then say that the **probability** of a head appearing is $\frac{1}{2}$.

DEFINITION   **Suppose an experiment is performed several times. The RELATIVE FREQUENCY of a possible outcome is the ratio**

$$\frac{\text{Number of times the outcome occurs}}{\text{Number of times the experiment is performed}}.$$

Let us agree to the following principle of relative frequency.

PRINCIPLE OF RELATIVE FREQUENCY   **Consider an experiment which is performed many times. As we increase the number of times the experiment is performed, the relative frequencies we obtain for a particular outcome will get close to some fixed number. This number is called the PROBABILITY of that outcome.**

In order to appreciate the above principle and the meaning of the probability of an outcome, you should perform the flipping-a-coin experiment. After each flip of the coin, include the result with those you have already tabulated and recompute the relative frequency of a head appearing. Even though these relative frequencies may not be close to $\frac{1}{2}$ when the number of flips is small, you should find that they eventually stay close to $\frac{1}{2}$, unless, of course, your coin or the way you flip it have some unusual characteristics.

For each outcome of an experiment which has been performed a large number of times, we could use the relative frequency of the outcome as its probability.

In addition to considering the probability of an individual outcome in an experiment, we will want to look at the probability of a set of outcomes. Let us determine what is meant by a question such as, "What is the probability of an even number appearing on the roll of a die?". The possibility set for this experiment is $S = \{1,2,3,4,5,6\}$. Corresponding to the sentence "the number is even" is a subset of S, namely, $E = \{2,4,6\}$.

We see that each statement about the outcomes of a given experiment defines a subset of the possibility set. Since this correspondence does exist between sentences and subsets, we could assign probabilities to the subsets. For example, since we know that the probability of obtaining a head is $\frac{1}{2}$, we could assign to the set $\{H\}$ the number $\frac{1}{2}$. Therefore for a particular experiment, if we know the numerical value assigned to each subset of the possibility set, then we would be able to answer any probability question concerning the experiment. We see now that it is desirable to have a procedure which assigns a probability to each subset of a possibility set S.

Since subsets of a possibility set are now of major importance, we give them a special name.

DEFINITION   **Given an experiment with possibility set S, an EVENT for the experiment is a subset of S.**

In order to *model* a particular experiment, we must give the possibility set and also a probability for each event. We shall denote the probability of an event E by $Pr(E)$.

We would like to determine the important properties of a **probability assignment**. To this end, consider the following situation.

An urn contains chips of three different colors—red, blue, and green; however, we do not know how many chips there are of each color. We select a chip from the urn, note its color, and return it to the urn. We see that the possibility set for this experiment is $S = \{r,b,g\}$. We perform the above experiment 100 times and the following results are obtained.

Color	Number of times drawn
Red	30
Blue	23
Green	47
	100

The probability we can assign to each outcome is its relative frequency in the 100 trials. Therefore we could say that the probability of obtaining a blue chip is $\frac{23}{100}$; that is, $Pr\{b\} = \frac{23}{100}$. The probabilities of the remaining two outcomes are $Pr\{g\} = \frac{47}{100}$ and $Pr\{r\} = \frac{30}{100}$. The total number of blue, green, and red chips appearing in the 100 trials must be 100; thus the sum of the relative frequencies and therefore the probabilities of the three outcomes must be 1.

Consider the following sentence: "The chip is white." We see that the subset of the possibility set S which this sentence defines is the empty set. Since the empty set is a subset of S, we would like to assign a probability to this event. Since no white chips appeared in any of the trials, it would only be reasonable to assign zero to this event.

Finally, consider the event $E = \{g,r\}$. Using set builder notation, we may write $E = \{x : x \text{ is green or red}\}$. The number of trials which yield an outcome belonging to E is 77. Therefore let us say that $Pr(E) = \frac{77}{100}$. We should observe that $\frac{77}{100}$ is the same as $Pr\{g\} + Pr\{r\} = \frac{47}{100} + \frac{30}{100}$.

With the above discussion in mind, we can state the basic assumptions for a probability assignment.

Consider an experiment with possibility set S. A probability assignment for S must satisfy the following: **PRINCIPLES OF PROBABILITY**

a) **The probability of the empty event is zero.**

b) **The probability of each outcome is greater than or equal to zero and less than or equal to one.**

c) **The sum of the probabilities of all the outcomes in S must be equal to one.**

d) **The probability of any nonempty event E is the sum of the probabilities of the outcomes which are contained in E.**

A PROBABILISTIC MODEL for a particular experiment consists of the possibility set S and a probability assignment for S which obeys the principles of probability.   **DEFINITION**

True or False? The probability that all the eggs will hatch is at least 1/6. (Photo by Peter Southwick, Stock Boston.)

When giving a probability assignment for S, it is enough just to assign a probability to each outcome. Once we have done this, we can find the probability of any nonempty event by adding the probabilities of the outcomes contained in it.

The probability that is assigned to each individual outcome can be obtained in one of many ways. For example, we could use the relative frequency for each outcome if the experiment had been performed many times. Sometimes probabilities are given for situations which occur only once. In this case, one may wish to assign probabilities based upon intuition. For example, you may feel that your chances of getting an A for this course are $\frac{9}{10}$. For this particular possible outcome, one may not be able to disagree with you. However, we would disagree with your assignment if you said at the same time that your chances of getting a B are also $\frac{9}{10}$. Your assignment now violates one of the principles of a probability assignment—namely, the sum of the probabilities of all the possible outcomes is 1.

Lastly, the assignment could be made based upon a knowledge of the physical characteristics of the experiment, as in selecting a card from a well-shuffled deck.

We shall consider these aspects of obtaining a probabilistic model in our next examples.

 # Probability is always between zero & one

■ Consider the experiment of rolling a fair die. Let us give a probabilistic model for this experiment.

The possibility set is $S = \{1,2,3,4,5,6\}$. Since we are assuming that each outcome has the same chance of occurring, our probability assignment for S is $Pr\{1\} = \frac{1}{6}$, $Pr\{2\} = \frac{1}{6}, \ldots, Pr\{6\} = \frac{1}{6}$. Observe that this assignment obeys the principles of probability; each number is greater than or equal to zero and the sum of all of them is 1.

Now, if we use this model to predict, for example, the chances of obtaining an even number, we can use the addition property (d). The event of an even number is $E = \{2,4,6\}$. Thus $Pr(E) = Pr\{2\} + Pr\{4\} + Pr\{6\} = \frac{1}{6} + \frac{1}{6} + \frac{1}{6} = \frac{3}{6}$. □

■ Consider the experiment where a mouse runs a T-maze. We shall give a probabilistic model for an experiment where it is assumed that the choice to go right or left is made randomly.

The possibility set for this experiment is $S = \{right, left\}$. To complete the model, we must assign a probability to each outcome. Since we are assuming that the chances of going to the right are the same as the chances of going left, then the probability assignment $Pr\{right\} = \frac{1}{2}$ and $Pr\{left\} = \frac{1}{2}$ would reflect this. □

■ Consider again an experiment where a mouse runs a T-maze. Suppose that this experiment was performed with 50 different mice and it was observed that 40 of the mice went right and 10 went left. Give a probabilistic model for this experiment.

The possibility set for the experiment is the same as in the previous example, namely, $S = \{right, left\}$. Suppose we complete the model by saying $Pr\{right\} = Pr\{left\}$ and both are $\frac{1}{2}$. The possibility set with this probability assignment would *not* be a realistic model for the experiment. Based upon relative frequencies, the chances of going right should not be the same as that of going left.

A model for the experiment, which more closely reflects the actual situation, is the possibility set S with the probabilities $Pr\{right\} = \frac{40}{50}$ and $Pr\{left\} = \frac{10}{50}$. □

■ Consider the situation in which three horses A, B, and C will race. From the results of prior races, a gambler feels that the chances of A winning are the same as for B, but twice those of C. Give a probabilistic model for the experiment and then, using this model, predict the chances of A or B winning.

We can let the possibility set be $S = \{A,B,C\}$. A probability assignment for S must be such that $Pr\{A\} = Pr\{B\}$ and $Pr\{A\} = 2 \cdot Pr\{C\}$. Therefore we have $Pr\{A\} + Pr\{B\} + Pr\{C\} = 2Pr\{C\} + 2Pr\{C\} + Pr\{C\} = 5Pr\{C\}$. Since this sum must be 1, we have $Pr\{C\} = \frac{1}{5}$. Knowing this, we can set $Pr\{A\} = \frac{2}{5}$ and $Pr\{B\} = \frac{2}{5}$.

Now the event of either A or B winning is $E = \{A,B\}$. Based on our model, we see that $Pr(E) = \frac{2}{5} + \frac{2}{5} = \frac{4}{5}$. □

■ Suppose for a certain experiment, the possibility set is $S = \{O_1,O_2,O_3,O_4\}$. Let us give a probability assignment for S so that

$$Pr\{O_1\} = Pr\{O_3\}, \qquad Pr\{O_2\} = 2 \cdot Pr\{O_1\}, \qquad \text{and} \qquad Pr\{O_4\} = 3 \cdot Pr\{O_1\}.$$

An experiment in which this could occur is the selecting of a chip from an urn. Each chip is one of four colors—red, blue, green, and yellow. The number of red chips equals the number of green chips; the number of blue chips is twice the number of red chips; and the number of yellow chips is three times the number of red chips. If we were to give the assignment based upon these physical characteristics, then we would have the above conditions.

From the principles of probability, we must give four numbers, each greater than or equal to zero, so that their sum is one. We can write

$$Pr\{O_1\} + Pr\{O_2\} + Pr\{O_3\} + Pr\{O_4\} = 1.$$

Furthermore, from the information given, we can rewrite this equation as

$$Pr\{O_1\} + 2Pr\{O_1\} + Pr\{O_1\} + 3Pr\{O_1\} = 1.$$

Now solving for $Pr\{O_1\}$, we obtain

$$7Pr\{O_1\} = 1 \qquad \text{and} \qquad Pr\{O_1\} = \tfrac{1}{7}.$$

For the other outcomes, we have

$$Pr\{O_2\} = \tfrac{2}{7}, \qquad Pr\{O_3\} = \tfrac{1}{7}, \qquad \text{and} \qquad Pr\{O_4\} = \tfrac{3}{7}.$$

Recall that it is sufficient to assign probabilities only to the outcomes in order to give a probability assignment for S. □

QUIZ YOURSELF*   As you do each of the following, be prepared to give explanations for your answer.

1. Let an experiment have the possibility set $S = \{O_1,O_2,O_3\}$. If $Pr\{O_1\} = \frac{1}{2}$, $Pr\{O_2\} = \frac{1}{4}$, and $Pr\{O_3\} = \frac{1}{4}$, will this give a probability assignment for S?

2. For the possibility set $S = \{O_1,O_2,O_3\}$, if $Pr\{O_1\} = \frac{1}{2}$, $Pr\{O_2\} = \frac{1}{4}$, and $Pr\{O_3\} = \frac{1}{2}$, will this give a probability assignment for S?

3. Suppose an experiment has three possible outcomes (a,b, or c) and each outcome has the same chance of occurring. Give a probabilistic model for the experiment.

4. Consider an urn which contains three black chips and six white ones. The experiment is to draw a chip from the urn and note its color. Is the possibility set $S = \{b,w\}$ with probabilities $Pr\{b\} = \frac{1}{2}$ and $Pr\{w\} = \frac{1}{2}$ a realistic model for the experiment? If not, what would be a model which reflects the situation?

***ANSWERS**   1. Yes   2. No   3. $S = \{a,b,c\}$; $Pr\{a\} = Pr\{b\} = Pr\{c\} = \frac{1}{3}$
4. No; $S = \{b,w\}$; $Pr\{b\} = \frac{1}{3}$; and $Pr\{w\} = \frac{2}{3}$.

We will now turn our attention to a special type of experiment, one in which the outcomes are **equally likely** to occur. In this case, whenever the outcomes have the same chance of occurring, we shall see that it is easy to give a probabilistic model for the experiment.

Consider the experiment of tossing a fair die. The possibility set is $S = \{1,2,3,4,5,6\}$. The probabilities we assign to these outcomes *must be the same*. Since the sum of these probabilities must be 1, we have that the probability of each outcome is $\frac{1}{6}$.

■ A coin is flipped three times. What is the probability of obtaining exactly   **EXAMPLE**
two heads?

First, consider as our possibility set $P = \{0H,1H,2H,3H\}$, the outcomes being the number of times heads appear. If we wish to give a model with equally likely outcomes, then this set will not do. There is only one way to obtain three heads. However, there are three ways in which two heads can appear; the two heads can occur on the first two flips, on the last two flips, or on the first and the last flip. Thus the outcomes in P do not have the same chance of occurring.

To give a model with equally likely outcomes, we must take for our possibility set $S = \{HHH,HHT,HTH,THH,TTH,HTT,THT,TTT\}$. Since each outcome in S has the same chance of occurring and since S contains eight outcomes, we assign to each outcome the probability $\frac{1}{8}$.

The event of interest is the set $E = \{HHT,HTH,THH\}$. By our principles of probability, $Pr(E) = \frac{1}{8} + \frac{1}{8} + \frac{1}{8} = \frac{3}{8}$. □

As we know from the principles of probability, we can use the additive property to find the probability of an event; the probability of a nonempty event is the sum of the probabilities of the outcomes in the event. In the case of equally likely outcomes, each outcome has the same probability and therefore we would be adding the same number repeatedly. Thus we would be able to express this sum as a multiplication.

With these comments in mind we should be able to accept the following principle.

Consider an experiment which has equally likely outcomes. A probabilistic   **PRINCIPLE FOR**
model consists of a possibility set S, with each outcome being assigned the   **EQUIPROBABLE**
probability $1/n(S)$. Furthermore, if E is an event in S, the probability of E is   **MODELS**
the number of outcomes in E divided by the total number of possible outcomes:

$$Pr(E) = n(E)/n(S).$$

■ Consider the experiment of randomly drawing 5 cards from a well-shuffled   **EXAMPLE**
deck of 52 cards. What is the probability of the five cards being all hearts?

The number of possible outcomes for the experiment is $C[52,5]$. Since these are equally likely outcomes, the probability of any outcome is $1/C[52,5]$.

Let E be the event of interest and we have $Pr(E) = n(E) \cdot (1/C[52,5])$. We should be able to convince ourselves that $n(E) = C[13,5]$. Therefore

$$Pr(E) = \frac{C[13,5]}{C[52,5]} = \frac{1287}{2598960} = .0005. \ \square$$

QUIZ YOURSELF*

1. Suppose the possibility set for an experiment contains 36 equally likely outcomes. What probability should be assigned to each of these outcomes?

2. Consider the experiment of randomly drawing a single card from a deck of 52 cards. What probability should be assigned to each possible outcome?

3. For the experiment in question 2 above, what is the probability of drawing an ace?

4. Two light bulbs are selected at random from a bin which contains ten good and five bad bulbs. What is the probability that both are good?

***ANSWERS**   1. $\frac{1}{36}$ 2. $\frac{1}{52}$ 3. $\frac{4}{52}$ 4. $\dfrac{C[10,2]}{C[15,2]} = \dfrac{45}{105}$

The following properties of probability will be useful at times.

PROPERTIES OF PROBABILITY   **Let S be the possibility set for a particular experiment. As usual, S is a finite set and is considered the universal set.**

a) Let E and F be two events which are disjoint. Then $Pr(E \cup F) = Pr(E) + Pr(F)$.

b) For any event E, $Pr(E) = 1 - Pr(E')$.

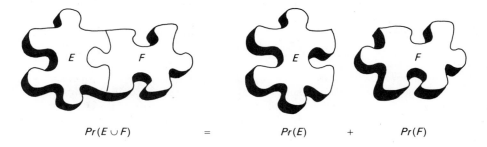

$Pr(E \cup F)$ = $Pr(E)$ + $Pr(F)$

The second formula of the above properties can be used when it is easier to compute $Pr(E')$ than it is to compute $Pr(E)$ directly. This is illustrated in the next example.

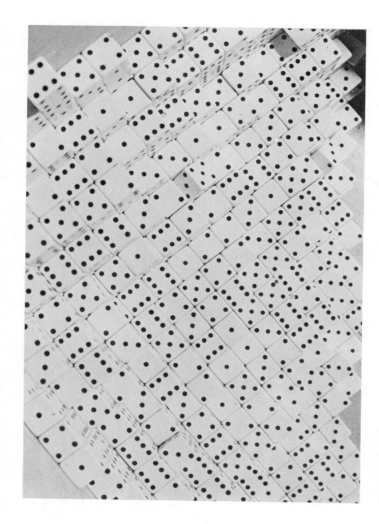

Although many people associate probability theory only with games of chance, it is an important area of mathematics that has applications in many areas of human endeavor.
(Photo by Frank Siteman, Stock Boston.)

■ A box contains 20 light bulbs; 15 are good and 5 are burned out. A sample of three is selected at random from the box and each bulb is tested. What is the probability that at least one of the three bulbs is good?

   The number of elements in the possibility set is $n(S) = C[20,3]$. Let E be the set of samples of three bulbs in which one, two, or three of the bulbs are good. We want to compute $Pr(E)$. We choose to use the formula $Pr(E) = 1 - Pr(E')$. Since E' is the set of samples of three bulbs where all the bulbs are burned out, $n(E') = C[5,3]$. Therefore $Pr(E') = C[5,3]/C[20,3]$ and $Pr(E) = 1 - C[5,3]/C[20,3] = 1 - \frac{10}{1140} = \frac{113}{114}$. □

EXAMPLE

As in the above example, when giving a model for an experiment in which the outcomes are equally likely, the probabilities can be determined by counting. However, we should keep in mind that not all experiments have equally likely outcomes. If the experiment is such that we cannot use an equiprobable model, then there are various ways to assign the probabilities to the outcomes, as we have already discussed. In any case, the model we use for the experiment must satisfy the principles of probability and must realistically reflect the experiment.

EXERCISES

1. An urn contains two red, one blue, and three white chips. An experiment is to draw a chip from the urn and observe its color. Give a probabilistic model for the experiment.

2. Let $S = \{O_1, O_2, O_3, O_4, O_5\}$ be the possibility set for a certain experiment. Give a probability assignment for S such that $Pr\{O_1\} = Pr\{O_3\} = 2 \cdot Pr\{O_5\}$, $Pr\{O_2\} = \frac{1}{2} \cdot Pr\{O_5\}$, and $Pr\{O_4\} = \frac{1}{4} \cdot Pr\{O_5\}$. Describe an experiment for which the above could be a probabilistic model.

3. There is a faulty candy machine in a student center. A student obtains the following data by observing others' experiences with the machine.

Outcome	Frequency of occurrence
Gives the candy only	34
Gives the candy and returns the money	5
Does not give the candy and keeps the money	11
	$\overline{50}$

Give a probabilistic model for the experiment of putting money into this machine. In giving your model, if the machine was tried a fifty-first time, what assumption are you making about the outcome? Based upon your model, what is the probability of obtaining candy from this machine?

4. Through observation, the following probabilities for the number of people waiting at a teller's window in the Farmers' Bank are determined.

Number waiting	0	1	2	3	4	5 or more
Probability	.05	.30	.35	.22	.05	.03

a) What is the probability of fewer than two people waiting in line?
b) What is the probability of at least two people waiting in line?
c) What is the probability that at most three people will be waiting in line?

5. In a raffle for a single prize, 500 tickets are sold. What is the probability that a person who buys one ticket will win? What is the probability of winning if five tickets are bought?

6. A mutual fund considering the purchase of stock will select from five bus lines and ten airlines. Assuming that each company has the same record of gain over the past months, the mutual fund decides to select three at random. What is the probability that the three selected are all airline stocks? What is the probability that two are airline stocks and one is a bus line?

7. What is the probability that in a family of four children all will be the same sex?

8. a) Consider the experiment of rolling a pair of fair dice and observing the sum of the dots on the upturned faces. The possibility set is $S = \{2,3,4,\ldots,12\}$. Explain why the outcomes in S are not equally likely.

   b) Consider the rolling of a pair of fair dice; what should be the possibility set so that the outcomes will be equally likely?

9. What is the probability of obtaining a seven on the roll of a pair of fair dice?

10. To fill four vacant seats on the student government board, a list of eight candidates is submitted to the student body and they are asked to vote for four. The four candidates receiving the largest number of votes will win the seats. If each student randomly chooses the four, then what is the probability that a particular candidate will win a seat?

11. Consider exercise 10 above where a list of seven candidates for the four seats is proposed. What is now the probability of a particular candidate winning a seat?

12. A five-question true-false test is given. A passing score on the test is having four out of five correct answers. What is the probability of passing the test by guessing?

13. Suppose in an experiment on ESP, a person is asked to concentrate on the arrangement of four cards, which are labeled 1, 2, 3, and 4. Without seeing the cards, a second person is asked to give the arrangement that he thinks he perceives. What is the probability that the second person will give the correct arrangement purely by guessing?

14. In poker, a flush (including a straight and royal flush) is five cards of the same suit. What is the probability of obtaining a flush?

15. What is the probability that a five-card poker hand contains at least one ace?

EXPECTED VALUE

Let us consider again the familiar flipping-a-coin experiment. You should recall from our discussion on relative frequency that if the coin is flipped 50 times, then the ratio of the number of heads that occur should be about $\frac{1}{2}$. Thus on the 50 flips, we could expect about $\frac{1}{2} \cdot 50 = 25$ heads. For any other experiment, we could say that the relative frequency of an outcome is nearly the same as its probability p. If the experiment is performed N times and the outcome occurs K times, then p would be about K/N. Thus K would be about $p \cdot N$; that is, we could expect the number of times the outcome occurs to be the probability of the outcome multiplied by the number of times the experiment is performed.

EXAMPLES   ■ In a board game, a die is rolled and a token is moved on a board as many places as the number appearing on the die. How many places can a player expect to move on 12 rolls of the die? What is the average number of places one can expect to move per roll?

To find the answer to the first question, we say that the expected number of times a 1 will show on the 12 rolls of the die is $\frac{1}{6} \cdot 12$. Each time a 1 appears on the die the token is moved 1 place. For each of the other possible outcomes on the die, the expected number of times they will appear on the 12 rolls is also $\frac{1}{6} \cdot 12$. Thus we can expect that the token would move $1 \cdot \frac{1}{6} \cdot 12 + 2 \cdot \frac{1}{6} \cdot 12 + \cdots + 6 \cdot \frac{1}{6} \cdot 12 = 42$ places on the 12 rolls.

In finding the average, we need only divide the above answer by 12 and we see that the 12 will divide out. Thus we obtain $1 \cdot \frac{1}{6} + 2 \cdot \frac{1}{6} + \cdots + 6 \cdot \frac{1}{6} = 3\frac{1}{2}$.

Even though $3\frac{1}{2}$ moves will not result from a single toss, it is reasonable for this number to be the average per toss.

We should realize that if we were to find the expected number of moves on 100 rolls of the die and then find the average per roll, we would obtain the same answer as in the above example, namely, $1 \cdot \frac{1}{6} + 2 \cdot \frac{1}{6} + \cdots + 6 \cdot \frac{1}{6}$. Thus the average number of moves per toss can be computed without regard to a total number of tosses. □

■ Consider the following board game. A single die is rolled and a token is moved backwards or forwards as many places as the number appearing on the die. If the number is even, then the token is moved forward and, if the number is odd, the token is moved backward. How many places can a player expect to move on 12 rolls of the die? What is the average number of places one can expect to move per roll?

It is easy enough to agree that a move backward can be designated by a negative number. Thus a move of one place backward is given by (-1).

Computing the answer to the first question, we have $(-1)\frac{1}{6}(12) + (2)\frac{1}{6}(12) + (-3)\frac{1}{6}(12) + \cdots + (6)\frac{1}{6}(12)$.

Now to find the average number of places, we divide the above answer by 12, giving us $(-1)\frac{1}{6} + (2)\frac{1}{6} + (-3)\frac{1}{6} + \cdots + (6)\frac{1}{6} = \frac{3}{6}$.

Again, we should realize that if we computed the average based on 100 tosses, we would still obtain the same answer, namely, $(-1)\frac{1}{6} + (2)\frac{1}{6} + (-3)\frac{1}{6} + \cdots + (6)\frac{1}{6}$. Thus the expected gain or loss can be computed knowing only the probability for each outcome and the amount of gain or loss associated with each outcome. □

In the above examples, the gain or loss for each outcome is a move of a token on a board. In other situations, the **payoff**, the gain or loss per outcome, could be money or one of many other things.

EXAMPLE   ■ The data in the following table give the number of minutes in a single hour that a toll collector at a turnpike entrance is idle—that is, no vehicles are passing through. The information is gathered over a 24-hour period. Based upon this data, how many minutes during a one-hour period can the toll collector expect to be idle?

Minutes idle during one hour	Number of one-hour periods this occurs
30	4
25	4
20	6
15	2
10	4
5	2
0	2
Total	24

(handwritten notes:)
$\frac{1}{4}(1.00) - \frac{3}{4}(.30) =$
$.25 - .225 = .025$

Heads vs. Tails $\$1.00$ person will win

$\frac{1}{4}(1.00) + \frac{3}{4}(.30) =$
$-.25 + .225 = -.025$

The experiment consists of seven outcomes—idle 30 minutes, idle 25 minutes, etc. The probability for each outcome can be taken to be the relative frequency of the outcome. For example, the probability of being idle 20 minutes in a one-hour period is $\frac{6}{24}$. Thus, the number of minutes idle per single hour that could be expected is $(30)\frac{4}{24} + (25)\frac{4}{24} + (20)\frac{6}{24} + (15)\frac{2}{24} + (10)\frac{4}{24} + (5)\frac{2}{24} + (0)\frac{2}{24} = 17\frac{1}{2}$ minutes. □

Consider an experiment with k possible outcomes having probabilities $P_1, P_2, \ldots, P_k$. For the outcomes there are PAYOFFS, $m_1, m_2, \ldots, m_k$; the outcome with probability P_1 having payoff m_1, etc. The EXPECTED VALUE is   **DEFINITION**

$$(P_1 \cdot m_1) + (P_2 \cdot m_2) + \cdots + (P_k \cdot m_k).$$

In the board-game examples, the expected value is the number of moves of the token per toss of the die. In the toll-collector example, the number of minutes idle per hour is the expected value.

---

1. An experiment has outcomes O_1, O_2, and O_3, where $Pr\{O_1\} = \frac{1}{3}$, $Pr\{O_2\} = \frac{1}{4}$, and   **QUIZ YOURSELF***
$Pr\{O_3\} = \frac{5}{12}$. Assume that O_1 has payoff 3, O_2 has payoff -4, and O_3 has payoff 6. For this experiment with these payoffs, what is the expected value?

2. In a game, a die is rolled and a token is moved forward as many places as the number showing if the number is odd. The token is moved backward $\frac{1}{2}$ the number shown if the number is even. What is the expected number of moves per toss of the die?

3. Suppose a company realizes a profit of $.50 on each good item produced by a certain machine and a loss of $.10 on each defective item produced by this machine. If the probability that the machine will produce a defective item is 0.1, then what is the expected amount of profit on each item produced by the machine?

---

1. $\frac{30}{12}$;   2. $\frac{1}{2}$;   3. $.44$

Whenever we associate a payoff (a positive or negative number or zero) with each outcome of an experiment, we can compute the expected value by using our definition. Since in games of chance there is usually a payoff in money, we can compute the expected value of the game and use this to evaluate the game. If the expected value of the game is positive, then the game is favorable to the player; if it is negative, then the game is unfavorable to the player.

EXAMPLE   ■ Consider the following game. A person randomly draws one card from an ordinary deck of 52 cards. If the card is an ace, the player wins $6. If the card is a King, Queen, or Jack, then the player wins $1. If any other card is drawn, the player loses $1. What is the expected value of this game?

The probability of obtaining an Ace is $\frac{4}{52}$. For a King, Queen, or Jack, the probability is $\frac{12}{52}$. For any other card, the probability is $\frac{36}{52}$. The payoffs for these three events are 6, 1, and -1, respectively. Therefore the expected value of the game for the player is

$$\tfrac{4}{52}(6) + \tfrac{12}{52}(1) + \tfrac{36}{52}(-1) = 0. \quad \square$$

In the above example we obtained an expected value of zero. Thus we say this game is **fair** for the player. We say a game is **fair** whenever the expected value is zero.

EXAMPLE   ■ How much should a raffle ticket cost for a prize of $100 if there are to be 100 tickets sold and the raffle is to be fair?

We shall assume that the group conducting the raffle is to make no profit. We therefore want the expected value for a single raffle-ticket buyer to be zero. The payoffs for this game are a loss of the price of the ticket and a gain of $100 minus the price of the ticket. Let us designate the price of the ticket with an x. Since the probability of winning is $\frac{1}{100}$ and the probability of losing is $\frac{99}{100}$, we wish to choose x such that $\frac{1}{100}(100 - x) + \frac{99}{100}(-x) = 0$. Therefore x should be $1.

In fact, for any raffle or lottery to be fair to the bettor, the price of the ticket must be the value of the prize divided by the number of tickets sold. However, if a lottery conducting organization is to make a profit, then more tickets must be sold or the price of the ticket must be increased. Thus no such lottery can be fair to the bettor. $\square$

All games of chance need not involve gambling. There are situations in business which can be considered games and we can therefore compute their expected values.

EXAMPLE   ■ The probability of a 20-year-old male living to age 21, based on mortality tables, is about .99. If he buys a one-year-term life-insurance policy valued at $1000 and pays a premium of $25, what is the expected value?

Based upon mortality tables as used by life insurance companies, the probability of a 20-year old male living to age 21 is .99. (Photo by John Running, Stock Boston.)

Handwritten margin notes:

1st prize = 200
2nd " = 150
3rd " = 100

raffle ticket
1 chance in 1000
spent 1 dollar so only
gains 199

$$\frac{1}{1000}(199) + \frac{1}{1000}(149) + \frac{2}{1000}(99) - \frac{996}{1000}(1)$$

$$\frac{199}{1000} + \frac{149}{1000} + \frac{198}{1000} - \frac{996}{1000} =$$

$$= \frac{-450}{1000}$$

$$= -.45$$

amount of money
lost per game.

$$\left(\frac{1}{1000}(200-x) + \frac{1}{1000}(150-x) + \frac{2}{1000}(100-x)\right.$$
$$- \frac{996}{1000}(x) = 0$$

or

for raffle ticket only

$$1000\,x = 550$$

$$x = .55¢$$

An individual buying a fixed-term insurance policy can be considered similar to a game of chance. There are two payoffs, loss in premium or gain to the beneficiary of the face value of the policy minus the premium. For the situation described, the expected value is

$$(.99)(-25) + (.01)(975) = -\$15.$$

As in a lottery, the expected value for the buyer will always be negative because there must be a profit to the insurance company. Since an expected value is an average per "play," we can interpret the answer $-\$15$ in our example as an average profit of $15 to the insurance company per policy sold. □

Expected value can also be used to compare alternatives and to give a basis for decision-making as discussed in the next examples.

EXAMPLE   ■ A company intends to buy one of two machines. Each machine produces the same item; however, they operate at slightly different rates. It would cost more for the slower machine to produce a single item because of the additional operator time required on the machine. But because it is slower, it is also more reliable; that is, the probability of the machine producing a bad item is lower. For either machine, each bad item produced results in a loss (cost of material and labor) and each good item yields a profit. Based upon the data in the following table, which machine would be the better buy for the company?

	Loss on bad item	Profit	Probability of producing bad item
Slower machine	$.07	$.28	.03
Faster machine	$.05	$.30	.09

For each machine we can compute the expected value, the average gain or loss per item produced on the machine. For the slower machine, the expected value is $(-.07)(.03) + (.28)(.97) = .2695$, while for the faster machine, the expected value is $(-.05)(.09) + (.30)(.91) = .2685$. The slower machine has a slightly higher expected value and is therefore the better buy for the company. □

In our next and last example, we will use expected value to determine the number of items a merchant should stock in order to realize the largest profit.

EXAMPLE   ■ Ned wants to determine how many copies of the *Daily Patriot* he should stock at his newsstands for a single day's sales. Ned has gathered the following statistics on the volume of sales over a 20-day period. The information is given in the table that follows.

Number of copies sold	1500	1400	1300	1200	1100	1000
Number of days with these sales	2	5	4	5	2	2

The *Daily Patriot* costs Ned $.06 and he sells it for $.10, which gives him a profit of $.04 on each copy sold. Ned loses $.06 on each unsold copy. Ned's **net profit** at the end of a day will be the profit on copies sold minus the loss on unsold copies. There are six different-size orders that he could place; 1500, 1400, 1200, etc. We can compute the expected net profit for each size of order. The order which gives the highest expected net profit would be the best one for Ned to choose.

The net profit for each of the six orders is given in the following table.

Table of net profit

Number of copies ordered	Number of copies in demand					
	1500	1400	1300	1200	1100	1000
1500	$60.00	$50.00	$40.00	$30.00	$20.00	$10.00
1400	56.00	56.00	46.00	36.00	26.00	16.00
1300	52.00	52.00	52.00	42.00	32.00	22.00
1200	48.00	48.00	48.00	48.00	38.00	28.00
1100	44.00	44.00	44.00	44.00	44.00	34.00
1000	40.00	40.00	40.00	40.00	40.00	40.00

To illustrate how the numbers are obtained for the net-profit table, let us assume that 1500 copies are ordered but only 1300 copies are sold. Ned will make a profit of ($.04) × 1300 = $52 on the copies sold; however, he will lose ($.06) × 200 = $12 on the unsold copies. Thus his net profit in this case is $40.

In order to obtain the expected net profit on each of the six orders, we must have the probabilities for each of the possible outcomes—sells 1500, sells 1400, etc. From the sales-volume table, we can make our probability assignment based upon relative frequency.

Sells	1500	1400	1300	1200	1100	1000
Probability	$\frac{1}{10}$	$\frac{1}{4}$	$\frac{1}{5}$	$\frac{1}{4}$	$\frac{1}{10}$	$\frac{1}{10}$

Now to obtain the expected net profit, let us suppose that Ned chooses to order 1500 copies of the *Daily Patriot*. If he sells all 1500 copies, he will have a net profit of $60. However, the probability that he will sell 1500 copies is $\frac{1}{10}$. If Ned sells only 1400 copies, his net profit will be $50. The probability of selling 1400 copies is $\frac{1}{4}$. Looking at the remaining possible outcomes and their respective payoffs, and *assuming that Ned orders 1500 copies*, we can compute the expected net profit to be

$$\$60 \cdot \tfrac{1}{10} + \$50 \cdot \tfrac{1}{4} + \$40 \cdot \tfrac{1}{5} + \$30 \cdot \tfrac{1}{4} + \$20 \cdot \tfrac{1}{10} + \$10 \cdot \tfrac{1}{10} = \$37.$$

The expected net profit for each of the remaining options on order size is given in the next table.

Table of expected net profit

Number of copies ordered	Expected net profit
1500	$37.00
1400	42.00
1300	44.50
1200	45.00
1100	43.00
1000	40.00

From these expected net profits, we see that the best choice is for Ned to order 1200 copies of the *Daily Patriot*. Recall how to interpret expected value. If Ned orders 1200 each day, then even though for some days he will have a net profit which is more than $45 and for some days less than $45, his net profit will average out to $45 per day. □

EXERCISES

1. The weather bureau reports a 75-percent chance of rain during the day and we assume this is an accurate forecast. A construction worker will lose $6 in transportation costs if he reports to work and it rains, or he will make $50 if he reports to work and it doesn't rain. What is the construction worker's expected loss or gain on one of these days?

2. If in exercise 1 the weather bureau forecasts a 90-percent chance of rain during the day, would it be best for the construction worker to stay home or go to work on one of these days?

3. When bidding on a particular contract valued at $50,000, a company estimates it has an 80-percent chance of receiving the contract. It will cost the company $5000 in consultant fees to prepare the bid. What is the expected gain or loss for the company if they submit a bid on this $50,000 contract?

4. In exercise 3, suppose that the company could bid on a second contract valued at $42,000; they estimate that they have a 90-percent chance of getting the contract. For this second contract, it will cost the company $2000 to prepare the bid. What is the expected gain for the company if they submit a bid on the $42,000 contract? Which of the two contracts would it be better for the company to bid on if they choose to bid on only one of them?

5. In the board-game Monopoly, if a player's token lands on the space marked Community Chest, then the player draws a card from a stack of 14 cards (this is assuming some of the cards have been lost from the set). Two of the cards instruct the player to collect $200; four of them instruct the player to collect $100; two of them instruct the player to collect $50; two of them instruct the player to pay $50; and, for the remaining four cards, there is no exchange of money. What is the expected value for the player who lands on Community Chest?

6. In a lottery where the winner receives $100, the number of tickets sold is 110 at $1 a piece. What is the expected value for a single-ticket buyer? Would the expected value change if the person bought two tickets?

7. During those years when financial responsibilities are greatest, such as when children are in school, there is a mortgage on the house, etc., a person may choose to buy a term life-insurance policy. The insurance company will pay the face value of the policy if the insured dies during the term of the policy. For how much should an insurance company sell a 10-year-term policy with a face value of $30,000 to a 30-year-old male in order for the company to make a profit? The probability of a 30-year-old male living to age 40 is .97.

8. If an insurance company sells a $30,000, 15-year-term life-insurance policy to 30-year-old males for $1500, what would be the average profit to the insurance company per policy sold? The probability of a 30-year-old male living to age 45 is .95.

9. A basketball team has scored the number of points given in the table below during the last 20 games. What is the expected number of points per game?

Number of points	60	65	70	75	80
Number of games	3	4	4	6	3

10. After determining the best order to place, Ned continues to gather information on the size of the demand for the *Daily Patriot*. He obtains the following information over a 20-day period.

Number of copies sold	1500	1400	1300	1200	1100	1000
Number of days with these sales	2	5	5	5	2	1

Using the net-profit table and the above information table, find the size of daily order which would be best for Ned to place.

MULTISTAGE EXPERIMENTS

In this section, we shall consider experiments which are composed of several processes performed in succession. We shall call such experiments **multi-stage**. An example of a two-stage experiment is the rolling of a die followed by the flipping of a coin. Let us develop a model for this type of experiment.

We know that in order to obtain a probabilistic model for a particular experiment, we must first determine the possibility set. For multistage experiments we can sometimes conveniently display the possible outcomes with a **tree-diagram**. To illustrate, consider the two-stage experiment of flipping a coin twice. The outcomes are displayed as branches on a tree.

The tree diagram displays all the outcomes in the possibility set. An outcome is found by beginning at the point *start* and reading along a branch of

The possibilities in a multistage experiment can be modeled by means of a tree diagram. (Photo by Rick Stafford, Stock Boston.)

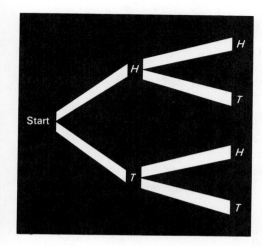

the tree to a terminal point. For example, the first branch of the tree displays the outcome *head* followed by *head*.

EXAMPLE

■ Consider an urn which contains one blue chip, one red chip, and one green chip. The experiment is to draw a chip from the urn, note its color, and without replacing the chip draw another chip until a blue one is drawn. We can display the possibility set for this multistage experiment using the tree diagram shown in Fig. 8.1. □

It is convenient to have a notation for writing an outcome of a multistage experiment. We can give an outcome by writing, within parentheses, the result at each stage in the order of occurrence. For example, the outcomes for the experiment in the above example can be written as (b), (r,b), (r,g,b), (g,r,b), and (g,b). Note that the outcome (r,g,b) is different from the outcome (g,r,b).

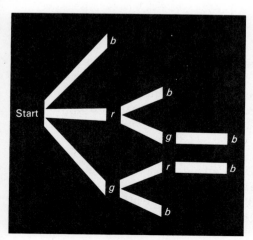

Figure 8.1

EXAMPLE

■ Consider the experiment of taking a three-question test, where the answer is either right or wrong. Let E be the set of outcomes when at least two questions are answered correctly. Using the listing method, give the set E.

If we let R indicate a correct answer and W an incorrect answer, then we can write the set as $E = \{(R,R,W),(R,W,R),(W,R,R),(R,R,R)\}$. □

QUIZ YOURSELF*

1. An urn contains three discs numbered 1, 2, and 3. A disc is drawn from the urn, its number noted, and without replacing it, a second disc is drawn and its number noted. Display the possibility set using a tree diagram.

2. Give by the listing method the possibility set for the experiment described in question 1.

3. Give by the listing method the possibility set for the experiment of tossing a coin two times.

***ANSWERS**

1.

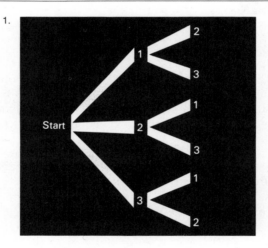

2. $\{(1,2),(1,3),(2,1),(2,3),(3,1),(3,2)\}$
3. $\{HH,HT,TH,TT\}$

In giving probabilistic models, the next step after determining the possibility set S for the experiment is to obtain a probability assignment for S. We saw how the possibility set can be displayed using a tree diagram for a multistage experiment. We will now give a procedure for obtaining the probability assignment for the set.

EXAMPLES

■ Consider the multistage experiment: A coin is flipped; if a head appears, then the coin is flipped again, but if a tail appears, then a die is rolled. The tree diagram of the possibility set is given in Fig. 8.2.

We should recall that the probability of any outcome is between 0 and 1 and that the sum of the probabilities of all the outcomes is 1. Now consider the two events E_H and E_T, where $E_H = \{(H,H),(H,T)\}$ and $E_T = \{(T,1),(T,2),\dots,(T,6)\}$. Since $S = E_H \cup E_T$ and $E_H \cap E_T = \varnothing$, we have $Pr(E_H) + Pr(E_T) = 1$.

Because the probability of obtaining a head or a tail on the flip of a coin is $\frac{1}{2}$, it would only be reasonable to set $Pr(E_H) = Pr(E_T) = \frac{1}{2}$. To obtain the probabilities of the outcomes, we now consider what can occur at the second stage of the experiment.

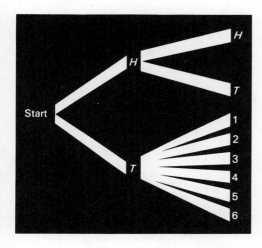

$$S = \left\{ \begin{array}{l} (H, H),\ (H, T),\ (T, 1), \\ (T, 2),\ (T, 3),\ (T, 4), \\ (T, 5),\ (T, 6) \end{array} \right\}$$

Figure 8.2

First, consider the outcomes (H,H) and (H,T) in the event E_H. The sum of the probabilities of these two outcomes must be the probability of the event E_H. Therefore each of these probabilities must be a *fractional part* of $Pr(E_H)$. Since the probability of obtaining a head (or a tail) on the second flip is $\frac{1}{2}$, the only reasonable assignment of a probability to the outcome (H,H) is $\frac{1}{2}$ of $Pr(E_H)$, and likewise for the outcome (H,T). Thus we can make the assignment $Pr(H,H) = Pr(H,T) = \frac{1}{2} \cdot \frac{1}{2} = \frac{1}{4}$.

Let us look at the case in which a *tail* occurs during the first stage. Again, the probabilities of the outcomes in the event E_T should each be a fractional part of $Pr(E_T)$. Furthermore, the outcomes in E_T are equally likely. Therefore the probability of each of these outcomes should be $\frac{1}{6}$ of $Pr(E_T)$. Thus we assign to each outcome in E_T the probability $\frac{1}{6} \cdot \frac{1}{2} = \frac{1}{12}$.

It is easy to verify that this assignment of probabilities satisfies the principles of probability, and from our discussion we should see that this assignment is reasonable. □

■ Consider an urn which contains two red chips and three blue ones. The experiment is to draw a chip from the urn, note its color, and, without returning it to the urn, draw a second chip. Let us give a probabilistic model for this two-stage experiment.

We see that the possibility set, displayed by the tree diagram in Fig. 8.3, shows the union of the two disjoint events, $E_r = \{(r,r),(r,b)\}$ and $E_b = \{(b,r),(b,b)\}$. Since E_r can be described as the event in which the first draw yields a red chip, its probability is the same as the probability of picking a red chip first; that is, $Pr(E_r) = \frac{2}{5}$. We can also agree that $Pr(E_b) = \frac{3}{5}$.

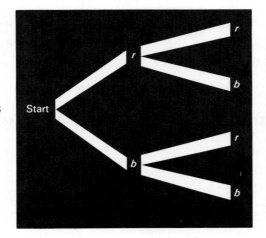

Figure 8.3

Let us now assign probabilities to the outcomes in E_r. It is reasonable to accept that the probability of each outcome in this event should be a *fractional part* of $Pr(E_r)$. To see what these fractional parts are, we observe that if the first chip drawn is red, then when the second chip is drawn the urn would have contained one red and three blue chips. In this case, the probability that the second chip drawn will be red is $\frac{1}{4}$. For a blue chip, the probability is $\frac{3}{4}$. Therefore the probability that we should assign to the outcome (r,r) is $\frac{1}{4}$ of $Pr(E_r)$. For the outcome (r,b), it should be $\frac{3}{4}$ of $Pr(E_r)$; that is, $Pr(r,r) = \frac{1}{4} \cdot \frac{2}{5} = \frac{2}{5} \cdot \frac{1}{4} = \frac{2}{20}$ and $Pr(r,b) = \frac{2}{5} \cdot \frac{3}{4} = \frac{6}{20}$.

Let us describe the probability we obtained for the outcome (r,b). $Pr(r,b)$ is "the probability of obtaining a red chip at the first stage" times "the probability of obtaining a blue chip at the second stage, knowing that a red chip was drawn first." We can also say that $Pr(r,r)$ is "the probability of obtaining a red chip at the first stage" times "the probability of obtaining a red chip at the second stage, knowing that a red chip was drawn first."

Now consider the outcomes in E_b. In this case when the second chip is drawn, the urn would have contained two red and two blue chips. Therefore the probability of drawing either a red or a blue chip at the second stage is $\frac{2}{4}$. Thus the probability of each of the outcomes in E_b should be $\frac{2}{4}$ of $Pr(E_b)$; that is, $Pr(b,r) = Pr(b,b) = \frac{3}{5} \cdot \frac{2}{4} = \frac{6}{20}$. If we were to describe the probability of (b,r), we would say it is "the probability of obtaining a blue chip at the first stage" times "the probability of obtaining a red chip at the second stage, knowing that a blue chip was drawn first."

For convenience, we can label the segments of the tree with their appropriate probabilities as shown in Fig. 8.4. Then, to find the probability of an outcome, we multiply the numbers which are encountered as we trace the branch of the tree for that outcome. Again (considering Fig. 8.4 this time) we see that $Pr(r,r) = \frac{2}{5} \cdot \frac{1}{4}$. $\square$

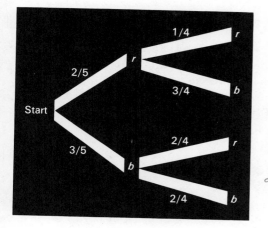

Figure 8.4

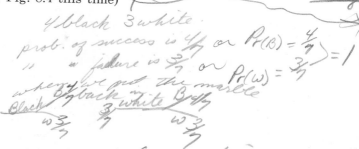

Consider an urn which contains two red and three blue chips. The experiment is to draw a chip from the urn, note its color, and, after replacing it, draw a second chip from the urn. The possibility set for this experiment is

$$S = \{(r,r),(r,b),(b,r),(b,b)\}.$$

1. Let $E = \{(r,r),(r,b)\}$. What is the probability of the event E?

2. The probability of the outcome (r,r) is what fractional part of $Pr(E)$?

3. What is the probability of the outcome (r,r)?

4. Let $F = \{(b,r),(b,b)\}$. The probability of the outcome (b,r) is what fractional part of $Pr(F)$?

5. What is the probability of the outcome (b,r)?

QUIZ YOURSELF*

---

1. $\frac{2}{5}$ 2. $\frac{2}{5}$ 3. $\frac{4}{25}$ 4. $\frac{2}{5}$ 5. $\frac{6}{25}$

***ANSWERS**

In the above examples we indicated a procedure for obtaining the probability assignment for the possibility set of a multistage experiment.

Let (x,y) be an outcome for a two-stage experiment. Then $Pr(x,y)$ is the probability of x occurring at the first stage times the probability of y occurring at the second stage, knowing that x occurred first. For an experiment with more than two stages, the assignment can be patterned after the two-stage experiment.

PRINCIPLE: PROBABILITY ASSIGNMENT FOR MULTISTAGE EXPERIMENT

As always, to find the probability of an event, we need only add the probabilities of the outcomes in the event, once a probabilistic model has been given for the experiment.

EXAMPLE   ■ A person plays a sequence of three games against two opponents, A and B; A is played first, then B, and then A again. The probability of winning against A, no matter when A is played, is $\frac{1}{3}$; against B it is $\frac{1}{4}$. In order for anyone to win the set, they must win at least two consecutive games. What is the probability of winning the set?

The event we are interested in for this multistage experiment is $E_W = \{(W,W,W),(W,W,L),(L,W,W)\}$. Consider the outcome (W,W,W). The probability of winning the first game is $\frac{1}{3}$. The probability of winning the second game is $\frac{1}{4}$ since at this stage opponent B is played. The probability of winning the third game is again $\frac{1}{3}$. Therefore $Pr(W,W,W) = (\frac{1}{3})(\frac{1}{4})(\frac{1}{3}) = \frac{1}{36}$.

It is not difficult to see that the probability of the outcome (W,W,L) is $(\frac{1}{3})(\frac{1}{4})(\frac{2}{3})$. Finally, $Pr(L,W,W) = (\frac{2}{3})(\frac{1}{4})(\frac{1}{3})$. Therefore we have $Pr(E_W) = (\frac{1}{3})(\frac{1}{4})(\frac{1}{3}) + (2)(\frac{1}{3})(\frac{1}{4})(\frac{2}{3}) = \frac{5}{36}$. □

EXERCISES

1. An urn contains three discs numbered 1, 2, and 3. A disc is drawn from the urn, its number is noted, and, without replacing it, a second disc is drawn and its number is noted. Give a probabilistic model for this experiment. Also, find the probability of obtaining the number three on either draw.

2. Consider two urns labeled A and B. Urn A contains two green chips and three blue chips, whereas urn B contains two blue chips and one red chip. The experiment is to draw a chip from urn A and, without looking at its color, place it in urn B. Next, a chip is drawn from urn B and its color is noted. Find the probability of drawing a green chip from urn B. Also find the probability of drawing a blue chip from urn B.

3. A bin contains five left-handed bolts and three right-handed bolts. Bolts are taken from the bin one at a time until a left-handed bolt is obtained. Find the probability that two bolts must be taken from the bin to obtain a left-handed bolt. Also find the probability that at least two bolts must be drawn from the bin before finding a left-handed bolt.

4. Two people, A and B, compete against each other in a best-of-five game tournament; that is, the first person to win three games wins the tournament. Suppose A's chances of beating B in any game of the tournament is 0.4. What is the probability that only three games will be played in the tournament? What is the probability that all five games must be played?

5. On a roulette wheel there are 37 equally-spaced slots, 18 red, 18 black, and 1 white. A gambler can bet on either red or black. Suppose he bets on red. The wheel is spun and if the ball comes to rest in a red slot, the gambler wins. If the ball comes to rest in a black slot, the gambler loses. If the ball drops into the white slot, then the wheel is spun again until the ball drops into either a red or black

slot. If on these further spins a black comes up first, then the player loses; if red comes up first, the player receives his bet back; that is, he breaks even. What are the probabilities of the three possibilities, the player wins, the player loses, and the player breaks even?

6. In pottery making, the molded, dried clay, called greenware, is fired, after which it is glazed and then fired a second time. After the first firing, the piece of pottery can be rejected or passed on to the next step. After the second firing, the piece can be rejected, sold as a second, or sold as first quality. Suppose a pottery company has obtained the following data from past performance. The probability that a piece will pass on to the next step after the first firing is 0.7. For a piece that has been given the second firing, the probability that it will be sold as first quality is 0.5 and the probability that it will be sold as a second is 0.3. What is the probability that a piece of greenware will be sold as first quality? What is the probability that a piece of greenware will be sold either as second or as first quality?

7. Consider the game of roulette described in exercise 5. If a player bets on red and wins, he receives $2, which is a gain of $1. Find the expected value for the player of this game. If a roulette player bets $.50 on red and $.50 on black, what is the expected value for this gambler?

As in the pottery question, we see that the probability of an outcome in a two-stage experiment is the probability of the outcome on the first stage times the probability of the second. (Photo by Sam Sweezy, Stock Boston.)

8. Suppose the pottery company described in exercise 6 sells its first-quality plates at a profit of $2 apiece and its seconds at a profit of $1 apiece. A piece of greenware which is rejected after the first firing will result in a loss of $.50 to the company and a loss of $.75 if rejected after the second firing. What is the expected profit for a piece of greenware? In order for the expected profit for a piece of greenware to be $.75, what must be the profit on a plate sold as first quality?

BINOMIAL EXPERIMENTS

We will consider a special type of multistage experiment, which we will call an **independent-trials process**. This type of multistage process is a single experiment performed repeatedly in succession, where an outcome at any stage of the process has no effect on the possible outcomes and their probabilities in any succeeding stage of the process. An example of an independent-trials process would be the rolling of a single die five times. Another example would be guessing on a ten-question true-false test.

Any independent-trials process is a multistage experiment, and we have already given a method in the previous section for determining a probabilistic model for the process. In this section, we will restrict our attention to independent-trials with two outcomes and obtain special formulas for this type of multistage experiment. This type of multistage experiment, an independent-trials process with two possible outcomes for each trial, will be called a **binomial experiment**. The flipping of a coin several times is one example of a binomial experiment.

EXAMPLES   ■ A mouse runs a T-maze twice. Assume that on the first run its choice to go right or left is made randomly. There is no stimulus present to cause learning and therefore we assume that its choice on the second run is also made randomly. We shall give a probabilistic model for this two-stage process.

We should first observe from the assumptions that the outcomes at the second stage are independent of what occurred on the first run; that is, the probability of the mouse going either right or left on the second run is not dependent on the outcome of the first run. Furthermore, each time the mouse runs the maze the same two possibilities are present, right or left. Therefore this two-stage process is a binomial experiment.

In giving the probability assignment, we can use the principle stated in the previous section for a multistage experiment. In particular, $Pr(R,R) = \frac{1}{2} \cdot \frac{1}{2} = \frac{1}{4}$. In fact, for each of the other three outcomes, (R,L), (L,R), and (L,L), the probability is also $\frac{1}{4}$. □

■ The probability of a 20-year-old woman living to age 30 is about 0.98. Suppose that three 20-year-old women each buy a ten-year-term life-insurance policy. Considering that each person could either live or die, give a probability assignment for the outcomes of this three-stage binomial experiment.

Let us first agree that we can model this as a binomial experiment. To this end, we need only observe that the probability of the second person living to age 30 is independent of the first person living to age 30. A similiar statement can also be made for the third person.

One outcome for the experiment is that all three women live to age 30. Since the chances of any one of the three women living to age 30 are 0.98, we have $Pr(L,L,L) = (0.98)(0.98)(0.98) = (0.98)^3$.

If we consider an outcome where one of the three dies, for example the outcome (L,L,D), its probability is $(0.98)(0.98)(0.02) = (0.98)^2(0.02)$.

The outcomes in which only one of the three lives to age 30 each have probability $(0.98)(0.02)^2$.

Finally, for the outcome (D,D,D), we have $Pr(D,D,D) = (0.02)^3$. □

With these examples in mind, we should realize how to determine the probability assignment for the outcomes of a binomial experiment and therefore we should easily accept the next principle.

Consider a binomial experiment where the two possibilities at each stage are denoted by s and f. Suppose that this multistage experiment has n stages and that at any stage $Pr(s) = p$ and $Pr(f) = 1 - p$. The probability of each outcome in which s occurs on k stages, no matter what the order, is $(p)^k(1 - p)^{n-k}$.

PRINCIPLE: PROBABIL- ITY ASSIGNMENT FOR A BINOMIAL EXPERIMENT

■ Suppose in a particular community with a large voting population 75 percent of the people are registered Democrats. If ten people are selected from the voters, what is the probability that the first seven selected are Democrats?

EXAMPLE

The process of selecting ten voters and observing whether each voter is a Democrat or not can be considered a binomial experiment. Of course, in doing so we are assuming that the voting population of the community is so large that after selecting a voter the percentage of Democrats has not significantly changed.

This binomial experiment consists of ten stages; the possibilities at any stage are that the person selected is a Democrat with probability 0.75 or the person selected is not a Democrat with probability 0.25. The probability that the first seven voters selected are Democrats is $(0.75)^7(0.25)^3$. In fact, each outcome for this binomial experiment in which seven of the ten are Democrats has probability $(0.75)^7(0.25)^3$. □

Consider the binomial experiment where a student guesses on a three-question multiple-choice test in which each question has four choices. Observe that for each question, the probability that the student will guess the correct choice is $\frac{1}{4}$ and the probability that he will get the question wrong is $\frac{3}{4}$.

QUIZ YOURSELF*

1. What is the probability that the student gets all three questions correct?

2. What is the probability that the student gets the first question right and the other two questions wrong?

3. What is the probability that the student gets the second question right and the first and last questions wrong?

4. What is the probability that the student guesses right on the last question but gets the first two questions wrong?

5. What is the probability for *each* outcome of this binomial experiment in which the student guesses two questions right and one question wrong?

---

*ANSWERS   1. $(\frac{1}{4})^3$ 2. $(\frac{1}{4})(\frac{3}{4})^2$ 3. $(\frac{1}{4})(\frac{3}{4})^2$ 4. $(\frac{1}{4})(\frac{3}{4})^2$ 5. $(\frac{1}{4})^2(\frac{3}{4})$

Since we now know how to compute the probability of any outcome of a binomial experiment, let us consider how to find the probability of an event.

EXAMPLE   ■ An urn contains three red chips, two blue chips, and three green chips. The experiment is to draw a chip from the urn, note its color, and return it to the urn. If this experiment is performed ten times, then what is the probability of drawing exactly four red chips on the ten trials?

Returning the chip to the urn before drawing the next one makes this an independent-trials process. Furthermore we can consider that one of two things will occur at each stage, namely, either drawing a red chip or drawing one of a different color. Thus in answer to our question, we can use a binomial model.

The event E, for which we want to compute the probability, is the set of outcomes in which four stages of this binomial experiment yield a red chip and six stages yield a chip of a different color. Since the probability of obtaining a red chip at any stage is $\frac{3}{8}$ and the probability of obtaining a different colored chip is $\frac{5}{8}$, the probability of any outcome in the event E is $(\frac{3}{8})^4(\frac{5}{8})^6$. Since each outcome in event E has the same probability, we realize that $Pr(E) = n(E) \cdot (\frac{3}{8})^4(\frac{5}{8})^6$.

Now we need only to determine the number of outcomes in event E. The outcome where the first four draws yield a red chip is one element in E. Another element in E is where the first three trials and the fifth trial yield a red chip. There are other outcomes in E, but we should realize that the number of elements in E is the same as the number of ways to choose from the ten trials, four of the trials on which a red chip is to appear. Therefore $n(E) = C[10,4]$ and we have $Pr(E) = C[10,4] \cdot (\frac{3}{8})^4(\frac{5}{8})^6$. □

In the above example, a typical question is asked with respect to a binomial experiment. In particular, for a binomial experiment of n stages with the possibilities at any stage being s or f, we may ask "What is the probability that exactly k of the n trials of the experiment result in the possibility s?" The procedure that we used to answer the question in the above example is the same for the general situation which we now give.

Consider a binomial experiment of n stages with the possibilities at any stage being s or f. Let $Pr(s) = p$ and $Pr(f) = 1 - p$. Then $C[n,k](p)^k(1 - p)^{n-k}$ is the probability that s will occur exactly k times in n repetitions of the experiment. We shall represent this number by the symbol $Pr(n, k; p)$.

PRINCIPLE: PROBABILITY OF EVENTS IN A BINOMIAL EXPERIMENT

In the appendix, a table of values for $Pr(n,k;p)$ is given for various values of n, k, and p. You could use this table when numerical values for $Pr(n,k;p)$ are desired instead of computing the value directly from the formula $C[n,k](p)^k(1 - p)^{n-k}$. For example, from the table we find the value of $Pr(2,0;\frac{1}{4})$ to be .5625. We shall refer to this table as the **binomial probability table**.

---

1. Using the binomial probability table, find the value of $Pr(3,1;\frac{1}{2})$. In order to see how this number was obtained for the table, compute it directly from the formula $C[3,1](\frac{1}{2})(\frac{1}{2})^2$.

2. Repeat question 1 for $Pr(3,2;\frac{1}{4})$.

3. Determine from the table the value of $C[10,3](0.25)^3(0.75)^7$.

QUIZ YOURSELF*

---

1. .3750   2. .1406   3. .2503

*****ANSWERS**

■ Suppose a new medicine is tested on 20 people and the drug company which produced the medicine claims that it is effective 90 percent of the time. What is the probability that exactly 15 of the 20 people will benefit from the medicine, assuming that the drug company's claim is correct?

EXAMPLE

Since the probability that one person will benefit from the medicine is 0.9, we obtain from our principle that the chances of exactly 15 people benefiting from the medicine are $C[20,15](0.9)^{15}(0.1)^5$. If you have a calculator available, compute the value of this number; the answer you should obtain is .0319.

Now let us use the binomial probability table to obtain the value of $C[20,15](0.9)^{15}(0.1)^5 = Pr(20,15;0.9)$. We see that there are no values listed for the probability 0.9. Observe that when exactly 15 of the 20 people benefit from the drug 5 people will not. Therefore $C[20,15] = C[20,5]$ and $C[20,15](0.9)^{15}(0.1)^5 = C[20,5](0.1)^5(0.9)^{15}$. Thus we can determine from the table the equivalent value $Pr(20,5;0.1) = .0319$. □

As in the above example, when we want to use the table to determine the value of $C[n,k](p)^k(1 - p)^{n-k}$ and p is larger than 0.5, we can look up $C[n,n - k](1 - p)^{n-k}(p)^k = Pr(n,n - k;1 - p)$ since it has the same value.

■ Consider the following binomial experiment. A student who has not studied is given a ten-question multiple-choice test in which each question has

EXAMPLE

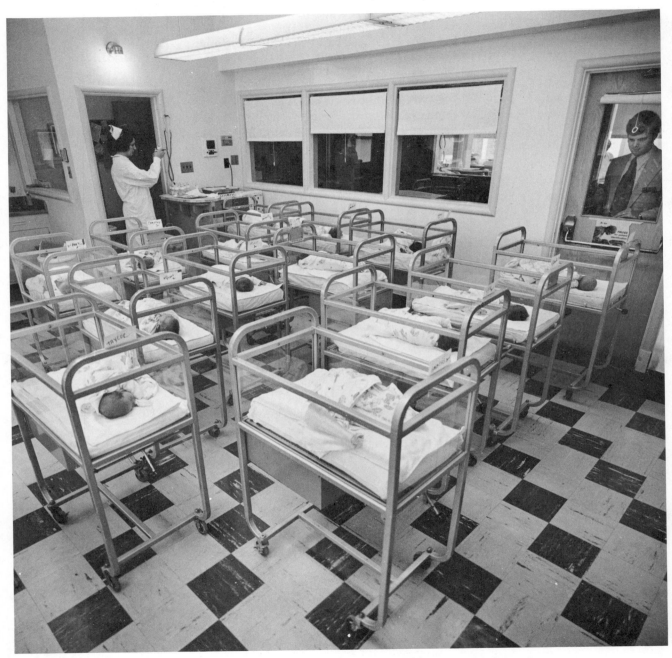

What is the probability that all fourteen babies are girls? (Photo by Frank Siteman, Stock Boston.)

four choices. Suppose that a passing grade is 70 percent or better. What is the probability of the student passing the test if he guesses on each question?

In this process, there are two outcomes for each stage. One possibility is that of obtaining a correct answer, which has probability $\frac{1}{4}$, and the other possibility is that of obtaining an incorrect answer, which has probability $\frac{3}{4}$. The answer to our question is $Pr(10,7;\frac{1}{4}) + Pr(10,8;\frac{1}{4}) + Pr(10,9;\frac{1}{4}) + Pr(10,10;\frac{1}{4})$. The value of this sum is found to be .0035, which means that this student has less than 4 chances in 1000 of passing this test. $\square$

EXERCISES

1. What is the probability of obtaining one head on the toss of three coins? What is the probability of obtaining one head on three tosses of a single coin?

2. From past game statistics, it is known that a certain basketball player makes 70 percent of his shots. If this player takes 20 shots during a game, what is the probability of him missing 6 of them? Out of the 20 shots taken, what is the probability of the player making at least 14 of the shots?

3. Consider the following binomial experiment. A student who has not studied is given a ten-question true-false test. Suppose that a passing grade is 70 percent or better. What are the chances that the student will pass the test if he guesses on each question?

4. a) If a coin is flipped ten times, then what is the probability that exactly five heads will appear?

   b) If a coin is flipped 20 times, then what is the probability that exactly 10 heads will appear? What is the probability that between 8 and 12 heads will appear on the 20 flips of a coin?

5. How many questions should be on a true-false test so that the chances of getting at least half of the questions correct, even when guessing, is greater than $\frac{1}{2}$?

6. A fraternity brother thinks he can tell the difference between beer and ale, but we know that he really can't. Suppose we test him by giving him four glasses of beer and one of ale and ask him to pick out the ale. If this test is given five times, then what is the probability that the brother will be right three out of the five tries, assuming, of course, that he only guesses?

7. In a very large organization, it is known that 60 percent of the members favor a particular issue; therefore if a vote was taken from the entire body, the issue would pass. However, ten representatives are selected at random from the members and their votes are taken. Explain why a binomial model can be used for this experiment. What is the probability that among the ten representatives there would be six or more "no" votes on the issue; that is, what are the chances that the issue would be defeated by the ten representatives?

NEW TERMS

Experiment

Outcome

Possibility set

Relative frequency

Probability

Event

Probability assignment

Probabilistic model

Equally likely

Expected value

Payoff

Fair game

Multistage experiment

Tree diagram

Independent trials process

Binomial experiment

MASTERY TEST: TOOLS OF PROBABILITY

1. Let $S = \{O_1, O_2, O_3, O_4\}$ be the possibility set for a certain experiment. Give a probability assignment for S such that $Pr\{O_2\} = 2 \cdot Pr\{O_1\}$, $Pr\{O_3\} = Pr\{O_1\}$, and $Pr\{O_4\} = Pr\{O_1\} + Pr\{O_2\}$.

2. Suppose the Weather Bureau compiled the following data over the past 50 years on the amount of rainfall during April.

Amount of rainfall in April	Frequency
Less than 1 inch	4
Between 1 and 2 inches	10
Between 2 and 3 inches	15
Between 3 and 4 inches	16
More than 4 inches	5

Use the above table to determine each of the following:
a) the probability of getting between 2 and 3 inches of rain in April;
b) the probability that there will be at least 2 inches of rain during April;
c) the probability that there will be less than 1 inch or more than 4 inches of rain during April.

3. An urn contains one red chip, two blue chips, and two white chips. Suppose that one chip is drawn from the urn and, without replacing it, a second chip is drawn. Give a probabilistic model for this experiment. Also find the probability of drawing one red chip and one blue chip.

4. Considering the urn in exercise 3, suppose that this time the first chip is put back into the urn before the second chip is drawn. Give a probabilistic model for this experiment.

5. A sample of 10 items is selected at random from a bin of 100 items. Suppose the bin contains 20 bad items and 80 good ones.
a) What is the probability that all the items selected are good?
b) What is the probability that at least one of the items selected is bad?

6. An organization is raffling a color TV set worth $600. The organization plans to sell 400 tickets and wants to make a profit of $200 on the raffle. How much should they charge for each ticket? What is the expected value for a person who buys one ticket?

7. An insurance company sells a $30,000 20-year term life-insurance policy to 25-year-old males for $2500. What would be the average profit to the insurance company per policy sold? The probability of a 25-year-old male living to age 45 is 0.94.

8. A company intends to buy one of two machines. Each machine produces the same item but they operate at slightly different rates. Since the slower machine is more reliable, the probability of it producing a defective item is smaller than

for the faster machine. Based upon the data in the following table, which machine is the better one for the company to buy?

	Loss on bad item	Profit on good item	Probability of producing bad item
Slower machine	$.12	$.42	.04
Faster machine	$.10	$.44	.08

9. Suppose that in making cut-glass plates two people work in sequence to produce a finished product and 95 percent of the first person's work is passed on to the second person for the next step. If 90 percent of the second person's work is considered first quality, what is the probability that a plate will end up as first quality?

10. A pollster claims that 25 percent of a city's population is opposed to a certain issue. If 12 people are selected at random, then what is the probability that exactly 3 of them will be opposed to the issue? What is the probability that either 2, 3, or 4 of the 12 will be opposed to the issue?

11. Assuming that the probability of a female answering a residential telephone during the day time is 0.75, what is the smallest number of telephone calls a pollster must make so that the probability of reaching at least one female is greater than 0.99?

III. THE DECISION-MAKING PROBLEM: A MODEL AND A SOLUTION

You will recall from Part I of this chapter that the Cure-All Pharmaceutical Company claims to have discovered a vaccine which would help protect a person from the common cold. We are interested in devising an experiment to test the validity of the company's claim that they can lower the probability of a person's catching a cold; that is, they can lower the infection rate.

As mentioned earlier, we are not going to try to find the exact probability of a person's catching a cold after being vaccinated. Instead we are going to design an experiment so that if we accept the company's claim, then the chances of our accepting a worthless vaccine are small.

Suppose we decide to do the following. We will vaccinate ten people. If four or fewer people catch a cold, then we will say that the company's claim is true. The experiment we are designing is a binomial experiment. There are ten stages in the process and the two outcomes at each stage are either the person catches a cold or does not catch a cold. Now to check the reliability of the experiment, we ask, "Suppose the vaccine is worthless, what is the probability of four or fewer people out of the ten catching a cold?" If indeed the vaccine is worthless, this will mean that the infection rate is still $\frac{1}{2}$. Therefore we wish to compute the following probability:

$$Pr(10,0;\tfrac{1}{2}) + Pr(10,1;\tfrac{1}{2}) + \cdots + Pr(10,4;\tfrac{1}{2}).$$

We can use the binomial probability table to determine the value of this sum. We find that the probability of four or fewer people in 10 catching a cold, if the infection rate were still $\frac{1}{2}$, is

$$.0010 + .0098 + .0439 + .1172 + .2051 = .3770.$$

We see that even if the vaccine is worthless the chances of our accepting it as being valuable are .3770. Since we may consider .3770 high, we should agree that our test is not reliable.

Let us consider a different criterion for accepting the claim of the company.

EXAMPLE   ■ Suppose we vaccinate ten people. If two or fewer of these people catch a cold, then we will accept the claim of the company. In this case, what is the probability of accepting a worthless vaccine?

Assuming that the vaccine is worthless is the same as saying the infection rate is still $\frac{1}{2}$ even when using the vaccine. Thus the probability of accepting a worthless vaccine is the sum of the three probabilities $Pr(10,k;\tfrac{1}{2})$, where $k = 0, 1, 2$. We find from the table that the answer is .0010 + .0098 + .0439 = .0547. □

The experiment and the criterion for acceptance in the example are more reliable than those of the previous case because there is less chance of accepting a worthless vaccine. Of course, one may say that if we vaccinate only ten people, then the most reliable test would be to require none of the ten to catch a cold. The chances of accepting a worthless vaccine in this case is .0010. However, there is a danger in having such a strict requirement. Even though we do not want to accept a worthless vaccine, at the same time we do not want to reject a valuable one.

■ Suppose the company claims that the vaccine lowers the infection rate to $\frac{1}{4}$. Considering the experiment and the acceptance criterion of the previous example, we will reject this claim if three or more people catch a cold. If the infection rate is indeed $\frac{1}{4}$ as the company claims, what is the probability of our rejecting this valuable vaccine?

EXAMPLES

In order to answer this question, we must compute the probabilities $Pr(10,k;\frac{1}{4})$ for $k = 3, 4, 5, \ldots, 10$ and add these numbers together. We obtain as our answer .2503 + .1460 + .0584 + .0162 + .0031 + .0004 = .4744. Thus we see that the probability of rejecting a valuable vaccine based upon our test is .4744.

In this test, we find that even though the chances of accepting a worthless vaccine are low, the chances of rejecting a valuable one is close to $\frac{1}{2}$, which may be considered high. □

■ Suppose the pharmaceutical company claims their vaccine lowers the infection rate for the common cold to $\frac{1}{4}$. Also suppose we wish to test this vaccine on only ten people. Is it possible to have a criterion so that the probability of accepting a worthless vaccine and the probability of rejecting a valuable vaccine are each less than .25?

The answer to our question is yes. Suppose we will accept the company's claim when three or fewer people out of the ten catch a cold. Then the probability of accepting a worthless vaccine is $Pr(10,0;\frac{1}{2}) + Pr(10,1;\frac{1}{2}) + Pr(10,2;\frac{1}{2}) + Pr(10,3;\frac{1}{2}) = .1719$, and the probability of rejecting a valuable vaccine is $Pr(10,4;\frac{1}{4}) + Pr(10,5;\frac{1}{4}) + \cdots + Pr(10,10;\frac{1}{4}) = .2241$. □

EXERCISES

1. Suppose the probability of a person's catching a cold is $\frac{2}{5}$. The Cure-All Company claims that their vaccine lowers this infection rate, and we decide to test this claim by vaccinating 15 people. If 4 or fewer of these people catch a cold, then we will say the company's claim is true. What is the probability that we are accepting a worthless vaccine?

2. Let us assume that the Cure-All Company claims that their cold vaccine lowers the infection rate from $\frac{2}{5}$ to $\frac{1}{5}$. In order to test their claim, we will inoculate 15 people and we will accept the company's claim if 4 or fewer people catch a cold. What is the probability that we may reject a valuable vaccine?

3. Assume that the infection rate is $\frac{2}{5}$ and the Cure-All Company claims their vaccine lowers this rate to $\frac{1}{5}$. If 15 people are vaccinated, is it possible to have a criterion

so that the probability of accepting a worthless vaccine and the probability of rejecting a valuable vaccine are each less than 0.1?

4. A college instructor wants to give a student either a 20-question true-false examination or a 5-question multiple-choice examination where each question has 4 choices. For either examination, the student will pass if 80 percent or better is scored. Which test should the instructor give if the probability of the student's passing the exam by purely guessing is to be the smaller?

5. Let us consider that the probability of a pepper seed sprouting is 0.7. The Big Grow Seed Company believes they have found a treatment process which will increase the germination rate of seeds. In order to test the company's claim, 20 pepper seeds which have been treated are planted and, if 17 or more of the seeds sprout, then the claim is accepted. What is the probability that with this test we will accept the claim of the Big Grow Seed Company even though the treatment process may be worthless?

6. Suppose that the Big Grow Seed Company claims that their process will increase the germination rate of pepper seeds to 0.9. We have planted 20 treated pepper seeds and have decided to accept the company's claim if 17 or more seeds sprout. What are the chances that we will reject the company's claim even though the treatment process may be valuable?

7. Assume again that the germination rate of pepper seeds is 0.7. The Big Grow Seed Company claims they can raise the rate to 0.9 with their process. If 18 seeds treated with the process are planted, is it possible to give a criterion so that the probability of accepting a worthless process and the probability of rejecting a valuable process are both less than 0.17?

8. The Tasty Coffee Company has a certain standard for the coffee beans they use. When receiving a shipment of coffee beans, they either accept or reject the entire shipment. Therefore they will allow up to 10 percent of the bags in a shipment to be less than their standard and since the company cannot test every bag in a shipment, they select 12 bags at random. If each of the 12 bags meet the good taste test, then the company will accept the shipment. What is the probability that the company will accept a shipment in which 10 percent (or more) of the bags fail to meet the company's standard of good quality?

SUGGESTED READINGS

GALE, D., "Optimal Strategy for Serving in Tennis." *Mathl Mag.*, September 1971, pp. 197–199.
Presents a simple mathematical model based upon probability.

GARDNER, M., "Mathematical Games: On the Meaning of Randomness and Some Ways to Achieve It." *Sci. Am.*, July 1968, pp. 116–121.
A discussion on the topic.

GARDNER, M., "Mathematical Games: On the Ancient Lore of Dice and the Odds Against Making a Point." *Sci. Am.*, November 1968, pp. 140–146.
An interesting explanation on how to construct dice with a bias, on what is craps, and on the odds in rolling the desired number to win.

NAHIKIAN, H. M., *A Modern Algebra for Biologists*. Chicago: University of Chicago Press, 1964.

Gives a brief but readable introduction to probability with applications to genetics.

NEWMAN, J. R., *The World of Mathematics*. Vol. II, New York: Simon & Schuster, 1956.

Part VII contains nontechnical essays on probability by Laplace, Pierce, Keynes, Poincaré, and Nagel.

WEAVER, W., "Probability," *Sci. Am.*, October 1950, pp. 44–47.

Brief introduction to probability; its meaning and role in the sciences.

Ring—a ring—of roses,*
A pocket full of posies;
Asha, Asha,
We all fall down.

Some feel that it was King Henry the VII's fear of the dreaded Black Plague that in 1532 prompted him to begin the registration of all deaths occurring in England. Later in the sixteenth century this practice evolved into the practice of publishing weekly Bills of Mortality, which informed the population of the number and causes of deaths that took place. As John Graunt, a wealthy English merchant, read these reports he noticed that the percentage of deaths due to accidents, suicides, and certain diseases did not vary greatly from bill to bill. His observations led him to conclude that social phenomena do not occur in a random fashion and, in 1662, he published his thoughts in the paper, "Natural and Political Observations . . . made upon the Bills of Mortality." King Charles II was so impressed with these "observations" that he nominated Graunt as one of the original members of the Royal Society of London, in spite of the fact that Graunt was only a shopkeeper and had few academic credentials.

Graunt's friend, Sir William Petty, felt that patterns of social behavior could be studied mathematically and gave the name "Political Arithmetic" to this new science, which is today known as **statistics**. The work of Graunt and Petty was continued by Edmund Halley, the noted English astronomer and mathematician, in a paper entitled "An Estimate of the Degrees of the Mortality of Mankind, Drawn from Curious Tables of the Births and Funerals

---

* Some analyze this rhyme as being descriptive of the symptoms and consequences of the Black Plague. If so, as statistics, it owes its origins to that horrible time in European history.

at the City of Breslaw; with an Attempt to Ascertain the Price of Annuities upon Lives." In this paper Halley studied human life expectancy mathematically and in doing so provided one of the mathematical foundations for the profitable operation of modern life-insurance companies.

The idea that social phenomena could be modeled mathematically was developed further in 1829 by the Belgian astronomer and statistician, Lambert A. J. Quetelet. After analyzing the results of the first Belgian census, he was struck by the way in which age, sex, climate, occupation, and economic status seemed to influence mortality. Quetelet's work further contributed to the beginnings of actuarial science, a branch of mathematics which not only has applications in the field of life insurance but also is used to determine how pension plans must operate to remain financially sound. Also through his studies, Quetelet was able to accurately predict patterns of crime which led him to conclude that social conditions influence the amount and nature of crime.

About the same time Quetelet was doing his pioneering work, Sir Francis Galton, a cousin of Charles Darwin, made an important contribution to the growing science of statistics. To investigate the question of whether abnormal height was inherited, Galton recorded the height of 1000 fathers and their sons. From his data he concluded that there was indeed a close relationship between these heights. Calling this relationship a correlation, he introduced a way of measuring it mathematically and then applied this new idea to studying other human characteristics which he felt could be inherited.

Many mathematicians from all over the world have contributed to the growth of statistics since its informal beginnings in the sixteenth and seventeenth centuries. In fact, the subject is now so vast that many universities have a department of statistics independent of their mathematics department. Applications of statistics can be found in most areas of endeavor. For example, statistical theory is used to improve crop yield and to analyze experimental results in biology, medicine, and other areas of science. It is also used in government to construct and analyze polls, in education to interpret the results of intelligence tests, and in industry to maintain effective quality control.

I. SHOULD THE GRADES BE CURVED?

Grades are a common concern to students. If a student does not do well on a test, it is hoped that there are many others in the same situation. After comparing one's performance against the performance of others in the class, the situation may not look so bad even though only 50 percent of the answers were correct. Students often ask, "Are you going to curve the grades?" There is no standard definition of what is meant by "curving" grades. However, to a student it generally means "Are you going to move my grade up?" To take a closer look at this, place yourself in the following situation.

You didn't do very well on Professor Erudite's midterm in Anthropology 100. In fact, you were lucky to squeak through with a 61. Nobody did very well; he said that there weren't any scores in the 90's. It's a good thing that Erudite's a decent guy; he's going to give you a break. He said that he'll curve the grades if the class wishes him to, and you're going to vote on it next Monday morning. That might help you. Maybe you'll even wind up with a *C*. But wait a minute . . . remember what happened to your roommate Charlie last term. When Professor Baker curved the final course averages in Music 105, Charlie's 91 turned out to be only a *B*. That could happen to you. Your 61 might become an *F*! Maybe you better ask Erudite what he has in mind.

(Photo by Jean-Claude Lejeune,
Stock Boston.)

"Uh. . . Excuse me. . . Uh. . . Professor. . . Sir. Exactly how are you going to curve the grades?"

"Well. . . Let's see. . . I'll first determine the mean of the raw scores and then I'll use that information to calculate the variance, which in turn would give me the standard deviation of the distribution. Those scores lying one half of a standard deviation on either side of the mean would correspond to C's; those scores an additional standard deviation away from the C's would be B's and D's, etc . . ."

What did he say? Mean. . .? Variance. . .? Standard deviation. . .? You are sorry that you asked. Maybe you can use a little common sense on this. It's a good thing you wrote those scores down—

$$14, \quad 32, \quad 61, \quad 72, \quad 77, \quad 78, \quad 79, \quad 82, \quad 83, \quad 85, \quad 88, \quad 89.$$

With all those scores above yours there is no way you are going to get a C. If he curves it you'll probably get an F. You've made up your mind . . . on Monday morning you are going to vote against curving.

Have you made the right decision? We will find out in Part III of this chapter.

II. TOOLS OF DESCRIPTIVE STATISTICS

DISTRIBUTIONS AND MEASURES OF CENTRAL TENDENCY

We live in an age of statistics. On any given day we may hear that the crime rate is up 17 percent; inflation is at an annual rate of 9.8 percent; the gross national product last year was $900 billion; there are 5,000,000 people unemployed; there are 3.6 million students in college; the incidence of leukemia for workers in a certain industry is twice the national average; there are fewer deaths per 100,000 women using the "pill" than there are per 100,000 women undergoing childbirth; the median income for black families is less than the median income for white families.

Hearing statements such as the above may lead you to conclude that the subject of statistics is concerned only with the gathering of numerical facts. There is an important area known as **descriptive statistics** which deals with the problem of summarizing and interpreting large quantities of numerical information on a given situation.

There is, however, another important area in statistics known as **inferential statistics** which is concerned with how we can draw conclusions about a large collection of objects although we only have complete information about a sample of that collection. We frequently see statements in the newspapers such as "57 percent approve of American foreign policy." How is such a figure obtained? Certainly it is not possible to interview every person in the United States to determine how they feel about our foreign policy. Instead, a sample of the population is polled and, using these results, statisticians try to deduce the feelings of the entire population on this issue.

Another interesting example of inferential statistics occurs every four years. On election night we sit before our TV sets and hear, "With only 1 percent of the vote counted, based on results from our sample precincts, XYZ News predicts that Senator Goodheart will carry the state of Ohio." Again we see that results from a sample of the population are used to predict the behavior of the whole population. Although it would be interesting to discuss this one aspect of statistics, the actual sampling process and the mathematical techniques involved are too complicated to be considered at this level.

We will now begin a discussion of descriptive statistics. A person working with statistics is often confronted with a large quantity of numerical data. For example, let us suppose that we have interviewed 20 inmates at the state prison to determine at what age they were first arrested. The ages are as follows: 9, 12, 18, 13, 13, 12, 18, 12, 9, 10, 11, 11, 10, 12, 14, 17, 16, 17, 10, 13. In a collection such as this, each of the numbers is called a **raw score** and the entire collection is called a **distribution**. Of course, a distribution of raw scores, particularly a very large distribution, may provide us with little information unless we organize it in some manner. For instance, we can present the above information more concisely in the following **frequency table**.

Age at first arrest	Frequency of occurrence
9	2
10	3
11	2
12	4
13	3
14	1
15	0
16	1
17	2
18	2

This frequency table now gives us a little better feeling for the information which has been gathered.

The cliché "a picture is worth a thousand words" certainly applies when working with large quantities of numerical data. Often a collection of figures becomes more meaningful when displayed visually. One of the most helpful ways of displaying a distribution of raw scores is by means of a **histogram**. A histogram of the distribution of prisoners' ages is shown in Fig. 9.1.

The histogram was constructed in the following manner:

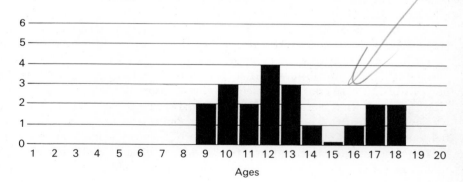

Figure 9.1

Ages

1) We located the various ages on a horizontal line.

2) We then centered a rectangle at each of the ages. The base of the rectangle has length 1 and the height of the rectangle is the number of times we observed that particular age in our distribution.

For example, since the age 12 occurs 4 times in our distribution, we draw a rectangle of height 4 centered over the 12 on the horizontal line. Note that since the age 15 does not occur in our distribution we place a "rectangle" of height zero over the 15.

Use the histogram given in Fig. 9.2 to answer the following questions. Assume that this histogram represents average daily temperatures, recorded over a certain time period.   **QUIZ YOURSELF***

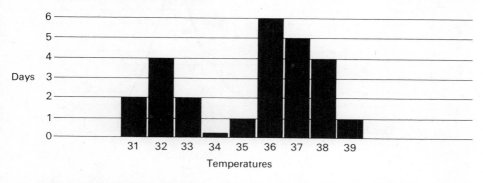

Figure 9.2

1. What was the lowest temperature recorded and how often did it occur?
2. What was the highest temperature recorded and how often did it occur?
3. Which temperature occurred most frequently?
4. For how many days were the temperatures recorded?
5. For what fractional part of the total number of days was the temperature above 37°F?

---

1. 31°F occurred twice   2. 39°F occurred once   3. 36°F   4. 25 days   5. $\frac{5}{25} = \frac{1}{5}$.   ***ANSWERS**

The next example shows how a distribution of raw scores can be effectively displayed using a figure known as a **frequency polygon**.

■ A representative of the City Bureau of Weights and Measures is checking   **EXAMPLE**
the scales in supermarkets throughout the city. He has weighed a 16-ounce object on 30 scales and obtained the results given in the frequency table below.

Weight	Frequency of occurrence
12	1
13	2
14	3
15	6
16	12
17	5
18	1

[handwritten margin notes: "always to look to find: 1. mean 2. median 3. mode"]

[handwritten top right: "Ex: 7500, 2000, 8000 — median, 100 000, 100 000, 13000, 57/132500 = mean, 26500 = average, mean"]

A frequency polygon for this distribution is given in Fig. 9.3.

[handwritten: "mode — the one that occurs most frequently."]

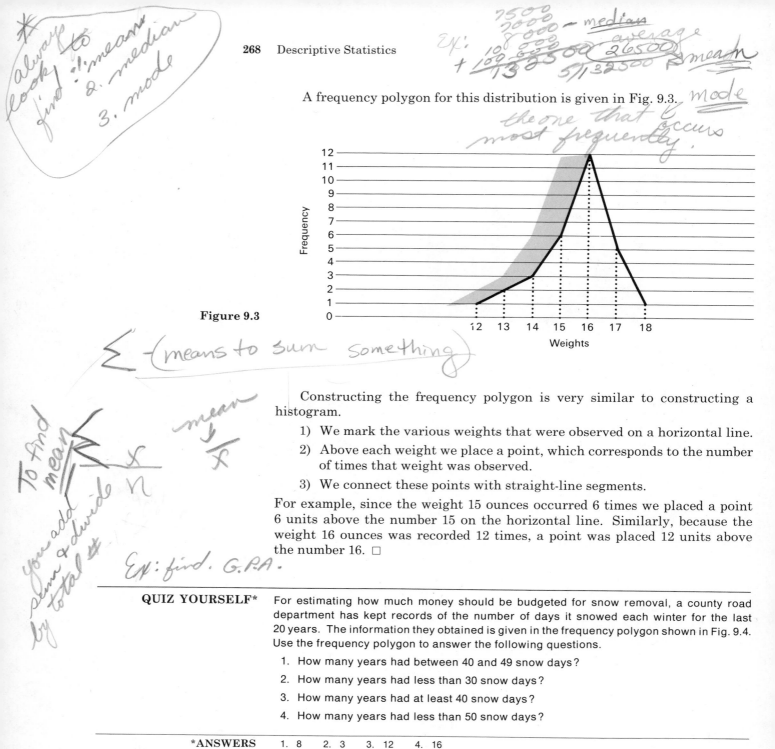

Figure 9.3

[handwritten: "∑ (means to sum something)"]

[handwritten: "mean ↓ x̄"]

[handwritten margin: "To find mean ∑x / n, you add plus divide by total #"]

Constructing the frequency polygon is very similar to constructing a histogram.

1) We mark the various weights that were observed on a horizontal line.

2) Above each weight we place a point, which corresponds to the number of times that weight was observed.

3) We connect these points with straight-line segments.

For example, since the weight 15 ounces occurred 6 times we placed a point 6 units above the number 15 on the horizontal line. Similarly, because the weight 16 ounces was recorded 12 times, a point was placed 12 units above the number 16. □

[handwritten: "Ex: find G.P.A."]

QUIZ YOURSELF*   For estimating how much money should be budgeted for snow removal, a county road department has kept records of the number of days it snowed each winter for the last 20 years. The information they obtained is given in the frequency polygon shown in Fig. 9.4. Use the frequency polygon to answer the following questions.

1. How many years had between 40 and 49 snow days?

2. How many years had less than 30 snow days?

3. How many years had at least 40 snow days?

4. How many years had less than 50 snow days?

*ANSWERS   1. 8   2. 3   3. 12   4. 16

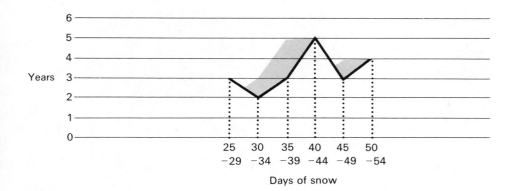

Figure 9.4

Organizing and displaying a collection of numerical data is generally not sufficient; we usually want more information about the nature of a total collection. We are probably interested in what should be considered a typical raw score for the distribution. For instance, a teacher may be interested in the class "average" on a test; a school board member may be concerned about the "average" class size at the local high school; a taxpayer might want to know the "average" amount paid yearly by his state for community health services. We have placed the word "average" in quotation marks to emphasize that asking "What is the *average*?" must be done carefully. We will see that there are at least three very different ways of explaining what could be meant by a "typical" score in a collection of data.

We now define what is usually meant by "average."

The MEAN of a distribution of raw scores is obtained by adding all of the raw scores together and then dividing this total by the number of raw scores in the collection.   DEFINITION

■ Ten students are asked how many hours they spend studying each week;   EXAMPLES
their replies are 12, 18, 13, 20, 18, 22, 20, 17, 17, and 21. What is the mean number of hours these students say they study each week?

To obtain the mean, we add these numbers and divide by 10.

$$\frac{12 + 18 + 13 + 20 + 18 + 22 + 20 + 17 + 17 + 21}{10} = \frac{178}{10} = 17.8 \ \square$$

■ A nuclear power plant is discharging hot water into a river. At a point two miles below the plant, the water temperature has been recorded in degrees Fahrenheit for each of the last 30 days. This information is summarized in the frequency table given below.

Temperature, degrees F	Frequency
50	1
51	1
52	3
53	5
54	3
55	8
56	4
57	3
58	2

We can simplify our calculations if we make some basic observations about the table. Note that the temperature 55°F occurred on 8 separate occasions, so when adding the temperatures we can calculate 8×55 rather than $55 + 55 + 55 + 55 + 55 + 55 + 55 + 55$. Treating the other lines of the table in a simlar way we obtain the sum 1635, which we divide by 30 to obtain the mean 54.5°F. □

True or false? The mean age of the children exhibiting vegetables in this display is thirteen. (Photo by Daniel S. Brody, Stock Boston.)

Before proceeding further with our discussion of statistics we wish to introduce some notation which will be useful throughout this chapter. Because we will often be computing sums of sets of numbers, it is convenient to use the Greek letter $\sum$ (sigma) to indicate sum. For example, we may wish to add all the numbers in the set $X = \{2,4,7,9,10\}$. Symbolically we would write $\sum X$, which would stand for the number $2 + 4 + 7 + 9 + 10 = 32$. The mean of the numbers in X could be expressed as $(\sum X)/5$, which of course means first calculate the sum $\sum X$ and then divide this number by 5.

The notation $\sum X^2$ would indicate that we should first square each member of the set X and then add these squares together. For the set X above, $\sum X^2 = 4 + 16 + 49 + 81 + 100 = 250$. Be careful not to mistakenly read $\sum X^2$ as $(\sum X)^2$; the latter notation means first add all members of X and then square.

The notation $\sum(X - 2)$ would tell us to subtract 2 from each member of the set X and then add the resulting numbers together. If X is as above, $\sum(X - 2) = (2 - 2) + (4 - 2) + (7 - 2) + (9 - 2) + (10 - 2) = 22$.

We can use the sigma notation to write the mean of an arbitrary set X as

$$\frac{\sum X}{n(X)}.$$

Consider the set $X = \{1,3,7,9\}$. Find:

1. $\sum X$;

2. $\sum X^2$;

3. $\sum(X - 5)^2$;

4. $\dfrac{\sum(X - 5)^2}{4}$;

5. the mean of the distribution of the numbers in X.

---

1. $\sum X = 20$

2. $\sum X^2 = 140$

3. $\sum(X - 5)^2 = 40$

4. $\dfrac{\sum(X - 5)^2}{4} = 10$

5. $\dfrac{\sum X}{n(X)} = 5$

The mean gives us some feeling for the general trend in a collection of raw scores. Another way of indicating a typical score in a distribution is given in the next definition.

The MEDIAN of an odd number of raw scores is the middle score when the scores are arranged in numerical order. When there is an even number of raw scores the median will be the mean of the two scores closest to the middle of the distribution.

DEFINITION

Some examples will make this notion clear.

■ Let us find the median of the distribution:

$$-4, \quad 11, \quad 25, \quad 8, \quad -3, \quad 62, \quad 21, \quad 13, \quad 2, \quad 8, \quad 61.$$

EXAMPLES

For some distributions the median is the middle score when all scores are arranged in numerical order. Do any utensils on this rack have median length?
(Photo by Fredrik D. Bodin, Stock Boston.)

We first arrange the scores in the following order:

$$-4, \quad -3, \quad 2, \quad 8, \quad 8, \quad 11, \quad 13, \quad 21, \quad 25, \quad 61, \quad 62.$$

The middle number in this list, namely 11, is the median of this distribution. □

The next example illustrates how we find the median when there is an even number of scores.

■ Assume that on a given day at the state welfare office, ten people apply for food stamps. Their annual family incomes, rounded off to the nearest thousand dollars are $7,000, $8,000, $6,000, $6,000, $4,000, $9,000, $4,000, $5,000, $7,000, $7,000. What is the mean family income?

We first arrange the incomes in ascending order. $4,000, $4,000, $5,000, $6,000, $6,000, $7,000, $7,000, $7,000, $8,000, $9,000.

Since we have an even number of incomes, we cannot have a unique number in the middle of this list; so to compute the median we find the mean of the two numbers nearest to the middle, namely \$6,000 and \$7,000, and obtain

$$\frac{\$6,000 + \$7,000}{2} = \$6,500$$

for the median family income. □

You can see from the last two examples that approximately half of the scores in a distribution lie below the median and the other half lie above it.

Here is a third way to indicate the trend in a collection of raw scores.

A **MODE** of a distribution of raw scores is a number which occurs most fre-   **DEFINITION**
quently. If there are no raw scores that occur more than once, then we say there is no mode.

■ The mode of the distribution 10, 7, 6, 12, 4, 10, 13, 10, 8, 7, 6, 9 is 10 because   **EXAMPLE**
it occurs more often than any of the other numbers. □

A distribution of numbers may have several modes, as we can see in the next example.

■ A representative from the bureau of consumer protection buys 10 boxes   **EXAMPLE**
of facial tissues which supposedly contain 200 tissues per box. The actual number of tissues in the 10 boxes was 202, 198, 198, 201, 200, 201, 198, 200, 201, 199. Since both 198 and 201 occur more frequently than any of the other numbers, 198 and 201 are modes for this distribution. □

Compute the mean, median, and mode for each of the following distributions: **QUIZ YOURSELF***

1. 7, 2, 8, 3, 4, 5, 6, 7;

2. 3, 3, 3, 9, 9, 9;

3. 2, 2, 68, 69, 70;

4. 98, 99, 100, 101, 102.

---

1. Mean $= \dfrac{42}{8} = 5.25$; median $= \dfrac{5 + 6}{2} = 5.5$; mode $= 7$.   ***ANSWERS**

2. Mean $= \dfrac{36}{6} = 6$; median $= \dfrac{3 + 9}{2} = 6$; mode $= 3$ and 9.

3. Mean $= \dfrac{211}{5} = 42.2$; median $= 68$; mode $= 2$.

4. Mean $= \dfrac{500}{5} = 100$; median $= 100$; no mode.

The mean, median, and mode all are measures of what is called the central tendency of a distribution of numbers. They tend to give us some information as to what a "typical" value in the distribution would be. We must be careful to realize that these three **measures of central tendency** do not tell us exactly the same things about a collection of data. In some circumstances one of the measures of central tendency may more accurately describe a *typical score* of the distribution than another would.

EXAMPLE   ■ The workers at the Magnum Industrial Corporation are on strike and a labor arbitrator has been called in to settle the dispute. There are 12 workers with salaries of $6000, $6000, $6000, $8000, $8000, $8000, $9000, $9000, $9000, $9000, $9000, $9000.

The arbitrator has gotten labor and management to agree that an across the board raise should be given to increase the "average" salary to $10,000. However, there is still an argument about what the "average" salary is. Management is insisting that since most workers earn $9000, the mode is the best measure of the "average." On the other hand, the union representatives maintain that the mean is the best indicator of a representative salary. Finally the arbitrator decides that the median is the best measure of what a typical salary is and decides that this figure should be raised to $10,000. Why did labor and management take the positions they did? What is the final raise granted to each employee?

In order to answer these questions, let us first calculate the mean, median, and mode of the distribution. If we call the distribution of salaries X, then the mean is

$$\frac{\sum X}{12} = \frac{96,000}{12} = \$8000.$$

Since there is no unique middle score, we average the two scores nearest the middle to get the median.

$6000, $6000, $6000, $8000, $8000, $\underline{$8000, $9000,}$ $9000, $9000, $9000, $9000, $9000

The median is

$$\frac{8000 + 9000}{2} = \$8500.$$

The mode is $9000.

It is now clear that management chose the mode to represent the "average" salary, because increasing the mode to $10,000 would result in a $1000 raise for each worker. Labor chose the mean for the "average" salary, because raising the mean to $10,000 gives each worker a $2000 pay boost. Since the arbitrator chose the median, which is $8500, raising that figure to $10,000 grants each worker a raise of $1500. □

Would the mean, median, or mode be the most accurate measure of what the average American family spends yearly on housing? (Photo by Ellis Herwig, Stock Boston.)

EXERCISES

1. Unemployment rates for the 12 months of 1975 were as follows:

   7%, 8%, 7.5%, 8%, 8.5%, 8.5%, 9%, 8.5%, 8%, 8.5%, 8.5%, 9%

   a) Draw a histogram for this distribution.
   b) Draw a frequency polygon for this distribution.

   mean = 8.25
   median = 8.5
   mode = 8.5

2. The EPA combined mileage ratings for 38 imported 1976 cars are given in the following frequency table.

Rating (mpg)	Frequency of occurrence
19	2
20	5
21	3
22	2
23	1
24	8
25	1
26	2
27	3
28	2
29	7
30	2

   a) Draw a histogram for this distribution.
   b) Draw a frequency polygon for this distribution.

3. What is the mean, median, and mode for the distribution given in exercise 1?

4. What is each measure of central tendency for the distribution given in exercise 2?

5. The following is the list of the scores for 20 students on a ten-point test.

   7, 8, 6, 5, 7, 10, 2, 7, 9, 5, 8, 8, 10, 9, 6, 5, 10, 7, 9, 8

   Give a frequency table for this distribution. Now determine each measure of central tendency for this distribution. What do you think is a "typical" score for this distribution?

6. The following is another list of scores for 20 students on a ten-point test.

   10, 5, 6, 8, 5, 9, 10, 5, 5, 9, 5, 10, 6, 8, 5, 9, 5, 10, 5, 5

   Give a frequency table for this distribution. Now determine each measure of central tendency for this distribution. What would you say is a "typical" score for this distribution?

7. The EPA combined-mileage ratings for fifty-eight 1976 American cars is given in the following frequency table.

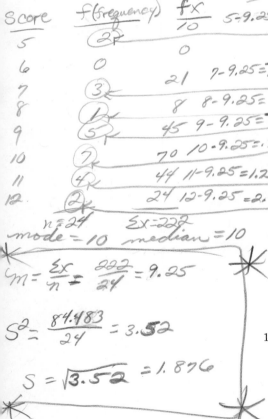

Rating (mpg)	Frequency of occurrence
12	1
13	9
14	13
15	20
16	1
17	4
18	0
19	0
20	10

   What is the mean, median, and mode for this distribution?

8. In a discussion of the fuel economy of imported cars vs. American cars, it was pointed out that there are two imported cars, Rolls-Royce and Jaguar XJ12, which have ratings of 11 mpg. Suppose this new score with its frequency is included in the distribution given in exercise 2. What is now the mean and median of the distribution?

9. The following two statements are somewhat common in sports discussions.

   "I read in Sunday's paper that the Golden Bears *averaged* 7 yards per carry in Saturday's game."

   "I have been attending all of our team's basketball games and I would say that Gunner *averages* 21 points per game."

   Which measure of central tendency do you think is being used in each statement?

10. Suppose there are nine scores from 1 to 10 in a distribution.
    a) Give an example of such a distribution in which the mean, median, and mode are each 5.
    b) Give an example of such a distribution in which the median is 3 and the mean is 5.
    c) Is it possible to have an example of such a distribution in which the median is 5 and the mean is less than 5?

MEASURES OF DISPERSION

We have seen that mean, median, and mode each provide us with some idea of what might be considered to be a *typical value* for a collection of data. You may feel that knowing these three numbers for a distribution gives a fairly accurate idea of the nature of the distribution. However, in the next example we will see that knowing the mean, median and mode does not give an indication of the spread of the raw scores.

■ Before dividing some revenue sharing funds between two cities of similar size, a federal administrator compiled some economic data on each city. He randomly selected 11 families from each city and recorded their annual income. In Urbanopolis, the mean, median, and mode of the incomes all were $10,000. Coincidentally, the mean, median, and mode of the incomes of Metrodelphia were also $10,000. Since the "average" family incomes in Urbanopolis and Metrodelphia were exactly the same, the administrator decided to split the funds equally between the two cities. The mayor of Metrodelphia immediately protested the decision stating that, although the sample of salaries accurately reflected the economic conditions of both cities, it was unfair to simply look at the mean, median, and mode of the distributions. She claimed that these measures of central tendency ignored important characteristics of the distributions, and in fact her city was far more deserving of federal funds than was Urbanopolis.

Here is the income distribution for the eleven families in each city.

Urbanopolis	Metrodelphia
$ 8,000	$ 5,000
9,000	5,000
9,000	6,000
10,000	6,000
10,000	6,000
10,000	10,000
10,000	10,000
10,000	10,000
11,000	10,000
11,000	10,000
12,000	32,000

Looking at the two distributions of income, we must agree with the mayor of Metrodelphia. We see that the incomes in Urbanopolis all tend to cluster around $10,000, whereas in Metrodelphia, more than half the incomes are quite a bit away from the "average" income of $10,000. In fact, we see that if we were to ignore the one income of $32,000 in Metrodelphia, the mean would drop to $7800, the median to $8000, and the mode would remain at $10,000. The three measures of central tendency do not indicate how much of a spread there is in the raw scores of the distributions. The simplest way to measure the spread or **dispersion** of a collection of numbers is given in the next definition. □

EXAMPLE

Although the range gives some indication of the dispersion of a distribution, it is not a very good measure, since it could be heavily influenced by a few extreme scores. (Photo by W. B. Finch, Stock Boston.)

DEFINITION The **RANGE** of a distribution is the difference between the largest and smallest scores in the distribution.

The range is very easy to compute, as we can see from our next examples.

EXAMPLE ■ In the Urbanopolis/Metrodelphia example mentioned earlier, the range of the Urbanopolis family incomes is $12,000 - $8,000 = $4,000. The range of the Metrodelphia family incomes is $32,000 - $5000 = $27,000. □

Our next example shows that range is not a very effective way to measure the dispersion of a distribution since it can be influenced too much by one number in the distribution.

EXAMPLE ■ A company has decided to work with the local parole board in providing jobs for exconvicts. Twelve parolees are hired, and after three months their work records show the following number of days missed for each parolee.

$$0, \quad 1, \quad 0, \quad 26, \quad 3, \quad 2, \quad 0, \quad 2, \quad 1, \quad 1, \quad 3, \quad 1$$

The range of this distribution is $26 - 0 = 26$, which seems to indicate that the scores are widely spread out, when in fact the number of absences tend to cluster around 1 or 2. If the 26 were omitted from this distribution, the range of the scores would be changed drastically to $3 - 0 = 3$. □

This last example shows the need for other ways of measuring the spread of a collection of raw scores. Let us return to our Urbanopolis/Metrodelphia example. Recall that the mean of the income distribution for each city is $10,000. We might feel that the reason for the small spread in the Urbanopolis distribution is that none of the raw scores is very far away from the mean. On the other hand the large spread of scores in the Metrodelphia incomes occurs because there are several scores, the $32,000 in particular, which are quite a distance from the mean.

This observation prompts us to make the following definition.

If x is a raw score in a distribution and the number m is the mean of that dis-   **DEFINITION**
tribution, we define the difference $x - m$ to be the DEVIATION OF x FROM
THE MEAN.

Let us calculate the deviation from the mean for each score in the above two distributions. We will denote the Urbanopolis distribution of income as X and the Metrodelphia distribution as Y.

Urbanopolis		Metrodelphia	
Score in thousands of dollars X	Deviation from the mean $X - 10$	Score in thousands of dollars Y	Deviation from the mean $Y - 10$
8	-2	5	-5
9	-1	5	-5
9	-1	6	-4
10	0	6	-4
10	0	6	-4
10	0	10	0
10	0	10	0
10	0	10	0
11	1	10	0
11	1	10	0
12	2	32	22

Comparing the incomes in the above table, we see that there is generally more deviation from the mean in the Metrodelphia distribution than there is in the Urbanopolis distribution. If fact, we might be tempted to find the mean of the deviations of the incomes for both cities and compare them.

The mean of the deviations of the Urbanopolis incomes is

$$\frac{\sum(X - 10)}{11} = \frac{0}{11} = 0.$$

And for Metrodelphia we have

$$\frac{\sum(Y - 10)}{11} = \frac{0}{11} = 0.$$

This, however, does not give us much information since the mean deviation is zero in both cases. Some of the deviations are negative and some are positive; by adding the deviations it is possible to obtain zero. In fact, *for any distribution the sum of the deviations will always be zero.* Symbolically, if X is any distribution with mean m, then $\sum(X - m) = 0$.

Let's try a different approach. Instead of finding the mean of the individual deviations, we will first square each deviation and then find the mean of these quantities. Note that squaring the deviations results in nonnegative numbers which will not cancel each other out when added together.

Urbanopolis		Metrodelphia	
Deviation $X - 10$	Deviation squared $(X - 10)^2$	Deviation $Y - 10$	Deviation squared $(Y - 10)^2$
-2	4	-5	25
-1	1	-5	25
-1	1	-4	16
0	0	-4	16
0	0	-4	16
0	0	0	0
0	0	0	0
0	0	0	0
1	1	0	0
1	1	0	0
2	4	22	484

If we now compute the mean of the squared deviations we get

Urbanopolis: $\quad \dfrac{\sum(X - 10)^2}{11} = \dfrac{12}{11} = 1.09;$

Metrodelphia: $\quad \dfrac{\sum(Y - 10)^2}{11} = \dfrac{582}{11} = 52.91.$

These calculations support our feeling that there is considerably more dispersion of the numbers in the Metrodelphia distribution than in the Urbanopolis distribution. The numbers we have just computed are given a special name.

The mean of the squared deviations of a distribution of numbers is called the   DEFINITION
VARIANCE of the distribution. If *m* is the mean of a distribution *X*, we can
express the variance of *X* as

$$\frac{\sum (X - m)^2}{n(X)}$$

We saw that the variance of the Urbanopolis incomes is 1.09 while the variance
of the Metrodelphia incomes is 52.91.

■ A consumer testing company has tested eight flashlight batteries of the   EXAMPLE
same brand to determine how long they last. The batteries worked for the
following number of hours:

104, 101, 97, 102, 104, 99, 102, 99.

To find the variance of this distribution, we first must compute the mean,
which is easily found to be 101.

Next we calculate the deviations from the mean and square them.

Score X	Deviation from the mean $(X - 101)$	Squared deviation $(X - 101)^2$
104	3	9
101	0	0
97	−4	16
102	1	1
104	3	9
99	−2	4
102	1	1
99	−2	4

Finally, we compute the mean of the squared deviations, which gives us
the variance

$$\frac{\sum (X - 101)^2}{8} = \frac{44}{8} = 5.5. \ \square$$

Compute the variance of the distribution 3, 4, 7, 1, 5. The mean for this distribution is 4.   **QUIZ YOURSELF***

$$\text{Variance} = \frac{(-1)^2 + (0)^2 + 3^2 + (-3)^2 + 1^2}{5} = \frac{1 + 0 + 9 + 9 + 1}{5} = \frac{20}{5} = 4.$$ ***ANSWER**

You may feel that in the last example the variance 5.5 is too large a number to represent the spread of the distribution because no score is more than
4 units away from the mean. Also, in the Metrodelphia example, the mean

Would you expect the group on the right or the one on the left to have the greater variance in its distribution of weights? (Photo by Rick Smolan, Stock Boston.)

is $10,000 and the variance is 52.91. Yet, if we consider $1000 as one unit of income, then the most extreme income, $32,000, is only 22 units away from the mean.

Recall that in calculating the variance, we have in some sense distorted this number by taking the individual deviations and squaring them. We compensate for this by taking the square root of the variance. In the flashlight-battery example, the square root of 5.5 is 2.345 which seems to be a more reasonable measure of how far away scores are from the mean. The square root of the variance 52.91 in the Metrodelphia distribution is 7.274, which again seems to be a more natural indicator of how far away incomes are from the mean.

This discussion leads to our next definition.

DEFINITION   **The nonnegative square root of the variance of a distribution is called the STANDARD DEVIATION of that distribution. Symbolically, if X is a distribution with mean m, then the standard deviation s can be given by**

$$s = \sqrt{\frac{\sum (X - m)^2}{n(X)}}.$$

By the definition, since the square root of the variance 52.91 is 7.274, this is the standard deviation of the Metrodelphia distribution.

■ A student interested in investigating her own longevity has compiled a list of the ages at which seven members of her family died.   **EXAMPLE**

Relative	Age at death
cousin	23
aunt	63
uncle	72
grandfather	68
grandmother	71
grandfather	69
grandmother	82

Compute the mean, variance, and standard deviation for this distribution, X. The mean is

$$\frac{\sum X}{7} = \frac{448}{7} = 64.$$

To find the variance we must first compute the deviations of each score from the mean and square them.

Score X	Deviation from the mean $(X - 64)$	Deviation squared $(X - 64)^2$
23	-41	1681
63	-1	1
72	8	64
68	4	16
71	7	49
69	5	25
82	18	324

The variance is the mean of these squared deviations which is

$$\frac{\sum(X - 64)^2}{7} = \frac{2160}{7} = 308.6.$$

The standard deviation is the positive square root of 308.6, which is approximately 17.567. □

EXERCISES

You should use the table of square roots given in the appendix when doing these exercises.

1. Consider that each of the following numbers is a variance for a distribution. Find the standard deviation for the distribution.
   a) 5   b) 50   c) .5   d) 23   e) 200

2. For each of the following distributions, compute the mean and the standard deviation.

   a) | Score | 18 | 19 | 20 | 21 | 22 |
      |-----------|----|----|----|----|----|
      | Frequency | 1 | 3 | 1 | 0 | 0 |

   b) | Score | 18 | 19 | 20 | 21 | 22 |
      |-----------|----|----|----|----|----|
      | Frequency | 0 | 1 | 3 | 1 | 0 |

3. For each of the following distributions, compute the mean and the standard deviation.

   a) | Score | 2 | 3 | 4 | 5 | 6 | 7 | 8 | 9 | 10 |
      |-----------|---|---|---|---|---|---|---|---|----|
      | Frequency | 1 | 1 | 0 | 2 | 3 | 4 | 4 | 3 | 2 |

   b) | Score | 2 | 3 | 4 | 5 | 6 | 7 | 8 | 9 | 10 |
      |-----------|---|---|---|---|---|---|---|---|----|
      | Frequency | 1 | 3 | 2 | 2 | 0 | 1 | 0 | 6 | 5 |

4. Consider the two distributions given by the frequency polygons in Fig. 9.5. Which distribution has the greater variance and hence the greater standard deviation?

Figure 9.5

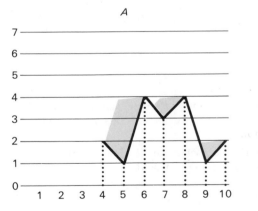
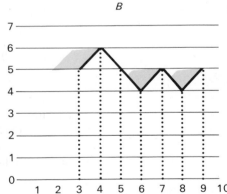

5. Which of the two distributions graphed in Fig. 9.6 has the greater variance and hence the greater standard deviation?

6. The number of people requiring treatment during a single day in the emergency room of a hospital was recorded over a 20-day period. The results are given in the following frequency table.

Number of people	17	18	19	20	21	22	23
Frequency (days)	2	3	5	6	0	2	2

What is the mean and the standard deviation for this distribution?

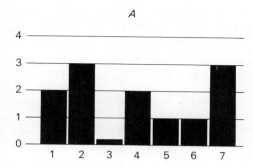
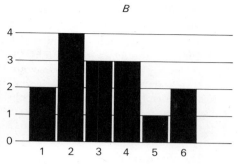

Figure 9.6

7. Suppose that in 1960 the mean family income in the United States was $8000 with a standard deviation of .5. In 1970, the mean family income was $11,000 with a standard deviation of 2. If a family earned $10,000 during 1960 and $14,000 during 1970, then in which of these two years was the family better off financially in relation to the rest of the population?

8. Below are the incomes, given in thousands of dollars and recorded in 1960 and 1970, of eight families:

	1960	1970
A	7	9
B	8	11
C	10	12
D	9	11
E	11	11
F	7	10
G	9	12
H	11	12

In terms of the number of standard deviations from the mean, which family made the greatest improvement in its economic condition during this 10-year period with respect to the rest of the population?

9. a) Flip four coins ten times and record the number of heads that appear on each flip. Find the mean and the standard deviation of the distribution of number of heads you obtain.
   b) Repeat (a) but this time flip the 4 coins 20 times.
   c) Repeat (a) but this time flip the 4 coins 30 times.

10. Pick any five numbers and let these be the five raw scores of a distribution. Compute the mean and the standard deviation of this distribution.
   a) Add 20 to each of the numbers in your original distribution and compute the mean and standard deviation for this new distribution.
   b) Subtract 5 from each number in your original distribution to form a new distribution of five scores. Compute the mean and standard deviation of this new distribution.
   c) Can you draw any conclusions about changes in the mean and the standard deviation when a number is added to or subtracted from each score in a distribution?

11. Use the conclusion reached in exercise 10 to simplify the calculation of the mean and standard deviation of the distribution 598, 597, 599, 596, 600, 601, 602, 603.

12. Pick any five numbers and let these be the five raw scores of a distribution. Compute the mean and standard deviation of this distribution.
   a) Multiply each number in your distribution by 4 and compute the mean and standard deviation of this new distribution.
   b) Multiply each number in your original distribution by 9 and compute the mean and standard deviation of this new distribution.
   c) Can you draw any conclusions about changes in the mean and standard deviation when each score in a distribution is multiplied by the same number?

13. The mean and standard deviation of the distribution 3, 4, 7, 1, 5 are 4 and 2, respectively. Consider your result for exercise 12(c) and compute the mean and standard deviation for the distribution 15, 20, 35, 5, 25.

CHEBYSHEV'S INEQUALITY

The standard deviation of a distribution in some way measures the spread of numbers in that distribution. In the Urbanopolis/Metrodelphia example we saw that since the Urbanopolis incomes were relatively close to the mean, we had the very small standard deviation of 1.044. On the other hand, the greater spread in the Metrodelphia incomes resulted in the considerably larger standard deviation of 7.274. In general if two distributions have the same mean but different standard deviations, then the distribution with the smaller standard deviation has its raw scores more tightly clustered about the mean.

We might wonder what can be said about a distribution if its mean and its standard deviation are known. In particular, can we make some estimate as to where the scores of the distribution will lie? Such estimates would be particularly important if we were dealing with information such as that reported by a government agency. We may be provided only with the mean and standard deviation of a distribution and may have no access to the raw scores. In this situation we may want to draw conclusions about the raw scores from the mean and standard deviation.

The Russian mathematician P. L. Chebyshev (1821–1894) gave a partial answer to this question in a famous theorem that bears his name.

PRINCIPLE:
CHEBYSHEV'S
INEQUALITY

Suppose that k is any real number which is greater than or equal to 1. If a score is chosen randomly from a distribution, the probability that it will lie less than k standard deviations away from the mean of the distribution is at least

$$1 - \frac{1}{k^2}.$$

A diagram makes this theorem a little clearer. Let us denote the mean of some distribution by the letter m and its standard deviation by s. Suppose x is a score selected randomly from the distribution.

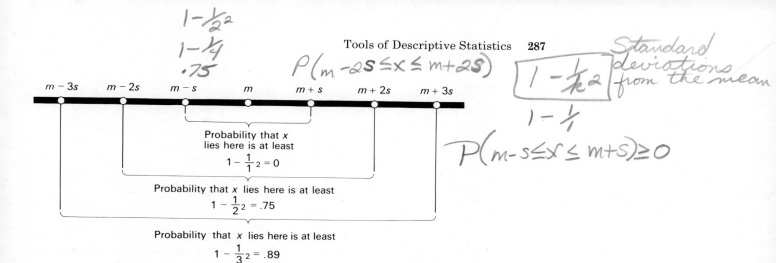

$$1 - \frac{1}{2^2}$$
$$1 - \frac{1}{4}$$
$$.75$$

$$P\left(m - 2s \leq x \leq m + 2s\right)$$

$$\boxed{1 - \frac{1}{k^2}}\ \text{Standard deviations from the mean}$$

$$1 - \frac{1}{1}$$

$$P\left(m - s \leq x \leq m + s\right) \geq 0$$

EXAMPLES

■ Suppose a distribution has a mean $m = 10$ and a standard deviation $s = 2$.

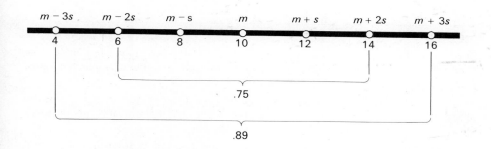

If a score is chosen randomly from this distribution, the probability that it will lie between 6 and 14 is at least .75, while the probability that the score lies between 4 and 16 is at least .89. □

■ A doctor who has installed a heart pacemaker in one of his patients would want to know when the pacemaker's battery should be replaced. Since an operation is required, it is important not to remove a good battery too soon, but on the other hand a delay in replacing a defective battery could be fatal.

By experimentation with a large number of batteries it has been found that the mean battery life is 2000 days and the standard deviation is 5 days. If one of these batteries is placed in a patient, when is the latest time it can be replaced and yet have a probability less than .0001 of being defective?

We can answer this question by finding an interval of time such that the probability of the battery becoming defective during this time period is at least .9999. If the battery is replaced prior to this period, the probability of it being defective will be less than .0001.

Chebyshev's inequality tells us that the probability of the battery becoming defective during a time interval extending k standard deviations on either side of the mean is at least $1 - (1/k)^2$. Therefore we should choose k so that

$$1 - \frac{1}{k^2} = .9999.$$

Solving for k^2 we get

$$k^2 = 10{,}000,$$

and therefore

$$k = 100.$$

We now know that the probability of a battery becoming defective during a time period extending 100 standard deviations on either side of the mean is at least .9999. Since the mean is 2000 days and the standard deviation is 5 days, this period extends from 1500 to 2500 days. Based upon our results, in order for the probability of being defective to be less than .0001, the battery must be replaced no later than 1500 days after it is first installed. □

EXERCISES

1. Suppose a distribution has a mean of 31 and a standard deviation of 3. Use Chebyshev's inequality to find the probability that a randomly selected score from this distribution lies between 25 and 37.

2. Suppose a distribution has a mean of 16 and a standard deviation of $\frac{1}{2}$. Find the probability that a score selected at random from this distribution lies between 11 and 21.

3. A distribution of raw scores has a mean of 50 and a standard deviation of 2. Find two numbers A and B such that if a score x is selected at random from the distribution, the probability that x lies between A and B is at least .96.

4. A distribution of raw scores has a mean of 67 and a standard deviation of 5. Find two numbers A and B such that if a score x is selected at random from the distribution, the probability that x lies between A and B is at least .99.

5. Ten thousand light bulbs have been placed in the score-board at a sports arena. If the mean lifetime of these bulbs is 500 days and the standard deviation is 20, use Chebyshev's inequality to estimate the probability that a randomly selected bulb will burn out between 450 and 550 days of use.

6. Reconsider exercise 5 with the mean now being 800 days and the standard deviation being 10. Estimate the probability that a randomly selected bulb will still be good on the 700th day of use.

7. The owner of a fleet of taxicabs is going to purchase new tires. He would like to be reasonably sure that the tires he buys will last 30,000 miles. More precisely, he wants the probability of a tire lasting 30,000 miles to be at least .96. Suppose that a tire manufacturer claims his tires have a mean life of 31,000 miles with a standard deviation of 800. Do these tires meet the taxicab owner's requirements?

8. Redo exercise 7 except assume that the mean life is now 30,500 miles and the standard deviation is 100.

9. A manufacturer is considering marketing one of two new models of microwave oven. From testing prototypes, it has been found that
   a) the mean amount of radiation emitted from type A is 14 units with a standard deviation of .05;
   b) the mean radiation emitted from type B is 12 units with a standard deviation of .4.

   Suppose there is a federal safety standard which requires that the probability of a model producing less than 15 units of radiation is at least .99. Is either model in compliance with this safety standard?

10. Suppose you read in a consumer magazine that television set A has a mean picture-tube life of 1900 days with a standard deviation of 10, while television set B has mean picture-tube life of 1950 days with a standard deviation of 25. Which set do you feel has the better chance of lasting 5 years?

MASTERY TEST: TOOLS OF DESCRIPTIVE STATISTICS

1. The histogram given in Fig. 9.7 represents monthly rainfall recorded over a certain period of time.
   a) What was the smallest amount of monthly rainfall and how often did it occur?
   b) What was the heaviest amount of monthly rainfall and how often did it occur?
   c) Which amount of monthly rainfall occurred most often?
   d) For how many months is the information given?
   e) For what fractional part of the total number of months was the rainfall below three inches?

NEW TERMS

Descriptive statistics
Inferential statistics
Raw score
Distribution
Frequency table
Histogram
Frequency polygon
Mean
Median
Mode
Measures of central tendency
Dispersion
Range
Deviation from the mean
Variance
Standard deviation
Chebyshev's inequality

Figure 9.7

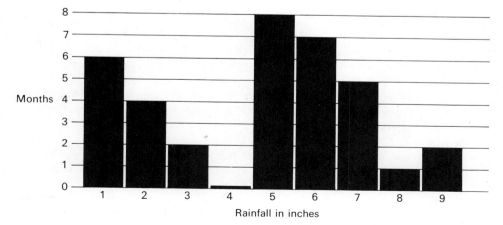

Rainfall in inches

2. A city police department has compiled the following information regarding the frequency of serious crime over a 100-day period.

Number of crimes	Frequency
0	3
1	2
2	7
3	12
4	15
5	12
6	13
7	14
8	10
9	8
10	4

a) Find the mean number of serious crimes per day.
b) Find the mode of this distribution.
c) Find the median number of serious crimes per day.

3. Let $X = \{1,4,6,5,12,2\}$. Find

a) $\dfrac{\sum X}{6}$ b) $\sum(X - 5)^2$ c) $\sqrt{\dfrac{\sum(X - 5)^2}{6}}$

4. Give a distribution of 7 numbers, each between 0 and 20, so that
a) the mean, median, and mode of the distribution are all 9.
b) the mean is 11 and the median is 12.

5. Find the mean, variance, and standard deviation for the distribution 1, 7, 14, 21, 2, 4, 7.

6. Suppose that in 1965 the mean family income in the United States was $10,000 with a standard deviation of .6 and in 1975, it was $12,000 with a standard deviation of 1.4. If a family earned $12,000 in 1965 and $15,000 in 1975, during which of these two years was the family better off financially in relation to the rest of the population?

7. Suppose a distribution has a mean of 70 and a standard deviation of 4. If a score is randomly selected from this distribution, what is the probability that it lies between 54 and 86?

8. A refrigerator manufacturer wants to guarantee his product. From past testing, it has been found that the cooling unit has a mean life expectancy of 2800 days with a standard deviation of 20 days. In order to avoid replacing too many units at no cost to the customer, the manufacturer wants to determine the latest date that a guarantee could still be in effect and yet have a probability of at least .99 that the cooling unit is still working properly. Determine this information for the manufacturer.

III. YOU WERE WRONG

We have developed enough of the "tools" of statistics to intelligently discuss the grading problem we posed in Part I of this chapter. You may recall that the test scores in Professor Erudite's anthropology class were 14, 32, 61, 72, 77, 78, 79, 82, 83, 85, 88, 89.

Professor Erudite is considering "curving" the grades by doing the following:

1) Scores which are less than $\frac{1}{2}$ of a standard deviation on either side of the mean will correspond with a C.

2) Scores between $\frac{1}{2}$ and $\frac{3}{2}$ of a standard deviation away from the mean would correspond with a B or a D.

3) Scores which are more than $\frac{3}{2}$ standard deviation away from the mean would correspond with an A or an F.

Your score is a 61 and you've decided to vote against curving the grades. Let's determine if you've made the right decision.

First we calculate the mean of this distribution X; in particular,

$$\frac{\sum X}{n(X)} = \frac{840}{12} = 70.$$

Next we compute the deviations from the mean of the individual scores.

Score X	Deviation from the mean $(X - 70)$	Deviation squared $(X - 70)^2$
14	-56	3136
32	-38	1444
61	-9	81
72	2	4
77	7	49
78	8	64
79	9	81
82	12	144
83	13	169
85	15	225
88	18	324
89	19	361

From the above table we see that the sum of the squared deviations is 6082. The variance is therefore 6082/12 = 506.8. The standard deviation is $\sqrt{506.8}$, which is approximately 22.51. If we denote the mean by m and the standard deviation by s, we can see in the following diagram how the letter grades would be assigned.

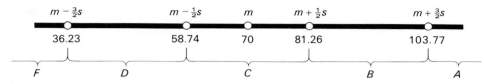

A would correspond with a score above 103.77.
B would correspond with a score between 81.26 and 103.77.
C would correspond with a score between 58.74 and 81.26.
D would correspond with a score between 36.23 and 58.74.
F would correspond with a score below 36.23.

If Professor Erudite "curves" the grades as indicated above, the following would be the 12 numerical grades and the corresponding letter grades:

$$\underbrace{14, 32,}_{F} \quad \underbrace{61, 72, 77, 78, 79}_{C} \quad \underbrace{82, 83, 85, 88, 89}_{B}$$

You see that if the grades were "curved" in this fashion, your 61 would correspond to a C.

Here is another example which should give you an appreciation of how the inclusion or omission of several scores can drastically change the mean and standard deviation of a distribution. Changing the mean and standard deviation can, of course, have a considerable effect on the way letter grades will be assigned to the numerical scores.

EXAMPLE   ■ Suppose the students with the two lowest scores drop the course. Professor Erudite decides that since these two students are no longer in the class, their scores should not be included in the distribution for the purpose of "curving" the grades.

Our distribution is now 61, 72, 77, 78, 79, 82, 83, 85, 88, 89. The mean is $\frac{794}{10} = 79.4$.

The deviations from the mean for this distribution Y are considered next.

Score Y	Deviation from the mean $(Y - 79.4)$	Deviation squared $(Y - 79.4)^2$
61	-18.4	338.56
72	-7.4	54.76
77	-2.4	5.76
78	-1.4	1.96
79	$-.4$	.16
82	2.6	6.76
83	3.6	12.96
85	5.6	31.36
88	8.6	73.96
89	9.6	92.16

We see that the variance is

$$\frac{\sum(Y - 79.4)^2}{n(Y)} = \frac{618.4}{10} = 61.84$$

and, hence, the standard deviation is $\sqrt{61.84} = 7.86$.

You can see from the following diagram how the letter grades would be assigned. As usual, m is the mean and s is the standard deviation.

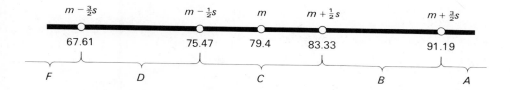

Below are the 10 numerical scores with their corresponding letter grade equivalents.

$$\underset{F}{61} \quad \underset{D}{72} \quad \underset{C}{77, 78, 79, 82, 83} \quad \underset{B}{85, 88, 89}$$

As you can see the score 61 is now an F, the 72 is a D, and 82 and 83 are C's. This is because of two things:

1) The mean has increased because we have omitted the two lowest scores.

2) Because the two extreme scores were omitted, the standard deviation has drastically changed. □

One of the reasons the mean is often preferred to the median or mode as a measure of the central tendency of a distribution is because it can be used to find the total of the individual scores in the distribution. The next example illustrates this.

■ In addition to your class, Professor Erudite has another large section of **EXAMPLE** Anthropology 100 with fifty students in it. If he curves the grades, he will do so by considering the set of all scores from the two classes to make up one distribution. He will then calculate the mean and standard deviation and assign letter grades as he did before.

Let's assume this time that you earned an 89 on the exam. Recall that the scores in your section are 14, 32, 61, 72, 77, 78, 79, 82, 83, 85, 88, 89. Your friend Lisa, a math major, is in Professor Erudite's other section of Anthropology 100. Although she does not remember the 50 individual scores in that section she has calculated the mean, 70, and the standard deviation, 6, for that class on her pocket calculator. Knowing this, should you vote to curve the grades?

In order to answer this question intelligently, you should know how many standard deviations your score is from the mean of the whole distribution. Therefore we want to first determine the mean of the distribution.

The mean m of the combined distribution of scores will be the total of all scores from both classes divided by the total number of scores, 62. We found earlier that the total of the scores in your class was 840. The total of the scores in Lisa's class can be easily calculated. Let us denote the distribution of scores in Lisa's class with a Y; then the mean in Lisa's class is

$$\frac{\sum Y}{50} = 70.$$

Thus $\sum Y = 50 \cdot 70 = 3500$. Therefore

$$m = \frac{840 + 3500}{62} = \frac{4340}{62} = 70.$$

Next we would like to find the standard deviation, s, for the combined distribution. This can be found by dividing the total of the squared deviations from the mean in both sections by 62. Do not be misled by what we are doing. We are fortunate the deviations from the mean were calculated with respect to the same mean for both sections. *If the mean were not the same in both sections, we could not add the squared deviations as we will do to obtain the variance.*

We have earlier calculated the total of the squared deviations for your class and found it to be 6082. The corresponding total for Lisa's class can be found without too much effort. The standard deviation in Lisa's class is 6. Thus we can write

$$\sqrt{\frac{\sum(Y - 70)^2}{50}} = 6.$$

This implies that

$$\frac{\sum(Y - 70)^2}{50} = 36,$$

and hence

$$\sum(Y - 70)^2 = 1800.$$

We should realize that $\sum(Y - 70)^2$ is the total of the squared deviations from the mean in Lisa's class. Hence the combined total of the squared deviations is $6082 + 1800 = 7882$. Dividing this by 62 we get 127.1, the variance of the entire distribution. Now we see that the standard deviation for this combined distribution is $s = \sqrt{127.1} = 11.28$.

The letter grades can be assigned according to the following diagram.

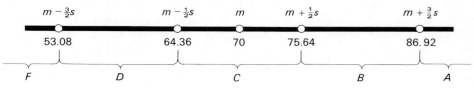

You see that it is now to your benefit to have the scores combined and the grades "curved" since your 89 would be an A. ☐

EXERCISES

In each of the exercises 1 to 8, a distribution of scores on a test is given. Consider yourself as one who has taken the test and your score is as indicated. First make an estimate based upon your intuition of the letter grade you would receive if the scores were "curved." Then using the method of curving described in this section, calculate the letter grade you would receive. Was your intuition correct?

1. 71, 76, 78, 78, 78, 78, 79, 80, 81, 81
    Your score: 81

2. Use the distribution of scores in exercise 1 and consider that your score is 71.

3. 50, 55, 57, 60, 60, 60, 63, 63, 65, 67
    Your score: 60

4. Use the distribution of scores in exercise 3 and consider that your score is 63.

5. 50, 50, 60, 60, 60, 60, 60, 60, 70, 70
    Your score: 70

6. 50, 60, 60, 60, 70, 70, 80, 80, 80, 90
    Your score: 90

7. 68, 69, 69, 69, 70, 70, 71, 71, 71, 72
    Your score: 70

8. Let us suppose that the scores on a particular test were 50, 59, 61, 70, and 100. The members of the class who received the four lower scores may feel that the student who received the 100 has hurt their chances of getting a better grade when the scores are curved; that is, the student who received the 100 "broke" the curve. If the 100 is dropped from the distribution, we see that the mean of the remaining four scores is 60. Suppose you are the one who received the 50. When the scores are curved, is it better for you if the 100 is included in the distribution or if it were replaced by another score of 60?

9. Redo exercise 8 but instead of the 50 assume that you received the 70.

10. a) Consider the distribution of the scores 56, 57, 58, 59, 60, 61, 62, 63, 64, 100. Assign letter grades by curving these scores. Which scores benefit from the curve?
    b) Consider the distribution of scores, 56, 57, 58, 59, 60, 61, 62, 63, 64, 60. Assign letter grades by curving these scores. Which scores benefit from the curve?

11. Suppose that the mean of the 10 scores in one class is 68 with a standard deviation of 6. In addition suppose that the mean of 12 scores in another class is also 68 but the standard deviation is 4. If the 22 scores from both classes are combined into one distribution, what is the standard deviation of this distribution?

12. Suppose that the mean of 20 scores for one class is 65 with a standard deviation of 5. For another class of 10 scores, the mean is also 65 but the standard deviation is 3. If the 30 scores are combined into one distribution, what is the mean and standard deviation of this distribution?

13. Suppose that you are in a class in which the mean of 10 scores on a test is 68 with a standard deviation of 8, and you got a 65. For another class, class *A*, the mean of the scores is also 68 but with a standard deviation of 1. For a third class, class *B*, the mean score is again 68 but with a standard deviation of 12. If the scores from your class are to be combined with the scores from either class *A* or *B* to form one distribution, which combination would you say is of more benefit to you?

14. Redo exercise 13 except assume now that your score is 81.

SUGGESTED READINGS

DALTON, A. G., "The Practice of Quality Control." *Sci. Am.*, March 1953, pp. 29–33.
A readable explanation, requiring little mathematics, of how statistical analysis is used in industry with respect to sampling techniques.

LICKERT, R., "Public Opinion Polls." *Sci. Am.*, December 1948, pp. 7–11.
Elementary discussion of the nonmathematical aspects of polling.

NEWMAN, J. R., *The World of Mathematics.* Vol. III, New York: Simon & Schuster, 1956.
Part VIII contains seven essays on statistics including the original papers of John Graunt and Edmund Halley.

WEAVER, W., "Statistics." *Sci. Am.*, January 1952, pp. 60–63.
A very brief discussion of descriptive statistics followed by an elementary introduction to inferential statistics.

WEINBERG, G. H., AND J. A. SCHUMAKER, *Statistics, an Intuitive Approach* (2nd ed.). Monterey, Calif.: Brooks/Cole, 1969.
The first three chapters cover the standard measures in descriptive statistics but with some additional commentary on their interpretations.

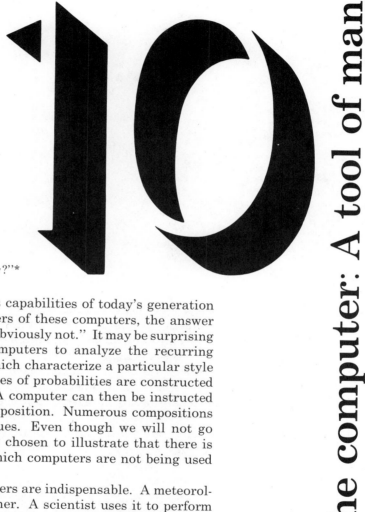

10

*"Can a computer be used to compose a symphony?"**

To one who is unaware of the tremendous capabilities of today's generation of computers and the creativity of the users of these computers, the answer to the above question may be a quick "No, obviously not." It may be surprising to learn that some musicologists use computers to analyze the recurring patterns of melody, harmony, and form which characterize a particular style such as Baroque. From this analysis, tables of probabilities are constructed which model the style being considered. A computer can then be instructed to use these tables to generate a new composition. Numerous compositions have been "written" using these techniques. Even though we will not go into any detail, this striking example was chosen to illustrate that there is scarcely an area of human endeavor in which computers are not being used in some way.

There are many areas in which computers are indispensable. A meteorologist uses a computer to predict the weather. A scientist uses it to perform the complex calculations necessary to design a nuclear reactor. The huge memory of a computer can aid a policeman in identifying a stolen car. A computer's fantastic ability to perform calculations very rapidly has helped the astronauts to reach the moon and return safely.

Computers have vast applications in industry. The auto industry, in particular, is a major user of the computer; in the early 1970's, General Motors

---

* "Computer Music," Lejaren A. Hiller, *Sci. Am.*, December, 1959.

was employing over 500 computers in its complex operations. In many manufacturing processes computers directly control other machines. For example,
they may control the rollers in a steel mill to ensure a uniform thickness in
the sheets being made. They may be used to control the temperature of a
solution during a complex chemical manufacturing process. Besides these
manufacturing applications, management uses computers for accounting,
maintaining inventories, and processing customer records.

Perhaps you are aware of some of the above-mentioned applications and
others, but you may not know that computers are also used to perform a
literary analysis to help identify the author of an anonymous work, to generate films, to play chess, to prove mathematical theorems and to perform
other acts which many would consider characteristic of "human intelligence."

Of course, a great many people have contributed to the development of
the modern computers existing today. One of the first to construct a computing machine was Blaise Pascal, the French mathematician and philosopher,
who in 1642 invented a machine which could add and subtract. He built this
machine to help his father perform the necessary calculations required in
his job as tax collector. More than 100 years later, another Frenchman,
Joseph Marie Jacquard, invented the punched card (which we have since
learned not to fold, spindle, or mutilate) to control the operation of a weaving
loom. Then in 1822 an Englishman named Charles Babbage designed an
elaborate machine called a difference engine, which among other things

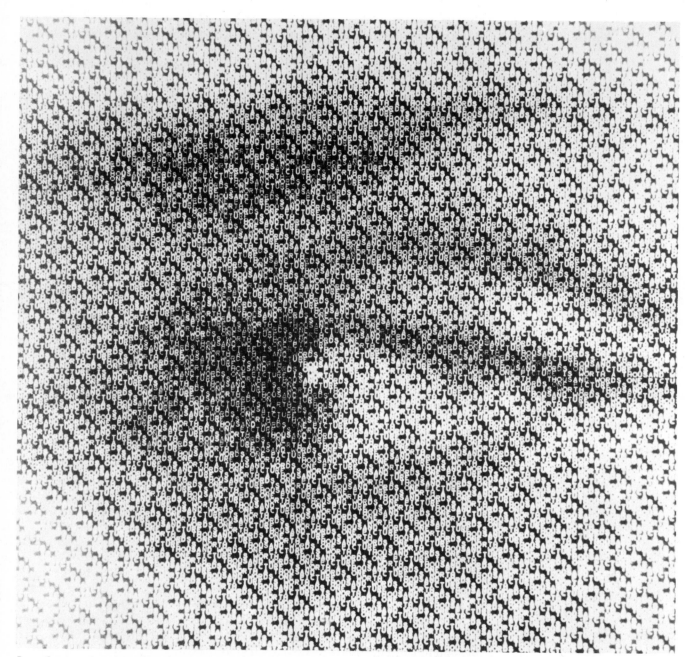

In order to appreciate this photograph you should view it from a distance of approximately ten feet. Gradations of dark and light have been assigned number values which were put into a computer. The computer then repeatedly spelled out "one picture is worth a thousand words" to create this image. (Photo by Bell Labs.)

could perform any arithmetical operation. However, his design was too ambitious and because of a lack of sufficient technology he was unable to actually build his invention. It was not until 1937 that Howard Aiken and George Stibitz developed machines that were capable of doing all that Babbage had envisioned.

In 1946 the era of the modern computer began, when J. Presper Eckert and John W. Mauchly invented the first truly electronic digital computer. This maze of vacuum tubes and electrical circuits was named ENIAC (*E*lectronic *N*umerical *I*ntegrator *A*nd *C*omputer). Although ENIAC was considered a big step forward, it had a serious drawback in that it had to be frequently rewired in order to perform various mathematical calculations. Research groups led by John von Neumann at Princeton and M. V. Wilkes at Cambridge, in attempting to avoid this constant rewiring, developed a method by which a computer could receive and store instructions. In 1949, the Cambridge group succeeded in their goal. If a different set of calculations was to be performed, a new set of instructions could be "fed" into the machine thus requiring no changes in the wiring.

The first computer for commercial use was placed on the market in 1951 and was immediately put to use in tabulating the results of the 1950 census. From that point on computers have progressed at a tremendous pace, aided in part by two important developments. First there was the invention of transistors and miniature printed circuits, which allowed the computers to do their calculations more rapidly. You must realize that a computer does its calculations as quickly as electricity moves through its internal circuits, which is roughly at the rate of 186,000 miles per second! As the length of connecting wires decreased, there was a corresponding increase in computer speed. A modern computer is now so fast that it can perform *millions* of operations per second. Assuming that you could do one addition every second, it would take you almost 12 full days to do 1,000,000 of them.

A second development which benefited computer users was the concept of time-sharing. While sitting at a teletype connected to a computer over telephone lines, a user can conduct a "dialogue" with the machine. If an error has been made in the instructions to the machine, it is possible for the machine to recognize the error and immediately ask for a correction.

Also with time-sharing it is possible for several different users, who may be hundreds of miles apart, to be connected to the same computer and for each of their problems to be worked on simultaneously. A musicologist could be analyzing a work of Bach; an English scholar could be making a study of a Shakespearean sonnet; a physician could be asking for the latest reference in cancer research; and a mathematician could be solving some complex problem. Each of these people could be working simultaneously with the same computer, without interfering with one another.

Often the fear is expressed that computers reduce us to numbers and in the process dehumanize us. This is simply not true! In fact, the computer in many ways has actually given us greater freedom and has contributed to our individuality. Certainly, the scientist, the engineer, or the businessman who deals with vast quantities of information has been freed from the tedious chore of performing lengthy calculations. The scholar who can walk up to a computer terminal, request a reference, and receive the desired information in seconds has certainly been saved hours of laborious searching through library shelves. The small child, whose teacher is using computer-aided instruction, does not have to proceed through the material at the same rate as the rest of the class. The course can be tailored to suit the student's abilities. And what other than a computer could tell you that you don't have to wait for that new car with red leather bucket seats, white vinyl roof, disc brakes, power steering, and AM-FM radio because another dealer, located 30 miles away, has one.

Computers have come a long way since Pascal invented this calculator in 1642. While Pascal's machine could only add and subtract, modern computers are so sophisticated that at times they appear to exhibit almost human intelligence. (Photo courtesy of IBM.)

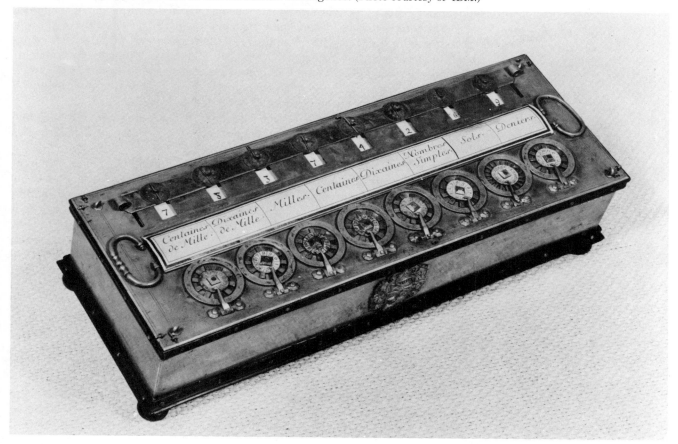

COMPUTERS AND ALGORITHMS

As we have previously discussed, computers are being used to help solve an unending list of problems. A computer is a very important "tool" of man. The advantage of this "tool" is that it can work with vast quantities of information at very high speeds and because computers have this capacity, there is more and more output from computers. However, the results may be at times a large bulk of useless or even incorrect information. Those in the computer field use the term GIGO—Garbage In, Garbage Out. This term, GIGO, points out that a computer will do no more or no less than it is instructed to do. If a computer is incorrectly instructed or is given the wrong information, then it will surely give back meaningless results. Therefore it is the responsibility of the computer user to know how to use and instruct this very powerful "tool" in solving problems.

In order to learn how to instruct a computer, we need a general understanding of its components. One of the computer's components consists of its **memory** cells. The computer can store **data** (information) in this memory.

A computer will also have an **input** and an **output** device for receiving data and giving back results. The output can be given on punch cards, magnetic tape, or as print on paper, or it can be displayed on a picture tube similar to the one used in television. The computer may receive input from punch cards, magnetic tape, or a keyboard similar to a typewriter's, or even by writing on the picture tube with what is called a light pencil.

We will not refer to any particular input or output device here. When we need to indicate an input of data, we will write "obtain . . ."; for an output of results, we will write "display . . .".

An important aspect of a computer, which distinguishes it from a pocket calculator, is its ability to process data according to a sequence of instructions which is also placed into its memory. The sequence of instructions which cause the computer to carry out a particular set of computations is called a **computer program**. The individual who writes this set of instructions is generally called a **computer programmer**.

The computer component which interprets the instructions of the program and controls their execution is called the **Central Processing Unit (CPU)**. The CPU can be considered the main component of the computer (Fig. 10.1); it controls the data flow between the input devices, the memory, and the output devices. The CPU contains the mechanism for performing the **arithmetic operations**—addition, subtraction, multiplication, and division. Since the CPU will do the arithmetic, data will move between it and the memory.

Besides performing arithmetic functions, the CPU can compare two numbers. For example, a program may require the CPU to test whether or not $A > B$. The result of this test or any comparison will be either yes or no (true or false). Based on this result, the CPU determines which instruction

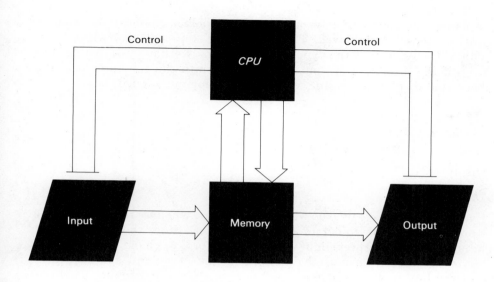

Figure 10.1

in the program it will perform next. If the result of the comparison is yes, the program may stipulate that the next instruction to be performed is a previous one on the list; otherwise, it will be the next instruction on the list.

The ability of a computer to do comparisons, **logical operations**, is very powerful and it allows the programmer to put **branches** in the program. The choice of which branch (path) is to be followed is determined when the computer executes the program.

After seeing the components of a computer, we can give a list of the type of instructions which can be used in a program:

1. *Input*: "Obtain . . ."
2. *Output*: "Display . . ."
3. *Arithmetic operations*: add, subtract, multiply, divide
4. *Logical operations*: compare two numbers and branch, based upon the result
5. *Move data between the CPU and memory*
6. *Start, Stop*

In using a computer to solve a particular problem, we must determine a sequence of steps to be followed. This sequence of steps, called an **algorithm**, can contain a variety of the type of instructions we have just listed. For example, if the problem is to put two numbers into the computer, add the numbers, and display the result, we may write for our algorithm:

1. Start.
2. Obtain A and B.
3. Let $S = A + B$.
4. Display S.
5. Stop.

Even though it is clear to us what must be done when *we* read these steps, the words would not be understood by a computer. Since the computer is a machine, it cannot interpret slight variations in words or changes in grammar. Therefore the algorithm must be written in one of several standard **programming languages** which are acceptable to the particular machine. Learning a computer language is less difficult than learning a "foreign" language; however, we will not consider this aspect of computer programming. We will write our algorithms using less formal language.

At this point we should have an understanding of the first four types of instruction. As we give examples of algorithms, they will become even more clear. Let us now discuss the fifth type of instruction, which moves data between the CPU and the memory. Each cell in memory which is required by the program for data storage is labeled with a name; these names are given in the program. The names that can be used are single letters of the alphabet, words, or any of these with a subscript such as A_1, A_2, etc. We can bring data out of a particular cell by using the appropriate name of that cell.

Consider again the algorithm:

1. Start.
2. Obtain A, B.
3. $S = A + B$.
4. Display S.
5. Stop.

This algorithm requires that three memory cells be labeled A, B, and S, respectively. When the step "Obtain A, B" is performed, the two numbers appearing in the input device will be placed in the two memory cells A and B, the first number going into cell A and the second into cell B. If there are any numbers already in these cells, they will be replaced by the two new numbers. When the step "$S = A + B$" is executed, the CPU will copy from memory the numbers appearing in cells A and B. The numbers in these two cells are not lost from the memory since only a copy of each is passed to the CPU. After

the CPU performs the addition, the sum is sent to the memory and placed in the cell labeled S. If there is a number already there, it will be replaced by this sum. Finally when the step "Display S" is performed, a copy of the number in memory cell S is sent to the output device for display.*

After seeing what happens when the computer performs "$S = A + B$," we find that a more descriptive way to write this step would be "$S \leftarrow A + B$." We use the arrow to indicate that the sum $A + B$ is placed into memory cell S.

Our first algorithm displaying the sum of two numbers can be simply given by listing the steps as we have done. However, as our problems become more extensive, we will find it convenient to organize the steps of the algorithm graphically in what is called a **flowchart**. The name is very appropriate since it is a chart (diagram) indicating the flow (path) in which the steps should be followed. The steps of the algorithm are written within various shaped "boxes," which are connected by flowlines. The shape of the box used

This two- by four-inch IBM data catridge can hold up to 100 million characters of information. Its memory capacity is equivalent to that of the much larger disk pack shown in the background. (Photo by IBM.)

---

* It would appear from our description that a computer operates very slowly; however, a computer is electronic and the movement and representation of data is accomplished by electrical impulses. The time it would take a computer to do the three steps of our algorithm could be measured in thousandths of a second.

for the step depends on the type of instruction given. A list of the geometric figures that are used is given in Fig. 10.2. We have already stated that we will use "←" to indicate that data is to be placed in a memory cell from the CPU.

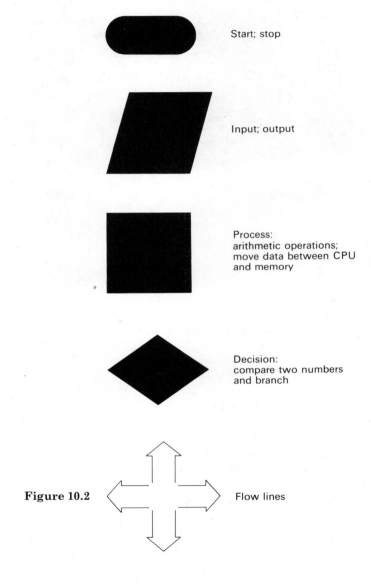

Start; stop

Input; output

Process: arithmetic operations; move data between CPU and memory

Decision: compare two numbers and branch

Figure 10.2

Flow lines

EXAMPLES   ■ A flowchart for the algorithm needed to put two numbers into the computer, to obtain the sum, and to display the result is given in Fig. 10.3. □

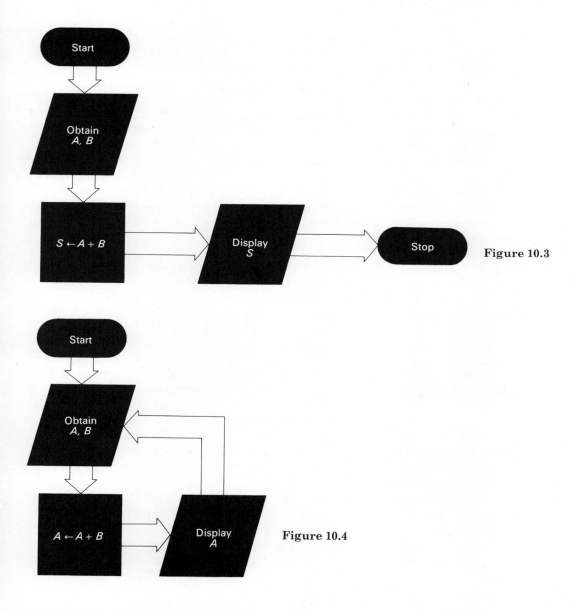

Figure 10.3

Figure 10.4

■ Consider the flowchart given in Fig. 10.4. Let us analyze what will occur. To begin, two numbers must be available for input into the computer; these numbers are placed into the memory cells which are labeled A and B. When the computer executes step "$A \leftarrow A + B$," the CPU receives from the memory copies of the numbers appearing in cells A and B. Next the sum of these two numbers is computed; then this sum is sent back to the memory and placed in the cell labeled A. The number that was in cell A is lost since it is replaced by the sum. For example, suppose 2 and 3 are available in that order for input. Upon performing the step "Obtain A, B," memory cell A will contain 2 and memory cell B will contain 3. After executing the step "$A \leftarrow A + B$," cell A will contain 5 and cell B will still contain 3.

The next step in the algorithm is "Display A." A copy of the number which is in cell A at the time this step is performed is sent to the output device. For our above illustration, since cell A would contain 5 when "Display A" is executed, the output device will receive a copy of 5 from memory.

In this flowchart, the flow is from "Display A" back to the input statement. Therefore after displaying the sum the computer is asked to return for *another* pair of numbers and to do the process over with this *new* pair of numbers. We see that we have not specified when the process should stop. Generally, it is poor practice not to specify the way in which the process terminates in an algorithm. □

■ For the flowchart in Fig. 10.5, let us analyze the result of performing the algorithm.

First we need a number available as input for storage in cell N. Next, 1 will be placed in memory cell K. When the computer comes to box 3, we will need to have a pair of numbers ready in the input device. The rest of the algorithm proceeds in the same manner as the previous example did up to the decision box, box 6. The decision that will be made is based on whether the number in cell K is the same as the number in cell N. If they are the same, the computer goes to the instruction "Stop" and terminates processing.

If the two numbers are not the same, then the next step to be done is "$K \leftarrow K + 1$." In this instruction, the number in cell K is increased by 1 and put back into cell K. We see that cell K will contain a count of the number of times steps 3 to 5 are done; therefore cell K is essentially used as a counter. After increasing this counter by 1, the computer is instructed to go back to the input step "Obtain A, B" and start the process over with a new pair of numbers, which are waiting in the input device. The computer will go through this cycle, box 3 to box 6, as many times as the number in cell N; the number put into cell N must be a positive integer.

Let us consider a specific example and go through the algorithm with this example. We give the input with the corresponding output as follows:

Counter	Input		Output	$K = N$?
K	N	A, B	S	
	3			
1		1, 3	4	No
2		2, 1	3	No
3		0, 5	5	Yes
				Stop

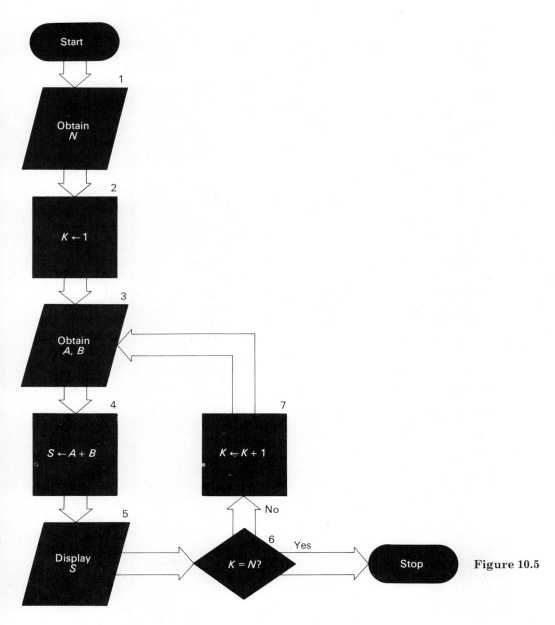

Figure 10.5

In summary, we see that in order to perform this algorithm the input data must be organized in the following manner. First a positive integer must be available, to be followed with as many pairs of numbers. For each pair of numbers put into the computer, we will receive back a single number, namely, the sum of the pair of numbers. □

When given a flowchart we should be able to read and analyze the result of performing the algorithm. Our analysis should include the following aspects:

1) Describe what is needed as input, if anything, and how it should be organized.

2) Describe what the output will be.

3) If input is needed, go through the algorithm with some trial data; give at least some of the output.

The third aspect of the analysis should be done. Unless the computer programmer knows what should be received back as output from the computer, there would be no way to recognize whether or not the algorithm is correctly written.

EXAMPLE

■ Let us interpret the result of performing the algorithm given in Fig. 10.6.

First we need a positive integer available for storage in cell N. The integer in cell N indicates the number of times the sequence of steps "Obtain A," "$X \leftarrow A \cdot A$," and "Display X" will be done. Thus there must also be available for input as many numbers as the integer in N. For example, if we use 4 as input for N, then four more numbers must be available for input, each to be placed one at a time in A.

After a number is read into A, its square is computed and displayed. Upon displaying this square, the computer will return for the next number provided that the counter K is still less than N; otherwise, the process terminates.

As trial data suppose 4, .5, 9, .33, and 11 are available for input in that order. The output will be .25, 81, .1089, and 121. □

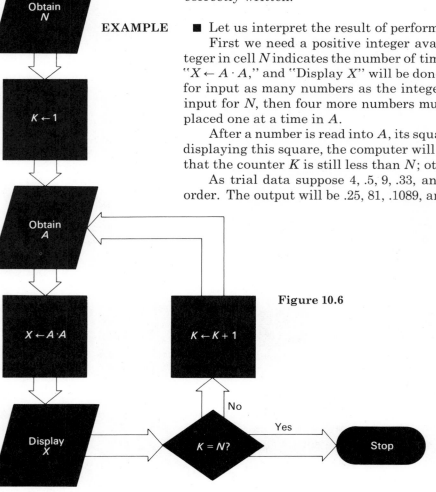

Figure 10.6

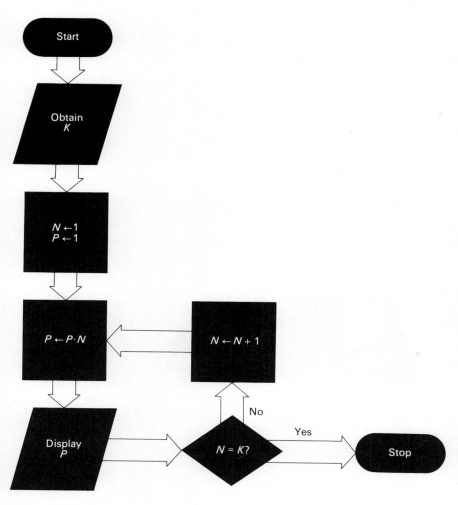

Figure 10.7

Consider the flowchart in Fig. 10.7.

1. Must there be any data available for input?

2. How many times will the step "Display P" be done?

3. What will be the output?

QUIZ YOURSELF*

---

1. Yes, one positive integer for K 2. K times   3. a table of factorials from 1! to K!   ***ANSWERS**

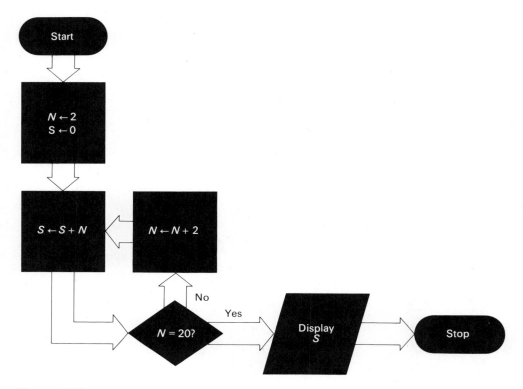

Figure 10.8

Upon analyzing a flowchart, if the algorithm does not give the desired results, we should then be able to correct it. We do this in the next example.

EXAMPLE   ■ Suppose we wish to obtain *each* of the numbers 2, 2 + 4, 2 + 4 + 6, up to 2 + 4 + 6 + $\cdots$ + 20. Does the algorithm in Fig. 10.8 give the desired result? If not, what can be changed in the flowchart to obtain the desired output?

We see that the step "Display S" will be done only once. The number which will be in cell S when the machine performs this step will be the sum 2 + 4 + 6 + $\cdots$ + 20; therefore we will receive only this number as output. In order to correct this flowchart to display each of the sums, the output box must be placed before the decision box as shown in Fig. 10.9. □

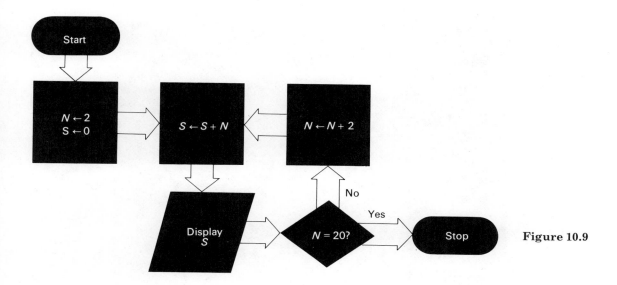

Figure 10.9

A replica of Charles Babbage's difference engine.
(Photo courtesy of IBM.)

Illustrating the phenomenon of field constriction, the computer plotted the electric field lines in and around a simple circular rod made of dielectric material. (Photo by Bell Labs.)

EXERCISES

For exercises 1 to 7, analyze the given algorithm. In your analysis, you should describe what is needed for input, if anything, and how it should be organized. You should also describe what the output will be. Finally if any input is needed, then, using an example of your own choosing, go through the algorithm and give the corresponding output.

1.

2.

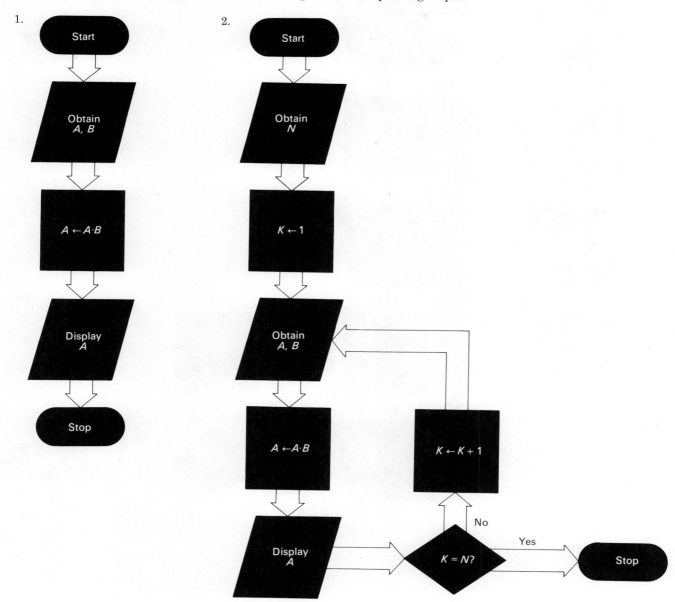

3.

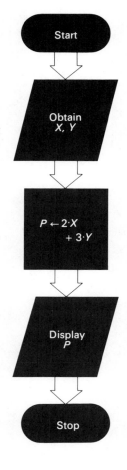

4.

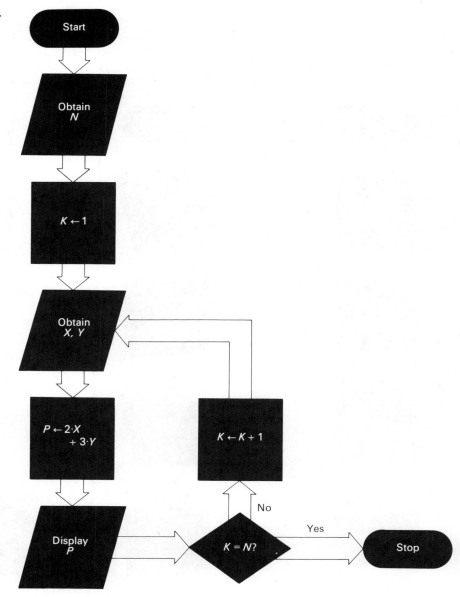

5.

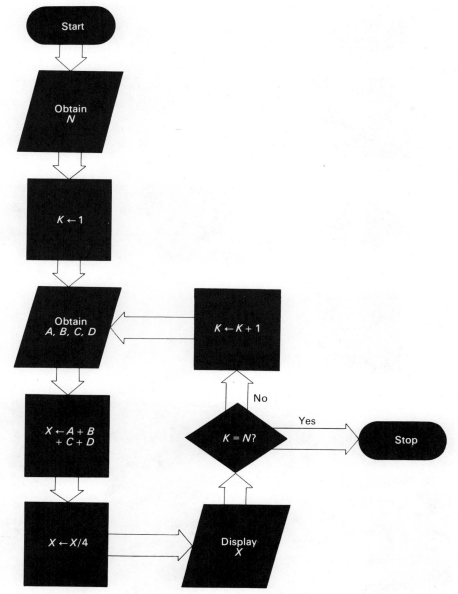

6.

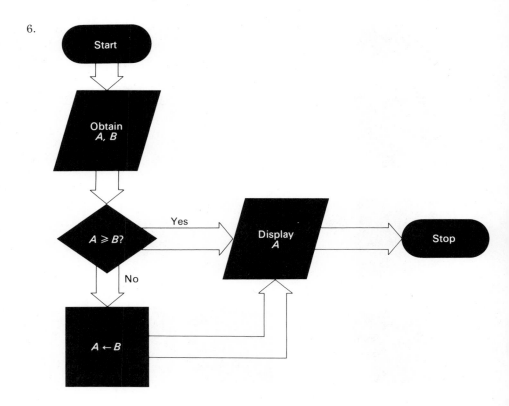

A computer's huge memory may provide assistance to any person who needs rapid access to large quantities of information. (Photo by Cary Wolinski, Stock Boston.)

7.

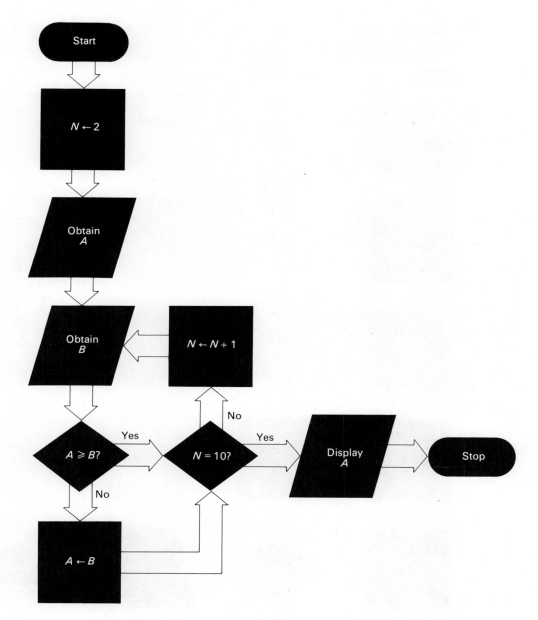

For exercises 8 to 12, analyze the suggested algorithm for the stated problem. If the algorithm does not give the desired output, then state how it may be changed to give the desired output. Also give a corrected flowchart.

8. Put 10 numbers into the computer; compute and display their sum. *Suggested Algorithm*:

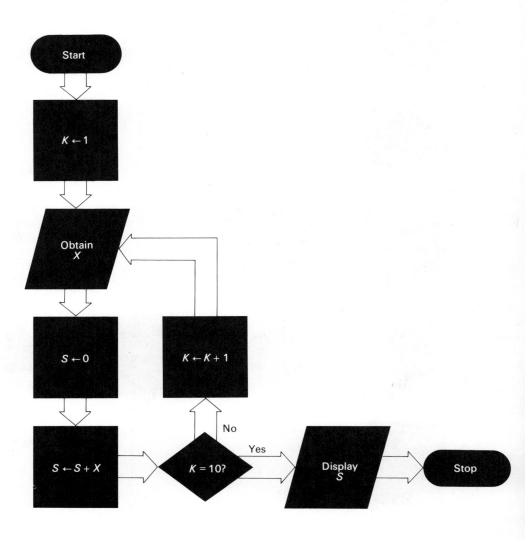

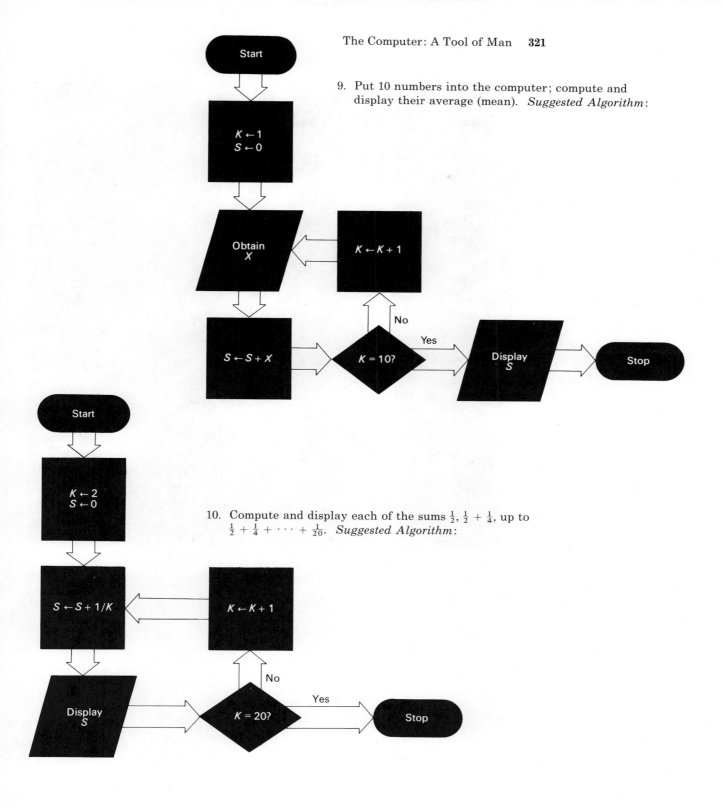

9. Put 10 numbers into the computer; compute and display their average (mean). *Suggested Algorithm*:

10. Compute and display each of the sums $\frac{1}{2}$, $\frac{1}{2} + \frac{1}{4}$, up to $\frac{1}{2} + \frac{1}{4} + \cdots + \frac{1}{20}$. *Suggested Algorithm*:

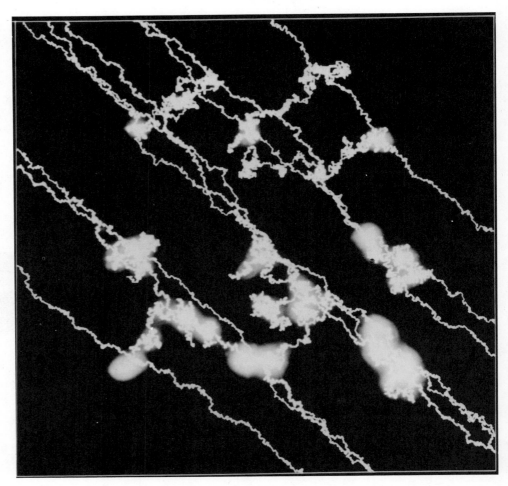

What appears to be lightning bolts are lines randomly drawn by a computer that had been given a minimum amount of instruction. (Photo by Bell Labs.)

Here is a scene from the Bell Laboratories film "A Computer Technique for the Production of Animated Movies" which was produced by a computer. The computer generates magnetic tape which is used to produce drawings on the face of a cathode ray tube. A camera photographs these drawings to produce the film. (Photo by Bell Labs.)

11. The Fibonacci sequence is a sequence in which the first term is 0, the second term is 1, and every term thereafter is the sum of the two previous terms. Obtain the first 15 terms in the Fibonacci sequence. For example, the first six terms are 0, 1, 1, 2, 3, 5. *Suggested Algorithm*:

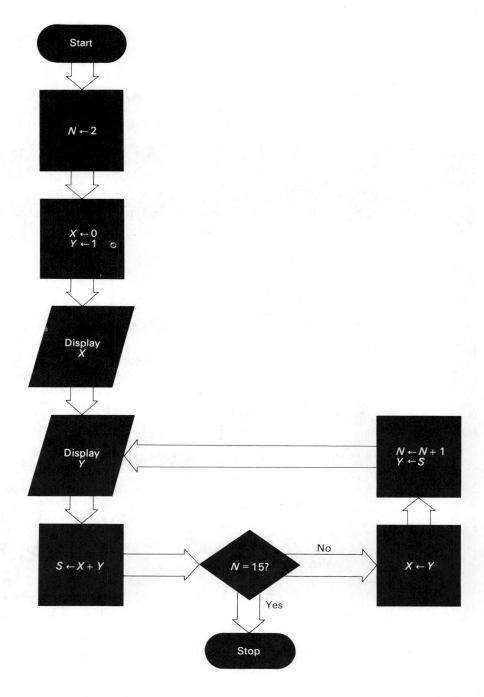

12. Put *K* positive numbers into the computer; obtain and display the largest of these numbers. *Suggested Algorithm*:

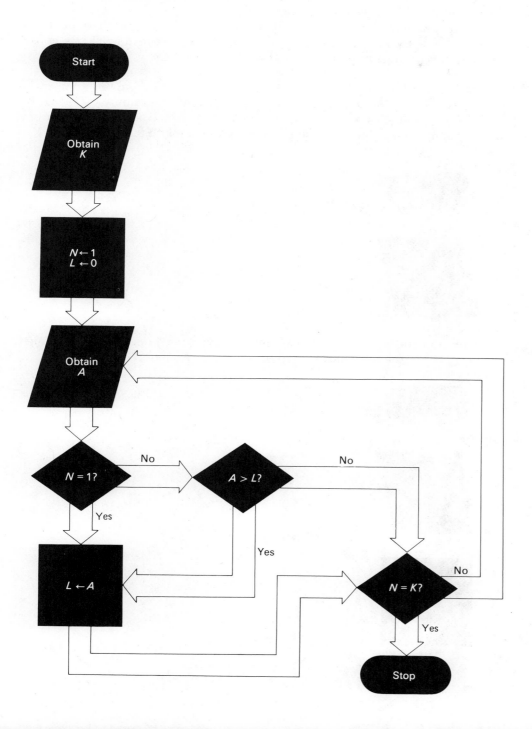

WRITING FLOWCHARTS

Now that we have acquired experience in reading and interpreting a flow-chart, we can proceed to writing an algorithm for a given problem.

In Chapter 6 we needed to compute the value of quantities of the type $P = Ax + By$ at specified points. In the next two examples we consider problems of this nature and write flowcharts for them.

EXAMPLES ■ We shall write a flowchart for the problem: Put a pair of numbers into the computer; for this pair compute and display the quantity $P = 4x + 5y$.

Let us label as X and Y the two memory cells needed for the input of the pair of numbers. Therefore the input step following "Start" is "Obtain X, Y."

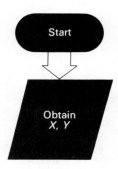

Now that the pair is in the machine the computation can be done. Thus the next step in the flowchart is the computation "$P \leftarrow 4 \cdot X + 5 \cdot Y$."

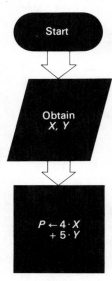

After computing the number P, we can display it and stop. The complete flowchart is given in Fig. 10.10. □

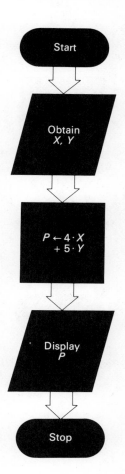

Figure 10.10

■ Let us draw a flowchart for the following: For each of six pairs of numbers, compute and display the quantity $P = 4x + 5y$.

The pairs of numbers are to be put into the computer one at a time. We see from the above example that the portion of the flowchart describing the input, computation, and output is the same (see Fig. 10.10).

Since this sequence of three steps will be done for each pair, the flow must return to the input step. But since there are only six pairs of numbers, a count must be taken of the numbers of times the input step is done. Therefore we use a memory cell, which we label K, to keep this count. Each time after the three steps are done, the number in K should be tested to see if it

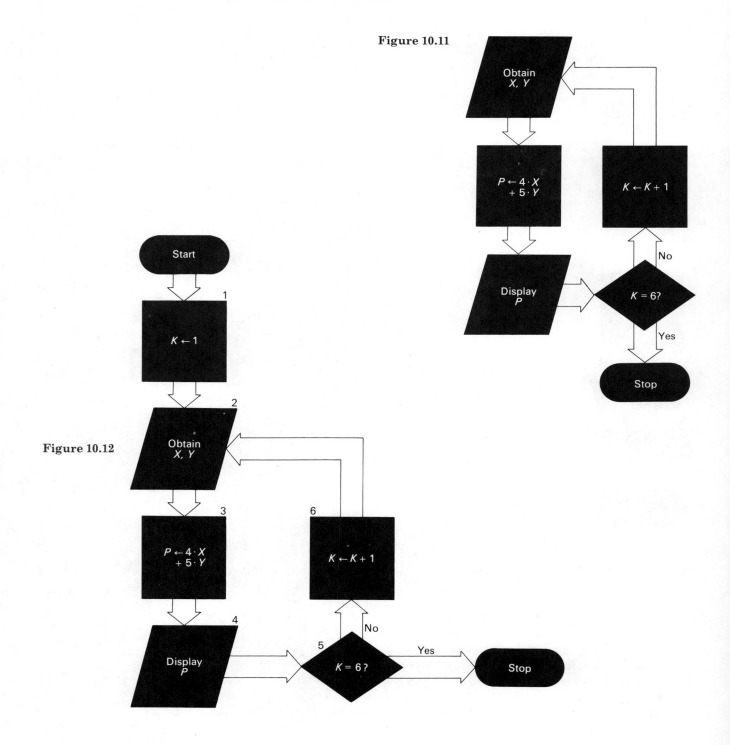

Figure 10.11

Figure 10.12

has reached 6. If K contains 6, we want the computer to stop. Otherwise the number in K is to be increased by 1 and the computer is to return to the input step for another pair of numbers. (See Fig. 10.11.)

Now to start the process the counter is set at 1 ($K \leftarrow 1$) and the computer proceeds to the input step for the first pair of numbers. We give the complete flowchart for this problem in Fig. 10.12. □

We see that in the above example there is a sequence of steps to be done several times. These steps are written in boxes 2, 3, and 4 of Fig. 10.12. After this sequence is performed with one pair of numbers, the computer is to return to the input step for a new pair. However, since this process is to be done only 6 times, we used a counter, cell K, and a test, box 5, to see when the counter reaches 6. The boxes 1 through 6 of the flowchart in Fig. 10.12 form what is called a **do loop**. Generally to write a do loop, we include the following parts in the flowchart (see Fig. 10.13).

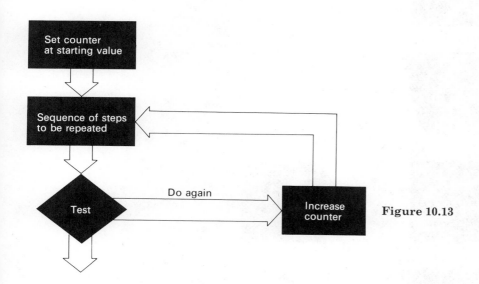

Figure 10.13

1) a step to place the starting value in the counter, box 1 of the example;

2) a sequence of steps to be done repeatedly, boxes 2, 3, and 4 of the example;

3) a test to determine when to stop going through the sequence, box 5 of the example;

4) a step to increase the counter before repeating the sequence, box 6 of the example.

QUIZ YOURSELF*
1. Fig. 10.14 is a partially completed flowchart for the problem: Put 10 numbers into the computer, one at a time; compute and display the square root of each.
   a) What question must be asked at the decision box 5?
   b) What step must be written in box 6?

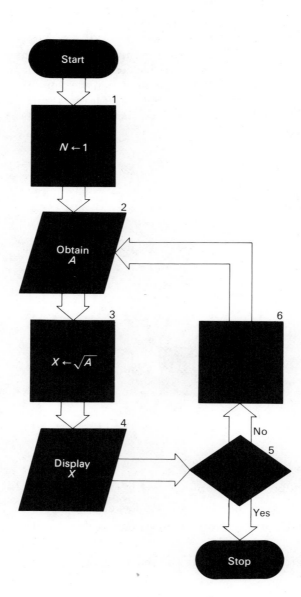

Figure 10.14

2. Figure 10.15 is a partially completed flowchart for the problem: Compute and display the list of numbers 2, 4, 6, 8, 10, 12.
   a) What question must be asked at the decision box 3?
   b) What step must be written in box 4?

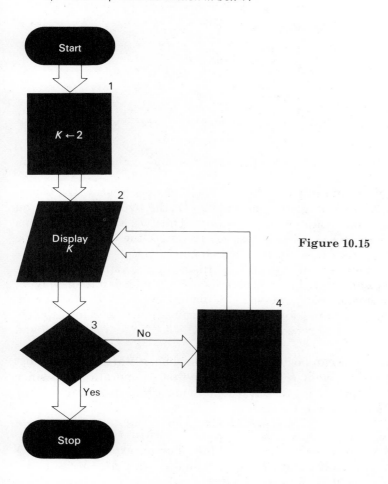

Figure 10.15

---

1. a) $N = 10$?   b) $N \leftarrow N + 1$

2. a) $K = 12$?   b) $K \leftarrow K + 2$

*ANSWERS

For some do loops, in the sequence of steps to be repeated, there may be variables in addition to the counter that must be given starting values before the sequence can be done for the first time. Therefore before beginning the do loop, a step must be performed which gives initial values to these variables. We will see this in our next example.

EXAMPLE   ■ In Chapter 4, we computed a table of numbers which was used to determine the apportionment of a faculty council. A portion of this table is given below.

$$
\begin{array}{c|l}
2 & \dfrac{(103)^2}{1 \cdot 2} = 5304.5 \\[2ex]
3 & \dfrac{(103)^2}{2 \cdot 3} = 1768.2 \\[1ex]
\vdots & \quad \vdots \\[1ex]
12 & \dfrac{(103)^2}{11 \cdot 12} = 80.4
\end{array}
$$

Let us write an algorithm for computing and displaying the entries in the above table.

The one computational step $(103)^2/(K \cdot N)$ will be done repeatedly, where K goes from 1 to 11 while N goes from 2 to 12. Of the two variables K and N, we will use N as our counter for a do loop. Thus we will initialize this counter at 2 and test to see when it reaches 12. Also, the counter must be increased by 1 before returning for the next computation.

Instead of recomputing $(103)^2$ each time the computational step is done, it can be found once and stored in a memory cell P. In addition to displaying the result of each computation, we wish to display N. Also K must be increased by 1 before returning for the next computation. Therefore the sequence of steps that will be repeated is "$K \leftarrow K + 1$," "$X \leftarrow P/(K \cdot N)$," and "Display N, X."

Before this sequence can be done for the first time, a starting value must be given to K. Since we want K to be 1 for the first computation and since the first step of the sequence is "$K \leftarrow K + 1$," we must initialize K at 0. Thus the step "$K \leftarrow 0$" must precede the do loop.

The complete flowchart for this problem is given in Fig. 10.16. □

In Chapter 8, we discussed expected value. For an experiment with N outcomes, having probabilities $P_1, P_2, \ldots, P_n$ and payoffs $M_1, M_2, \ldots, M_n$, respectively, the expected value is

$$E = P_1 M_1 + P_2 M_2 + \cdots + P_n M_n.$$

Generally, the letters with subscripts P_1, P_2, etc. would be written in programming language as $P(1)$, $P(2)$, etc. We will therefore do the same. Similarly, we will label the memory cells for the payoffs as $M(1)$, $M(2)$, etc. In the next example we give an algorithm for computing the expected value.

EXAMPLE   ■ Write a flowchart for putting into a computer the probabilities and payoffs for N outcomes of an experiment. Then compute and display the expected value.

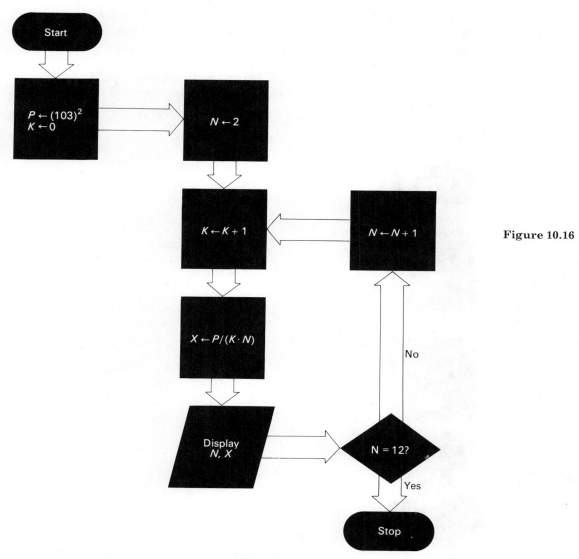

Figure 10.16

The probabilities with their corresponding payoffs will be put into the computer as pairs of numbers. Therefore we might be tempted to write a list of input statements

"Obtain $P(1)$, $M(1)$"

"Obtain $P(2)$, $M(2)$"

$\vdots$

However, we can use a loop with the one input statement "Obtain $P(I)$, $M(I)$." The loop will keep returning to this input statement as many times as there are outcomes in the experiment. Thus before entering the loop the number of outcomes for the experiment must be put into the computer; this can be stored in cell N. The input portion of our flowchart appears in Fig. 10.17.

We need to add to this input segment a segment for computing the expected value. In order to avoid rewriting the input segment, we can use in our flowchart the **connector** symbol. This flowchart symbol is a circle with a number written inside it and it is used to join segments of the flowchart.

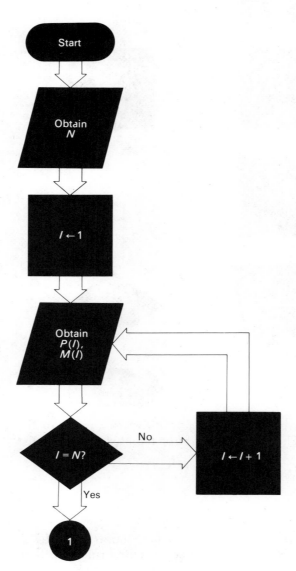

Fig. 10.17

Input segment

We now write the computational segment of the flowchart. The expected value can be stored in a memory cell labeled as E. If there is only one outcome, then the product $P(1) \cdot M(1)$ would be put into E. If there are two outcomes, then $P(2) \cdot M(2)$ would be added to the number in E and put back into E. We would repeat this process for all N outcomes. Thus we see that the computational portion of the algorithm can be given with the step "$E \leftarrow E + P(I) \cdot M(I)$" written within a do loop. For this loop, we can use I as the counter, where it goes from 1 to N. Also, before entering this loop, we must have 0 in E.

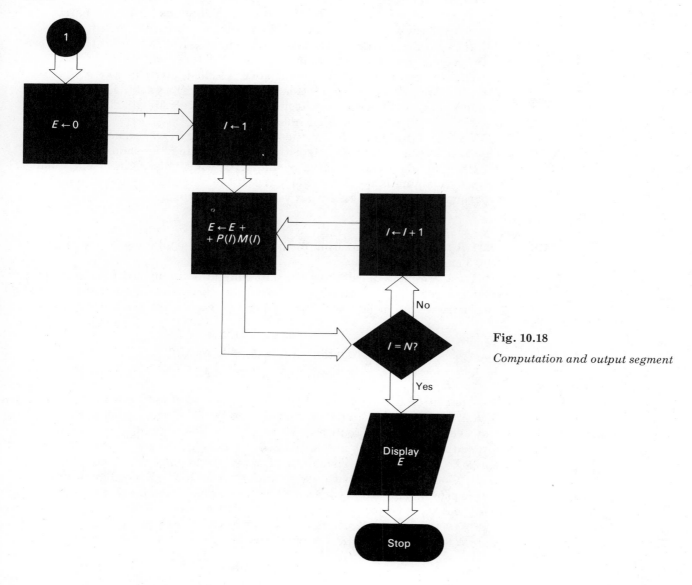

Fig. 10.18

Computation and output segment

Upon completing the computation, we wish to display the result and stop. The computational and output segments are given together in Fig. 10.18.

The complete flowchart for this problem of computing expected value is obtained by joining the input segment with the computation and output segment, as indicated by the connector symbol. □

The flowchart for the next example and the last one we give includes a loop within a loop. This example involves working with a matrix.

Recall from Chapter 3 that a matrix is a table of numbers. The size of the matrix is given by the number of rows and columns. A 3 by 3 matrix is a table with 3 rows and 3 columns. A particular entry can be identified by the row and column in which it appears. For example, the 2, 3-entry is the number in the second row and third column.

To store a matrix in computer memory, we must have a cell for *each* entry. The cells can be labeled by a letter with two subscripts such as $M(I,J)$. The memory cell $M(2,3)$ is different from the memory cell $M(3,2)$. The first holds the 2, 3-entry while the second holds the 3, 2-entry.

We found in Chapter 3 that the square of a matrix was sometimes needed. The next example gives an algorithm for computing the square of a 3 by 3 matrix. The algorithm we give is similar to what must be written if using the programming language, FORTRAN. If the algorithm were written in the programming language BASIC, it would be much simpler.

EXAMPLE   ■ Let us write an algorithm for computing and displaying the square of a 3 by 3 matrix.

First, we must decide how the data should be organized for input. It is easy to agree that the matrix would be put into the computer one row at a time. Therefore we might be tempted to use the following list of input statements.

"Obtain $M(1,1)$, $M(1,2)$, $M(1,3)$"

"Obtain $M(2,1)$, $M(2,2)$, $M(2,3)$"

"Obtain $M(3,1)$, $M(3,2)$, $M(3,3)$"

Instead of having this list we could use the input statement "Obtain $M(I,1)$, $M(I,2)$, $M(I,3)$" written within a do loop, where I goes from 1 to 3. However, we should realize that, especially when dealing with a matrix larger than 3 by 3, it is not convenient to write one input statement for the entire row. Therefore we should go further and replace this input statement with "Obtain $M(I,J)$" written within another do loop. Even though this is a better programming procedure, we prefer to leave it as an exercise and write our input segment for now as given in Fig. 10.19.

Now that we have the matrix in memory, we can proceed to obtain its square. Recall that the I, J-entry of the square is obtained by "multiplying"

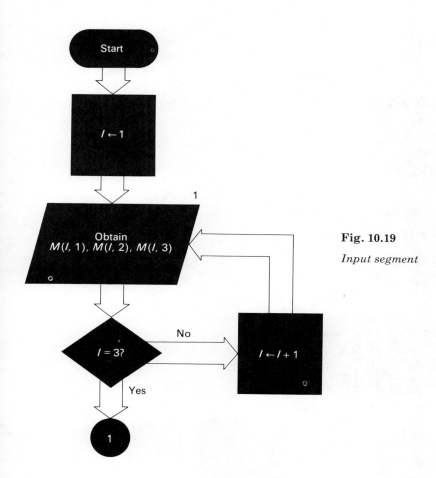

Fig. 10.19

Input segment

the Ith row by the Jth column. For example, the 1, 2-entry of the square is

$$M(1,1) \cdot M(1,2) + M(1,2) \cdot M(2,2) + M(1,3) \cdot M(3,2).$$

Since to carry out the entire matrix computation the entries of the original matrix must be saved, each entry of the squared matrix must be stored in a new memory cell as it is computed. We can label these cells $MS(I,J)$, where I goes from 1 to 3 and J goes from 1 to 3. The I, J-entry of this computed matrix can be obtained by performing the step

$$\text{``}MS(I,J) \leftarrow M(I,1) \cdot M(1,J) + M(I,2) \cdot M(2,J) + M(I,3) \cdot M(3,J)\text{''}.$$

Therefore this step written within a double do loop, I goes from 1 to 3, and J goes from 1 to 3, will give all the entries of the squared matrix. Again it should

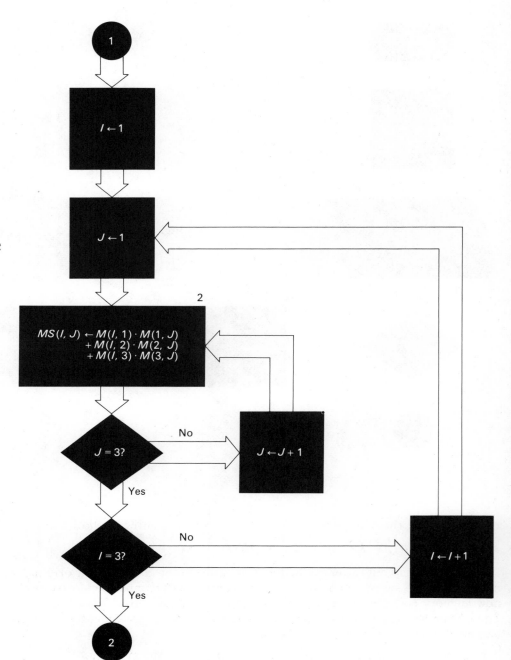

Fig. 10.20

Computation segment

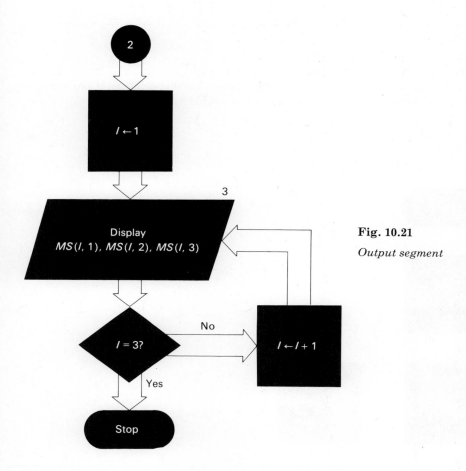

Fig. 10.21

Output segment

be noted that a statement similar to what we give for the computation is inconvenient to write, especially with a large matrix. Better programming procedure would be to rewrite this computational step with another do loop. However, we leave it as an exercise and give for now the computational segment of our flowchart as in Fig. 10.20.

Upon computing the square of the matrix, we wish to display it. We can bring out from memory the entries one row at a time. The procedure for accomplishing this is similar to the input segment. We give this output segment of the flowchart in Fig. 10.21.

The flowchart for our matrix multiplication is the above three segments, input, computation, and output, joined as indicated by the connectors. □

EXERCISES

Exercises 1 to 3 are of the general type.

1. Compute and display each of the numbers 1^2, $1^2 + 2^2$, up to $1^2 + 2^2 + 3^2 + \cdots + 10^2$. A partially completed flowchart is given below for this problem. Complete the flowchart by writing in the empty boxes the needed instructions.

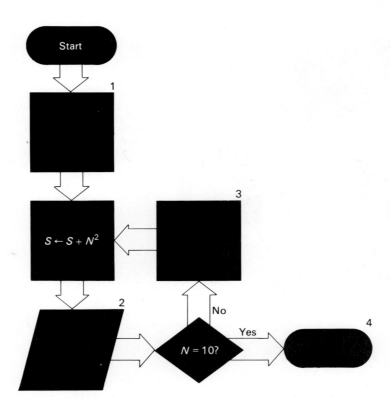

2. Put 11 numbers into the computer one at a time and display every other one starting with the first. A partially completed flowchart is given below. Complete the flowchart for this problem by filling in the empty boxes with the needed instructions.

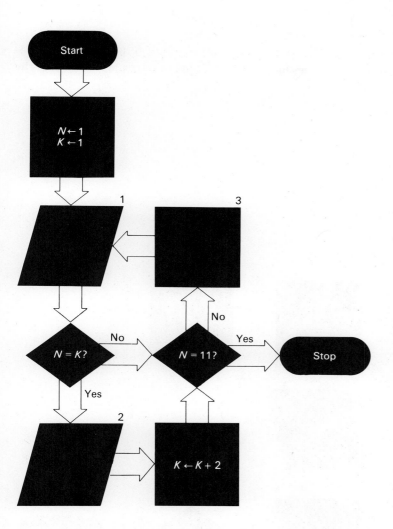

3. The following table gives the amount to which $1 will accumulate at the end of each year where interest is compounded annually at 6%.

Year	Amount at end of year
1	$(1 + .06)$
2	$(1 + .06)^2$
⋮	⋮
20	$(1 + .06)^{20}$

Draw a flowchart to compute and display the entries of this table.

Exercises 4, 5, and 6 come from Chapters 7 and 8.

4. Compute and display each of the numbers $P[10,1]$, $P[10,2]$, up to $P[10,10]$. For this problem complete the flowchart below by filling in the empty boxes with the needed instructions.

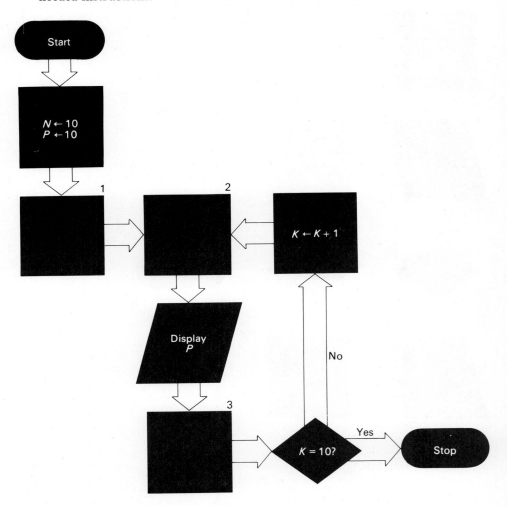

5. Draw a flowchart to compute and display each of the numbers $C[10,0]$, $C[10,1]$, up to $C[10,10]$.

6. Draw a flowchart to compute and display each of the binomial probabilities $Pr(10,0;.25)$, $Pr(10,1;.25)$ up to $Pr(10,10;.25)$.

Exercises 7 and 8 are taken from Chapter 9.

7. Draw a flowchart for the following. Put into the computer the N scores of a distribution; compute and display the mean.

8. Draw a flowchart for the following. Put into the computer the N scores of a distribution; compute and display the standard deviation for the distribution.

Exercises 9 and 10 are taken from Chapter 4.

9. Put into the computer the population and the number of representatives for each of two states A and B; compute and display the relative unfairness. For this problem complete the flowchart given below.

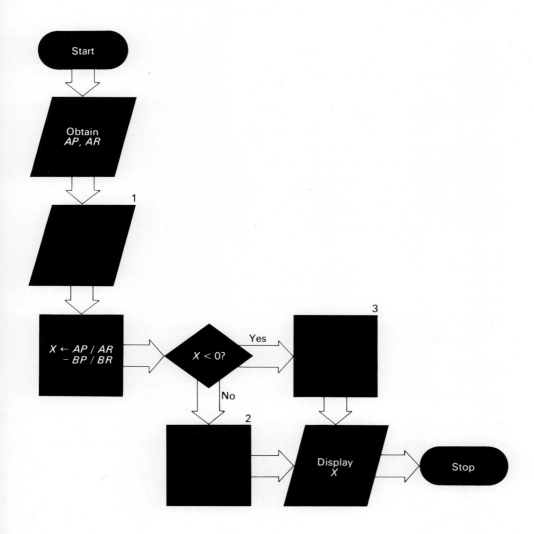

10. Draw a flowchart to compute and display the entries of Table 4.1 for the science division.

Exercises 11, 12, and 13 are taken from Chapter 5. For each of these exercises draw a flowchart to obtain the desired result.

11. Put into the computer the coordinates of two points; compute and display the slope, if one exists, of the line going through these two points.

12. Put into the computer three numbers A, B, and C; compute and display the coordinates of two points on the line $Ax + By = C$.

13. Put into the computer six numbers, $A(1)$, $A(2)$, . . . , $A(6)$; compute and display the solution, if one exists, to the system of equations

$$A(1)x + A(2)y = A(3),$$
$$A(4)x + A(5)y = A(6).$$

Exercises 14, 15, 16, and 17 are based upon Chapter 3. For each of these exercises draw a flowchart.

14. a) Replace the input step of the matrix multiplication of our example (box 1 of Fig. 10.19) with a second do loop.
   b) Replace the computation step, box 2 of Fig. 10.20, with a third do loop.
   c) Replace the output step, box 3 of Fig. 10.21, with a second do loop.

15. Suppose we already have in memory the entries of two 4 by 4 matrices, M and MS; the entries of the matrix M are in cells $M(1,1)$, $M(1,2)$, etc. and the entries of the matrix MS are in cells $MS(1,1)$, $MS(1,2)$, etc. Compute and display the entries of the matrix T, where $T = M + MS$.

16. Suppose we already have in memory the entries of a 4 by 4 matrix T; the entries are in cells $T(1,1)$, $T(1,2)$, etc. Compute and display for each row of T the sum of the entries in that row.

17. The flowchart needed for this exercise is very involved; however, if you have done exercises 15 and 16, then you have most of the segments. Recall that knowing the total one- and two-stage preference could be used to determine a ranking order in a paired-comparison test. Therefore draw a flowchart to put into the computer, a 4 by 4 incidence matrix M; compute and display the sum of the entries in each of the rows of the matrix $T = M + M^2$.

SUGGESTED READINGS

FORSYTHE, A. I., *et al.*, *Computer Science, a First Course* (2nd ed.). New York: John Wiley, 1975.

Chapter 1 contains a good description on how data is handled within a computer.

MORRISON, P. AND E., "The Strange Life of Charles Babbage." *Sci. Am.*, April 1952, pp. 66–73.

An interesting biography of C. Babbage including an explanation of how his difference engine worked.

Readings From Sci. Am.: Computers and Computation. San Francisco, Calif. W. H. Freeman Co., 1971.

A collection of 26 papers on the subject which appeared in Sci. Am. The topics covered range from computer hardware to computer models of the real world, chess-playing machines, and artificial intelligence.

Appendix

Appendix

n	k	0.10	0.20	0.25	0.30	0.40	0.50
1	0	0.9000	0.8000	0.7500	0.7000	0.6000	0.5000
	1	0.1000	0.2000	0.2500	0.3000	0.4000	0.5000
2	0	0.8100	0.6400	0.5625	0.4900	0.3600	0.2500
	1	0.1800	0.3200	0.3750	0.4200	0.4800	0.5000
	2	0.0100	0.0400	0.0625	0.0900	0.1600	0.2500
3	0	0.7290	0.5120	0.4219	0.3430	0.2160	0.1250
	1	0.2430	0.3840	0.4219	0.4410	0.4320	0.3750
	2	0.0270	0.0960	0.1406	0.1890	0.2880	0.3750
	3	0.0010	0.0080	0.0156	0.0270	0.0640	0.1250
4	0	0.6561	0.4096	0.3164	0.2401	0.1296	0.0625
	1	0.2916	0.4096	0.4219	0.4116	0.3456	0.2500
	2	0.0486	0.1536	0.2109	0.2646	0.3456	0.3750
	3	0.0036	0.0256	0.0469	0.0756	0.1536	0.2500
	4	0.0001	0.0016	0.0039	0.0081	0.0256	0.0625
5	0	0.5905	0.3277	0.2373	0.1681	0.0778	0.0313
	1	0.3280	0.4096	0.3955	0.3601	0.2592	0.1563
	2	0.0729	0.2048	0.2637	0.3087	0.3456	0.3125
	3	0.0081	0.0512	0.0879	0.1323	0.2304	0.3125
	4	0.0004	0.0064	0.0146	0.0283	0.0768	0.1563
	5	0.0000	0.0003	0.0010	0.0024	0.0102	0.0313
6	0	0.5314	0.2621	0.1780	0.1176	0.0467	0.0156
	1	0.3543	0.3932	0.3560	0.3025	0.1866	0.0938
	2	0.0984	0.2458	0.2966	0.3241	0.3110	0.2344
	3	0.0146	0.0819	0.1318	0.1852	0.2765	0.3125
	4	0.0012	0.0154	0.0330	0.0595	0.1382	0.2344
	5	0.0001	0.0015	0.0044	0.0102	0.0369	0.0938
	6	0.0000	0.0001	0.0002	0.0007	0.0041	0.0156

Table A *Binomial Probability Pr(n,k;p) (continued)*

n	k	0.10	0.20	0.25	0.30	0.40	0.50
7	0	0.4783	0.2097	0.1335	0.0824	0.0280	0.0078
	1	0.3720	0.3670	0.3115	0.2471	0.1306	0.0547
	2	0.1240	0.2753	0.3115	0.3177	0.2613	0.1641
	3	0.0230	0.1147	0.1730	0.2269	0.2903	0.2734
	4	0.0026	0.0287	0.0577	0.0972	0.1935	0.2734
	5	0.0002	0.0043	0.0115	0.0250	0.0774	0.1641
	6	0.0000	0.0004	0.0013	0.0036	0.0172	0.0547
	7	0.0000	0.0000	0.0001	0.0002	0.0016	0.0078
8	0	0.4305	0.1678	0.1001	0.0576	0.0168	0.0039
	1	0.3826	0.3355	0.2670	0.1977	0.0896	0.0313
	2	0.1488	0.2936	0.3115	0.2965	0.2090	0.1094
	3	0.0331	0.1468	0.2076	0.2541	0.2787	0.2188
	4	0.0046	0.0459	0.0865	0.1361	0.2322	0.2734
	5	0.0004	0.0092	0.0231	0.0467	0.1239	0.2188
	6	0.0000	0.0011	0.0038	0.0100	0.0413	0.1094
	7	0.0000	0.0001	0.0004	0.0012	0.0079	0.0313
	8	0.0000	0.0000	0.0000	0.0001	0.0007	0.0039
9	0	0.3874	0.1342	0.0751	0.0404	0.0101	0.0020
	1	0.3874	0.3020	0.2253	0.1556	0.0605	0.0176
	2	0.1722	0.3020	0.3003	0.2668	0.1612	0.0703
	3	0.0446	0.1762	0.2336	0.2668	0.2508	0.1641
	4	0.0074	0.0661	0.1168	0.1715	0.2508	0.2461
	5	0.0008	0.0165	0.0389	0.0735	0.1672	0.2461
	6	0.0001	0.0028	0.0087	0.0210	0.0743	0.1641
	7	0.0000	0.0003	0.0012	0.0039	0.0212	0.0703
	8	0.0000	0.0000	0.0001	0.0004	0.0035	0.0176
	9	0.0000	0.0000	0.0000	0.0000	0.0003	0.0020
10	0	0.3487	0.1074	0.0563	0.0282	0.0060	0.0010
	1	0.3874	0.2684	0.1877	0.1211	0.0403	0.0098
	2	0.1937	0.3020	0.2816	0.2335	0.1209	0.0439
	3	0.0574	0.2013	0.2503	0.2668	0.2150	0.1172
	4	0.0112	0.0881	0.1460	0.2001	0.2508	0.2051
	5	0.0015	0.0264	0.0584	0.1029	0.2007	0.2461
	6	0.0001	0.0055	0.0162	0.0368	0.1115	0.2051
	7	0.0000	0.0008	0.0031	0.0090	0.0425	0.1172
	8	0.0000	0.0001	0.0004	0.0014	0.0106	0.0439
	9	0.0000	0.0000	0.0000	0.0001	0.0016	0.0098
	10	0.0000	0.0000	0.0000	0.0000	0.0001	0.0010

Table A *Binomial Probability Pr(n,k;p) (continued)*

n	k	0.10	0.20	0.25	0.30	0.40	0.50
11	0	0.3138	0.0859	0.0422	0.0198	0.0036	0.0005
	1	0.3835	0.2362	0.1549	0.0932	0.0266	0.0054
	2	0.2131	0.2953	0.2581	0.1998	0.0887	0.0269
	3	0.0710	0.2215	0.2581	0.2568	0.1774	0.0806
	4	0.0158	0.1107	0.1721	0.2201	0.2365	0.1611
	5	0.0025	0.0388	0.0803	0.1321	0.2207	0.2256
	6	0.0003	0.0097	0.0268	0.0566	0.1471	0.2256
	7	0.0000	0.0017	0.0064	0.0173	0.0701	0.1611
	8	0.0000	0.0002	0.0011	0.0037	0.0234	0.0806
	9	0.0000	0.0000	0.0001	0.0005	0.0052	0.0269
	10	0.0000	0.0000	0.0000	0.0000	0.0007	0.0054
	11	0.0000	0.0000	0.0000	0.0000	0.0000	0.0005
12	0	0.2824	0.0687	0.0317	0.0138	0.0022	0.0002
	1	0.3766	0.2062	0.1267	0.0712	0.0174	0.0029
	2	0.2301	0.2835	0.2323	0.1678	0.0639	0.0161
	3	0.0852	0.2362	0.2581	0.2397	0.1419	0.0537
	4	0.0213	0.1329	0.1936	0.2311	0.2128	0.1208
	5	0.0038	0.0532	0.1032	0.1585	0.2270	0.1934
	6	0.0005	0.0155	0.0401	0.0792	0.1766	0.2256
	7	0.0000	0.0033	0.0115	0.0291	0.1009	0.1934
	8	0.0000	0.0005	0.0024	0.0078	0.0420	0.1208
	9	0.0000	0.0001	0.0004	0.0015	0.0125	0.0537
	10	0.0000	0.0000	0.0000	0.0002	0.0025	0.0161
	11	0.0000	0.0000	0.0000	0.0000	0.0003	0.0029
	12	0.0000	0.0000	0.0000	0.0000	0.0000	0.0002
13	0	0.2542	0.0550	0.0238	0.0097	0.0013	0.0001
	1	0.3672	0.1787	0.1029	0.0540	0.0113	0.0016
	2	0.2448	0.2680	0.2059	0.1388	0.0453	0.0095
	3	0.0997	0.2457	0.2517	0.2181	0.1107	0.0349
	4	0.0277	0.1535	0.2097	0.2337	0.1845	0.0873
	5	0.0055	0.0691	0.1258	0.1803	0.2214	0.1571
	6	0.0008	0.0230	0.0559	0.1030	0.1968	0.2095
	7	0.0001	0.0058	0.0186	0.0442	0.1312	0.2095
	8	0.0000	0.0011	0.0047	0.0142	0.0656	0.1571
	9	0.0000	0.0001	0.0009	0.0034	0.0243	0.0873
	10	0.0000	0.0000	0.0001	0.0006	0.0065	0.0349
	11	0.0000	0.0000	0.0000	0.0001	0.0012	0.0095
	12	0.0000	0.0000	0.0000	0.0000	0.0001	0.0016
	13	0.0000	0.0000	0.0000	0.0000	0.0000	0.0001

Table A   *Binomial Probability Pr(n,k;p) (continued)*

n	k	0.10	0.20	0.25	0.30	0.40	0.50
14	0	0.2288	0.0440	0.0178	0.0068	0.0008	0.0001
	1	0.3559	0.1539	0.0832	0.0407	0.0073	0.0009
	2	0.2570	0.2501	0.1802	0.1134	0.0317	0.0056
	3	0.1142	0.2501	0.2402	0.1943	0.0845	0.0222
	4	0.0349	0.1720	0.2202	0.2290	0.1549	0.0611
	5	0.0078	0.0860	0.1468	0.1963	0.2066	0.1222
	6	0.0013	0.0322	0.0734	0.1262	0.2066	0.1833
	7	0.0002	0.0092	0.0280	0.0618	0.1574	0.2095
	8	0.0000	0.0020	0.0082	0.0232	0.0918	0.1833
	9	0.0000	0.0003	0.0018	0.0066	0.0408	0.1222
	10	0.0000	0.0000	0.0003	0.0014	0.0136	0.0611
	11	0.0000	0.0000	0.0000	0.0002	0.0033	0.0222
	12	0.0000	0.0000	0.0000	0.0000	0.0005	0.0056
	13	0.0000	0.0000	0.0000	0.0000	0.0001	0.0009
	14	0.0000	0.0000	0.0000	0.0000	0.0000	0.0001
15	0	0.2059	0.0352	0.0134	0.0047	0.0005	0.0000
	1	0.3432	0.1319	0.0668	0.0305	0.0047	0.0005
	2	0.2669	0.2309	0.1559	0.0916	0.0219	0.0032
	3	0.1285	0.2501	0.2252	0.1700	0.0634	0.0139
	4	0.0428	0.1876	0.2252	0.2186	0.1268	0.0417
	5	0.0105	0.1032	0.1651	0.2061	0.1859	0.0916
	6	0.0019	0.0430	0.0917	0.1472	0.2066	0.1527
	7	0.0003	0.0138	0.0393	0.0811	0.1771	0.1964
	8	0.0000	0.0035	0.0131	0.0348	0.1181	0.1964
	9	0.0000	0.0007	0.0034	0.0116	0.0612	0.1527
	10	0.0000	0.0001	0.0007	0.0030	0.0245	0.0916
	11	0.0000	0.0000	0.0001	0.0006	0.0074	0.0417
	12	0.0000	0.0000	0.0000	0.0001	0.0016	0.0139
	13	0.0000	0.0000	0.0000	0.0000	0.0003	0.0032
	14	0.0000	0.0000	0.0000	0.0000	0.0000	0.0005
	15	0.0000	0.0000	0.0000	0.0000	0.0000	0.0000
16	0	0.1853	0.0281	0.0100	0.0033	0.0003	0.0000
	1	0.3294	0.1126	0.0535	0.0228	0.0030	0.0002
	2	0.2745	0.2111	0.1336	0.0732	0.0150	0.0018
	3	0.1423	0.2463	0.2079	0.1465	0.0468	0.0085
	4	0.0514	0.2001	0.2252	0.2040	0.1014	0.0278
	5	0.0137	0.1201	0.1802	0.2099	0.1623	0.0667
	6	0.0028	0.0550	0.1101	0.1649	0.1983	0.1222
	7	0.0004	0.0197	0.0524	0.1010	0.1889	0.1746
	8	0.0001	0.0055	0.0197	0.0487	0.1417	0.1964
	9	0.0000	0.0012	0.0058	0.0185	0.0840	0.1746
	10	0.0000	0.0002	0.0014	0.0056	0.0392	0.1222
	11	0.0000	0.0000	0.0002	0.0013	0.0142	0.0667
	12	0.0000	0.0000	0.0000	0.0002	0.0040	0.0278
	13	0.0000	0.0000	0.0000	0.0000	0.0008	0.0085
	14	0.0000	0.0000	0.0000	0.0000	0.0001	0.0018
	15	0.0000	0.0000	0.0000	0.0000	0.0000	0.0002
	16	0.0000	0.0000	0.0000	0.0000	0.0000	0.0000

Table A *Binomial Probability* $Pr(n,k;p)$ *(continued)*

n	k	0.10	0.20	0.25	0.30	0.40	0.50
17	0	0.1668	0.0225	0.0075	0.0023	0.0002	0.0000
	1	0.3150	0.0957	0.0426	0.0169	0.0019	0.0001
	2	0.2800	0.1914	0.1136	0 0581	0.0102	0.0010
	3	0.1556	0.2393	0.1893	0.1245	0.0341	0.0052
	4	0.0605	0.2093	0.2209	0.1868	0.0796	0.0182
	5	0.0175	0.1361	0.1914	0.2081	0.1379	0.0472
	6	0.0039	0.0680	0.1276	0.1784	0.1839	0.0944
	7	0.0007	0.0267	0.0668	0.1201	0.1927	0.1484
	8	0.0001	0.0084	0.0279	0.0644	0.1606	0.1855
	9	0.0000	0.0021	0.0093	0.0276	0.1070	0.1855
	10	0.0000	0.0004	0.0025	0.0095	0.0571	0.1484
	11	0.0000	0.0001	0.0005	0.0026	0.0242	0.0944
	12	0.0000	0.0000	0.0001	0.0006	0.0081	0.0472
	13	0.0000	0.0000	0.0000	0.0001	0.0021	0.0182
	14	0.0000	0.0000	0.0000	0.0000	0.0004	0.0052
	15	0.0000	0.0000	0.0000	0.0000	0.0001	0.0010
	16	0.0000	0.0000	0.0000	0.0000	0.0000	0.0001
	17	0.0000	0.0000	0.0000	0.0000	0.0000	0.0000
18	0	0.1501	0.0180	0.0056	0.0016	0.0001	0.0000
	1	0.3002	0.0811	0.0338	0.0126	0.0012	0.0001
	2	0.2835	0.1723	0.0958	0.0458	0.0069	0.0006
	3	0.1680	0.2297	0.1704	0.1046	0.0246	0.0031
	4	0.0700	0.2153	0.2130	0.1681	0.0614	0.0117
	5	0.0218	0.1507	0.1988	0.2017	0.1146	0.0327
	6	0.0052	0.0816	0.1436	0.1873	0.1655	0.0708
	7	0.0010	0.0350	0.0820	0.1376	0.1892	0.1214
	8	0.0002	0.0120	0.0376	0.0811	0.1734	0.1669
	9	0.0000	0.0033	0.0139	0.0386	0.1284	0.1855
	10	0.0000	0.0008	0.0042	0.0149	0.0771	0.1669
	11	0.0000	0.0001	0.0010	0.0046	0.0374	0.1214
	12	0.0000	0.0000	0.0002	0.0012	0.0145	0.0708
	13	0.0000	0.0000	0.0000	0.0002	0.0045	0.0327
	14	0.0000	0.0000	0.0000	0.0000	0.0011	0.0117
	15	0.0000	0.0000	0.0000	0.0000	0.0002	0.0031
	16	0.0000	0.0000	0.0000	0.0000	0.0000	0.0006
	17	0.0000	0.0000	0.0000	0.0000	0.0000	0.0001
	18	0.0000	0.0000	0.0000	0.0000	0.0000	0.0000

Table A *Binomial Probability Pr(n,k;p) (continued)*

n	k	0.10	0.20	0.25	0.30	0.40	0.50
19	0	0.1351	0.0144	0.0042	0.0011	0.0001	0.0000
	1	0.2852	0.0685	0.0268	0.0093	0.0008	0.0000
	2	0.2852	0.1540	0.0803	0.0358	0.0046	0.0003
	3	0.1796	0.2182	0.1517	0.0869	0.0175	0.0018
	4	0.0798	0.2182	0.2023	0.1491	0.0467	0.0074
	5	0.0266	0.1636	0.2023	0.1916	0.0933	0.0222
	6	0.0069	0.0955	0.1574	0.1916	0.1451	0.0518
	7	0.0014	0.0443	0.0974	0.1525	0.1797	0.0961
	8	0.0002	0.0166	0.0487	0.0981	0.1797	0.1442
	9	0.0000	0.0051	0.0198	0.0514	0.1464	0.1762
	10	0.0000	0.0013	0.0066	0.0220	0.0976	0.1762
	11	0.0000	0.0003	0.0018	0.0077	0.0532	0.1442
	12	0.0000	0.0000	0.0004	0.0022	0.0237	0.0961
	13	0.0000	0.0000	0.0001	0.0005	0.0085	0.0518
	14	0.0000	0.0000	0.0000	0.0001	0.0024	0.0222
	15	0.0000	0.0000	0.0000	0.0000	0.0005	0.0074
	16	0.0000	0.0000	0.0000	0.0000	0.0001	0.0018
	17	0.0000	0.0000	0.0000	0.0000	0.0000	0.0003
	18	0.0000	0.0000	0.0000	0.0000	0.0000	0.0000
	19	0.0000	0.0000	0.0000	0.0000	0.0000	0.0000
20	0	0.1216	0.0115	0.0032	0.0008	0.0000	0.0000
	1	0.2702	0.0576	0.0211	0.0068	0.0005	0.0000
	2	0.2852	0.1369	0.0669	0.0278	0.0031	0.0002
	3	0.1901	0.2054	0.1339	0.0716	0.0123	0.0011
	4	0.0898	0.2182	0.1897	0.1304	0.0350	0.0046
	5	0.0319	0.1746	0.2023	0.1789	0.0746	0.0148
	6	0.0089	0.1091	0.1686	0.1916	0.1244	0.0370
	7	0.0020	0.0545	0.1124	0.1643	0.1659	0.0739
	8	0.0004	0.0222	0.0609	0.1144	0.1797	0.1201
	9	0.0001	0.0074	0.0271	0.0654	0.1597	0.1602
	10	0.0000	0.0020	0.0099	0.0308	0.1171	0.1762
	11	0.0000	0.0005	0.0030	0.0120	0.0710	0.1602
	12	0.0000	0.0001	0.0008	0.0039	0.0355	0.1201
	13	0.0000	0.0000	0.0002	0.0010	0.0146	0.0739
	14	0.0000	0.0000	0.0000	0.0002	0.0049	0.0370
	15	0.0000	0.0000	0.0000	0.0000	0.0013	0.0148
	16	0.0000	0.0000	0.0000	0.0000	0.0003	0.0046
	17	0.0000	0.0000	0.0000	0.0000	0.0000	0.0011
	18	0.0000	0.0000	0.0000	0.0000	0.0000	0.0002
	19	0.0000	0.0000	0.0000	0.0000	0.0000	0.0000
	20	0.0000	0.0000	0.0000	0.0000	0.0000	0.0000

TABLE **B** | **Square Root**

N	$\sqrt{N}$	$\sqrt{10N}$	N	$\sqrt{N}$	$\sqrt{10N}$	N	$\sqrt{N}$	$\sqrt{10N}$	N	$\sqrt{N}$	$\sqrt{10N}$
1.00	1.000	3.162	1.40	1.183	3.742	1.80	1.342	4.243	2.20	1.483	4.690
1.01	1.005	3.178	1.41	1.187	3.755	1.81	1.345	4.254	2.21	1.487	4.701
1.02	1.010	3.194	1.42	1.192	3.768	1.82	1.349	4.266	2.22	1.490	4.712
1.03	1.015	3.209	1.43	1.196	3.782	1.83	1.353	4.278	2.23	1.493	4.722
1.04	1.020	3.225	1.44	1.200	3.795	1.84	1.356	4.290	2.24	1.497	4.733
1.05	1.025	3.240	1.45	1.204	3.808	1.85	1.360	4.301	2.25	1.500	4.743
1.06	1.030	3.256	1.46	1.208	3.821	1.86	1.364	4.313	2.26	1.503	4.754
1.07	1.034	3.271	1.47	1.212	3.834	1.87	1.367	4.324	2.27	1.507	4.764
1.08	1.039	3.286	1.48	1.217	3.847	1.88	1.371	4.336	2.28	1.510	4.775
1.09	1.044	3.302	1.49	1.221	3.860	1.89	1.375	4.347	2.29	1.513	4.785
1.10	1.049	3.317	1.50	1.225	3.873	1.90	1.378	4.359	2.30	1.517	4.796
1.11	1.054	3.332	1.51	1.229	3.886	1.91	1.382	4.370	2.31	1.520	4.806
1.12	1.058	3.347	1.52	1.233	3.899	1.92	1.386	4.382	2.32	1.523	4.817
1.13	1.063	3.362	1.53	1.237	3.912	1.93	1.389	4.393	2.33	1.526	4.827
1.14	1.068	3.376	1.54	1.241	3.924	1.94	1.393	4.405	2.34	1.530	4.837
1.15	1.072	3.391	1.55	1.245	3.937	1.95	1.396	4.416	2.35	1.533	4.848
1.16	1.077	3.406	1.56	1.249	3.950	1.96	1.400	4.427	2.36	1.536	4.858
1.17	1.082	3.421	1.57	1.253	3.962	1.97	1.404	4.438	2.37	1.539	4.868
1.18	1.086	3.435	1.58	1.257	3.975	1.98	1.407	4.450	2.38	1.543	4.879
1.19	1.091	3.450	1.59	1.261	3.987	1.99	1.411	4.461	2.39	1.546	4.889
1.20	1.095	3.464	1.60	1.265	4.000	2.00	1.414	4.472	2.40	1.549	4.899
1.21	1.100	3.479	1.61	1.269	4.012	2.01	1.418	4.483	2.41	1.552	4.909
1.22	1.105	3.493	1.62	1.273	4.025	2.02	1.421	4.494	2.42	1.556	4.919
1.23	1.109	3.507	1.63	1.277	4.037	2.03	1.425	4.506	2.43	1.559	4.930
1.24	1.114	3.521	1.64	1.281	4.050	2.04	1.428	4.517	2.44	1.562	4.940
1.25	1.118	3.536	1.65	1.285	4.062	2.05	1.432	4.528	2.45	1.565	4.950
1.26	1.122	3.550	1.66	1.288	4.074	2.06	1.435	4.539	2.46	1.568	4.960
1.27	1.127	3.564	1.67	1.292	4.087	2.07	1.439	4.550	2.47	1.572	4.970
1.28	1.131	3.578	1.68	1.296	4.099	2.08	1.442	4.561	2.48	1.575	4.980
1.29	1.136	3.592	1.69	1.300	4.111	2.09	1.446	4.572	2.49	1.578	4.990
1.30	1.140	3.606	1.70	1.304	4.123	2.10	1.449	4.583	2.50	1.581	5.000
1.31	1.145	3.619	1.71	1.308	4.135	2.11	1.453	4.593	2.51	1.584	5.010
1.32	1.149	3.633	1.72	1.311	4.147	2.12	1.456	4.604	2.52	1.587	5.020
1.33	1.153	3.647	1.73	1.315	4.159	2.13	1.459	4.615	2.53	1.591	5.030
1.34	1.158	3.661	1.74	1.319	4.171	2.14	1.463	4.626	2.54	1.594	5.040
1.35	1.162	3.674	1.75	1.323	4.183	2.15	1.466	4.637	2.55	1.597	5.050
1.36	1.166	3.688	1.76	1.327	4.195	2.16	1.470	4.648	2.56	1.600	5.060
1.37	1.170	3.701	1.77	1.330	4.207	2.17	1.473	4.658	2.57	1.603	5.070
1.38	1.175	3.715	1.78	1.334	4.219	2.18	1.476	4.669	2.58	1.606	5.079
1.39	1.179	3.728	1.79	1.338	4.231	2.19	1.480	4.680	2.59	1.609	5.089

Table B *Square Root (continued)*

N	$\sqrt{N}$	$\sqrt{10N}$	N	$\sqrt{N}$	$\sqrt{10N}$	N	$\sqrt{N}$	$\sqrt{10N}$	N	$\sqrt{N}$	$\sqrt{10N}$
2.60	1.612	5.099	3.10	1.761	5.568	3.60	1.897	6.000	4.10	2.025	6.403
2.61	1.616	5.109	3.11	1.764	5.577	3.61	1.900	6.008	4.11	2.027	6.411
2.62	1.619	5.119	3.12	1.766	5.586	3.62	1.903	6.017	4.12	2.030	6.419
2.63	1.622	5.128	3.13	1.769	5.595	3.63	1.905	6.025	4.13	2.032	6.427
2.64	1.625	5.138	3.14	1.772	5.604	3.64	1.908	6.033	4.14	2.035	6.434
2.65	1.628	5.148	3.15	1.775	5.612	3.65	1.910	6.042	4.15	2.037	6.442
2.66	1.631	5.158	3.16	1.778	5.621	3.66	1.913	6.050	4.16	2.040	6.450
2.67	1.634	5.167	3.17	1.780	5.630	3.67	1.916	6.058	4.17	2.042	6.458
2.68	1.637	5.177	3.18	1.783	5.639	3.68	1.918	6.066	4.18	2.045	6.465
2.69	1.640	5.187	3.19	1.786	5.648	3.69	1.921	6.075	4.19	2.047	6.473
2.70	1.643	5.196	3.20	1.789	5.657	3.70	1.924	6.083	4.20	2.049	6.481
2.71	1.646	5.206	3.21	1.792	5.666	3.71	1.926	6.091	4.21	2.052	6.488
2.72	1.649	5.215	3.22	1.794	5.675	3.72	1.929	6.099	4.22	2.054	6.496
2.73	1.652	5.225	3.23	1.797	5.683	3.73	1.931	6.107	4.23	2.057	6.504
2.74	1.655	5.234	3.24	1.800	5.692	3.74	1.934	6.116	4.24	2.059	6.512
2.75	1.658	5.244	3.25	1.803	5.701	3.75	1.936	6.124	4.25	2.062	6.519
2.76	1.661	5.254	3.26	1.806	5.710	3.76	1.939	6.132	4.26	2.064	6.527
2.77	1.664	5.263	3.27	1.808	5.718	3.77	1.942	6.140	4.27	2.066	6.535
2.78	1.667	5.273	3.28	1.811	5.727	3.78	1.944	6.148	4.28	2.069	6.542
2.79	1.670	5.282	3.29	1.814	5.736	3.79	1.947	6.156	4.29	2.071	6.550
2.80	1.673	5.292	3.30	1.817	5.745	3.80	1.949	6.164	4.30	2.074	6.557
2.81	1.676	5.301	3.31	1.819	5.753	3.81	1.952	6.173	4.31	2.076	6.565
2.82	1.679	5.310	3.32	1.822	5.762	3.82	1.954	6.181	4.32	2.078	6.573
2.83	1.682	5.320	3.33	1.825	5.771	3.83	1.957	6.189	4.33	2.081	6.580
2.84	1.685	5.329	3.34	1.828	5.779	3.84	1.960	6.197	4.34	2.083	6.588
2.85	1.688	5.339	3.35	1.830	5.788	3.85	1.962	6.205	4.35	2.086	6.595
2.86	1.691	5.348	3.36	1.833	5.797	3.86	1.965	6.213	4.36	2.088	6.603
2.87	1.694	5.357	3.37	1.836	5.805	3.87	1.967	6.221	4.37	2.090	6.611
2.88	1.697	5.367	3.38	1.838	5.814	3.88	1.970	6.229	4.38	2.093	6.618
2.89	1.700	5.376	3.39	1.841	5.822	3.89	1.972	6.237	4.39	2.095	6.626
2.90	1.703	5.385	3.40	1.844	5.831	3.90	1.975	6.245	4.40	2.098	6.633
2.91	1.706	5.394	3.41	1.847	5.840	3.91	1.977	6.253	4.41	2.100	6.641
2.92	1.709	5.404	3.42	1.849	5.848	3.92	1.980	6.261	4.42	2.102	6.648
2.93	1.712	5.413	3.43	1.852	5.857	3.93	1.982	6.269	4.43	2.105	6.656
2.94	1.715	5.422	3.44	1.855	5.865	3.94	1.985	6.277	4.44	2.107	6.663
2.95	1.718	5.431	3.45	1.857	5.874	3.95	1.987	6.285	4.45	2.110	6.671
2.96	1.720	5.441	3.46	1.860	5.882	3.96	1.990	6.293	4.46	2.112	6.678
2.97	1.723	5.450	3.47	1.863	5.891	3.97	1.992	6.301	4.47	2.114	6.686
2.98	1.726	5.459	3.48	1.865	5.899	3.98	1.995	6.309	4.48	2.117	6.693
2.99	1.729	5.468	3.49	1.868	5.908	3.99	1.997	6.317	4.49	2.119	6.701
3.00	1.732	5.477	3.50	1.871	5.916	4.00	2.000	6.325	4.50	2.121	6.708
3.01	1.735	5.486	3.51	1.873	5.925	4.01	2.002	6.332	4.51	2.124	6.716
3.02	1.738	5.495	3.52	1.876	5.933	4.02	2.005	6.340	4.52	2.126	6.723
3.03	1.741	5.505	3.53	1.879	5.941	4.03	2.007	6.348	4.53	2.128	6.731
3.04	1.744	5.514	3.54	1.881	5.950	4.04	2.010	6.356	4.54	2.131	6.738
3.05	1.746	5.523	3.55	1.884	5.958	4.05	2.012	6.364	4.55	2.133	6.745
3.06	1.749	5.532	3.56	1.887	5.967	4.06	2.015	6.372	4.56	2.135	6.753
3.07	1.752	5.541	3.57	1.889	5.975	4.07	2.017	6.380	4.57	2.138	6.760
3.08	1.755	5.550	3.58	1.892	5.983	4.08	2.020	6.387	4.58	2.140	6.768
3.09	1.758	5.559	3.59	1.895	5.992	4.09	2.022	6.395	4.59	2.142	6.775

Table B *Square Root (continued)*

N	√N	√10N	N	√N	√10N	N	√N	√10N	N	√N	√10N
4.60	2.145	6.782	5.10	2.258	7.141	5.60	2.366	7.483	6.10	2.470	7.810
4.61	2.147	6.790	5.11	2.261	7.148	5.61	2.369	7.490	6.11	2.472	7.817
4.52	2.149	6.797	5.12	2.263	7.155	5.62	2.371	7.497	6.12	2.474	7.823
4.63	2.152	6.804	5.13	2.265	7.162	5.63	2.373	7.503	6.13	2.476	7.829
4.64	2.154	6.812	5.14	2.267	7.169	5.64	2.375	7.510	6.14	2.478	7.836
4.65	2.156	6.819	5.15	2.269	7.176	5.65	2.377	7.517	6.15	2.480	7.842
4.66	2.159	6.826	5.16	2.272	7.183	5.66	2.379	7.523	6.16	2.482	7.849
4.67	2.161	6.834	5.17	2.274	7.190	5.67	2.381	7.530	6.17	2.484	7.855
4.68	2.163	6.841	5.18	2.276	7.197	5.68	2.383	7.537	6.18	2.486	7.861
4.69	2.166	6.848	5.19	2.278	7.204	5.69	2.385	7.543	6.19	2.488	7.868
4.70	2.168	6.856	5.20	2.280	7.211	5.70	2.387	7.550	6.20	2.490	7.874
4.71	2.170	6.863	5.21	2.283	7.218	5.71	2.390	7.556	6.21	2.492	7.880
4.72	2.173	6.870	5.22	2.285	7.225	5.72	2.392	7.563	6.22	2.494	7.887
4.73	2.175	6.877	5.23	2.287	7.232	5.73	2.394	7.570	6.23	2.496	7.893
4.74	2.177	6.885	5.24	2.289	7.239	5.74	2.396	7.576	6.24	2.498	7.899
4.75	2.179	6.892	5.25	2.291	7.246	5.75	2.398	7.583	6.25	2.500	7.906
4.76	2.182	6.899	5.26	2.293	7.253	5.76	2.400	7.589	6.26	2.502	7.912
4.77	2.184	6.907	5.27	2.296	7.259	5.77	2.402	7.596	6.27	2.504	7.918
4.78	2.186	6.914	5.28	2.298	7.266	5.78	2.404	7.603	6.28	2.506	7.925
4.79	2.189	6.921	5.29	2.300	7.273	5.79	2.406	7.609	6.29	2.508	7.931
4.80	2.191	6.928	5.30	2.302	7.280	5.80	2.408	7.616	6.30	2.510	7.937
4.81	2.193	6.935	5.31	2.304	7.287	5.81	2.410	7.622	6.31	2.512	7.944
4.82	2.195	6.943	5.32	2.307	7.294	5.82	2.412	7.629	6.32	2.514	7.950
4.83	2.198	6.950	5.33	2.309	7.301	5.83	2.415	7.635	6.33	2.516	7.956
4.84	2.200	6.957	5.34	2.311	7.308	5.84	2.417	7.642	6.34	2.518	7.962
4.85	2.202	6.964	5.35	2.313	7.314	5.85	2.419	7.649	6.35	2.520	7.969
4.86	2.205	6.971	5.36	2.315	7.321	5.86	2.421	7.655	6.36	2.522	7.975
4.87	2.207	6.979	5.37	2.317	7.328	5.87	2.423	7.662	6.37	2.524	7.981
4.88	2.209	6.986	5.38	2.319	7.335	5.88	2.425	7.668	6.38	2.526	7.987
4.89	2.211	6.993	5.39	2.322	7.342	5.89	2.427	7.675	6.39	2.528	7.994
4.90	2.214	7.000	5.40	2.324	7.348	5.90	2.429	7.681	6.40	2.530	8.000
4.91	2.216	7.007	5.41	2.326	7.355	5.91	2.431	7.688	6.41	2.532	8.006
4.92	2.218	7.014	5.42	2.328	7.362	5.92	2.433	7.694	6.42	2.534	8.012
4.93	2.220	7.021	5.43	2.330	7.369	5.93	2.435	7.701	6.43	2.536	8.019
4.94	2.223	7.029	5.44	2.332	7.376	5.94	2.437	7.707	6.44	2.538	8.025
4.95	2.225	7.036	5.45	2.335	7.382	5.95	2.439	7.714	6.45	2.540	8.031
4.96	2.227	7.043	5.46	2.337	7.389	5.96	2.441	7.720	6.46	2.542	8.037
4.97	2.229	7.050	5.47	2.339	7.396	5.97	2.443	7.727	6.47	2.544	8.044
4.98	2.232	7.057	5.48	2.341	7.403	5.98	2.445	7.733	6.48	2.546	8.050
4.99	2.234	7.064	5.49	2.343	7.409	5.99	2.447	7.740	6.49	2.548	8.056
5.00	2.236	7.071	5.50	2.345	7.416	6.00	2.449	7.746	6.50	2.550	8.062
5.01	2.238	7.078	5.51	2.347	7.423	6.01	2.452	7.752	6.51	2.551	8.068
5.02	2.241	7.085	5.52	2.349	7.430	6.02	2.454	7.759	6.52	2.553	8.075
5.03	2.243	7.092	5.53	2.352	7.436	6.03	2.456	7.765	6.53	2.555	8.081
5.04	2.245	7.099	5.54	2.354	7.443	6.04	2.458	7.772	6.54	2.557	8.087
5.05	2.247	7.106	5.55	2.356	7.450	6.05	2.460	7.778	6.55	2.559	8.093
5.06	2.249	7.113	5.56	2.358	7.457	6.06	2.462	7.785	6.56	2.561	8.099
5.07	2.252	7.120	5.57	2.360	7.463	6.07	2.464	7.791	6.57	2.563	8.106
5.08	2.254	7.127	5.58	2.362	7.470	6.08	2.466	7.797	6.58	2.565	8.112
5.09	2.256	7.134	5.59	2.364	7.477	6.09	2.468	7.804	6.59	2.567	8.118

Table B *Square Root* (continued)

N	$\sqrt{N}$	$\sqrt{10N}$	N	$\sqrt{N}$	$\sqrt{10N}$	N	$\sqrt{N}$	$\sqrt{10N}$	N	$\sqrt{N}$	$\sqrt{10N}$
6.60	2.569	8.124	7.10	2.665	8.426	7.60	2.757	8.718	8.10	2.846	9.000
6.61	2.571	8.130	7.11	2.666	8.432	7.61	2.759	8.724	8.11	2.848	9.006
6.62	2.573	8.136	7.12	2.668	8.438	7.62	2.760	8.729	8.12	2.850	9.011
6.63	2.575	8.142	7.13	2.670	8.444	7.63	2.762	8.735	8.13	2.851	9.017
6.64	2.577	8.149	7.14	2.672	8.450	7.64	2.764	8.741	8.14	2.853	9.022
6.65	2.579	8.155	7.15	2.674	8.456	7.65	2.766	8.746	8.15	2.855	9.028
6.66	2.581	8.161	7.16	2.676	8.462	7.66	2.768	8.752	8.16	2.857	9.033
6.67	2.583	8.167	7.17	2.678	8.468	7.67	2.769	8.758	8.17	2.858	9.039
6.68	2.585	8.173	7.18	2.680	8.473	7.68	2.771	8.764	8.18	2.860	9.044
6.69	2.587	8.179	7.19	2.681	8.479	7.69	2.773	8.769	8.19	2.862	9.050
6.70	2.588	8.185	7.20	2.683	8.485	7.70	2.775	8.775	8.20	2.864	9.055
6.71	2.590	8.191	7.21	2.685	8.491	7.71	2.777	8.781	8.21	2.865	9.061
6.72	2.592	8.198	7.22	2.687	8.497	7.72	2.778	8.786	8.22	2.867	9.066
6.73	2.594	8.204	7.23	2.689	8.503	7.73	2.780	8.792	8.23	2.869	9.072
6.74	2.596	8.210	7.24	2.691	8.509	7.74	2.782	8.798	8.24	2.871	9.077
6.75	2.598	8.216	7.25	2.693	8.515	7.75	2.784	8.803	8.25	2.872	9.083
6.76	2.600	8.222	7.26	2.694	8.521	7.76	2.786	8.809	8.26	2.874	9.088
6.77	2.602	8.228	7.27	2.696	8.526	7.77	2.787	8.815	8.27	2.876	9.094
6.78	2.604	8.234	7.28	2.698	8.532	7.78	2.789	8.820	8.28	2.877	9.099
6.79	2.606	8.240	7.29	2.700	8.538	7.79	2.791	8.826	8.29	2.879	9.105
6.80	2.608	8.246	7.30	2.702	8.544	7.80	2.793	8.832	8.30	2.881	9.110
6.81	2.610	8.252	7.31	2.704	8.550	7.81	2.795	8.837	8.31	2.883	9.116
6.82	2.612	8.258	7.32	2.706	8.556	7.82	2.796	8.843	8.32	2.884	9.121
6.83	2.613	8.264	7.33	2.707	8.562	7.83	2.798	8.849	8.33	2.886	9.127
6.84	2.615	8.270	7.34	2.709	8.567	7.84	2.800	8.854	8.34	2.888	9.132
6.85	2.617	8.276	7.35	2.711	8.573	7.85	2.802	8.860	8.35	2.890	9.138
6.86	2.619	8.283	7.36	2.713	8.579	7.86	2.804	8.866	8.36	2.891	9.143
6.87	2.621	8.289	7.37	2.715	8.585	7.87	2.805	8.871	8.37	2.893	9.149
6.88	2.623	8.295	7.38	2.717	8.591	7.88	2.807	8.877	8.38	2.895	9.154
6.89	2.625	8.301	7.39	2.718	8.597	7.89	2.809	8.883	8.39	2.897	9.160
6.90	2.627	8.307	7.40	2.720	8.602	7.90	2.811	8.888	8.40	2.898	9.165
6.91	2.629	8.313	7.41	2.722	8.608	7.91	2.812	8.894	8.41	2.900	9.171
6.92	2.631	8.319	7.42	2.724	8.614	7.92	2.814	8.899	8.42	2.902	9.176
6.93	2.632	8.325	7.43	2.726	8.620	7.93	2.816	8.905	8.43	2.903	9.182
6.94	2.634	8.331	7.44	2.728	8.626	7.94	2.818	8.911	8.44	2.905	9.187
6.95	2.636	8.337	7.45	2.729	8.631	7.95	2.820	8.916	8.45	2.907	9.192
6.96	2.638	8.343	7.46	2.731	8.637	7.96	2.821	8.922	8.46	2.909	9.198
6.97	2.640	8.349	7.47	2.733	8.643	7.97	2.823	8.927	8.47	2.910	9.203
6.98	2.642	8.355	7.48	2.735	8.649	7.98	2.825	8.933	8.48	2.912	9.209
6.99	2.644	8.361	7.49	2.737	8.654	7.99	2.827	8.939	8.49	2.914	9.214
7.00	2.646	8.367	7.50	2.739	8.660	8.00	2.828	8.944	8.50	2.915	9.220
7.01	2.648	8.373	7.51	2.740	8.666	8.01	2.830	8.950	8.51	2.917	9.225
7.02	2.650	8.379	7.52	2.742	8.672	8.02	2.832	8.955	8.52	2.919	9.230
7.03	2.651	8.385	7.53	2.744	8.678	8.03	2.834	8.961	8.53	2.921	9.236
7.04	2.653	8.390	7.54	2.746	8.683	8.04	2.835	8.967	8.54	2.922	9.241
7.05	2.655	8.396	7.55	2.748	8.689	8.05	2.837	8.972	8.55	2.924	9.247
7.06	2.657	8.402	7.56	2.750	8.695	8.06	2.839	8.978	8.56	2.926	9.252
7.07	2.659	8.408	7.57	2.751	8.701	8.07	2.841	8.983	8.57	2.927	9.257
7.08	2.661	8.414	7.58	2.753	8.706	8.08	2.843	8.989	8.58	2.929	9.263
7.09	2.663	8.420	7.59	2.755	8.712	8.09	2.844	8.994	8.59	2.931	9.268

Table B *Square Root (continued)*

N	√N	√10N	N	√N	√10N	N	√N	√10N
8.60	2.933	9.274	9.10	3.017	9.539	9.60	3.098	9.798
8.61	2.934	9.279	9.11	3.018	9.545	9.61	3.100	9.803
8.62	2.936	9.284	9.12	3.020	9.550	9.62	3.102	9.808
8.63	2.938	9.290	9.13	3.022	9.555	9.63	3.103	9.813
8.64	2.939	9.295	9.14	3.023	9.560	9.64	3.105	9.818
8.65	2.941	9.301	9.15	3.025	9.566	9.65	3.106	9.823
8.66	2.943	9.306	9.16	3.027	9.571	9.66	3.108	9.829
8.67	2.944	9.311	9.17	3.028	9.576	9.67	3.110	9.834
8.68	2.946	9.317	9.18	3.030	9.581	9.68	3.111	9.839
8.69	2.948	9.322	9.19	3.031	9.586	9.69	3.113	9.844
8.70	2.950	9.327	9.20	3.033	9.592	9.70	3.114	9.849
8.71	2.951	9.333	9.21	3.035	9.597	9.71	3.116	9.854
8.72	2.953	9.338	9.22	3.036	9.602	9.72	3.118	9.859
8.73	2.955	9.343	9.23	3.038	9.607	9.73	3.119	9.864
8.74	2.956	9.349	9.24	3.040	9.612	9.74	3.121	9.869
8.75	2.958	9.354	9.25	3.041	9.618	9.75	3.122	9.874
8.76	2.960	9.359	9.26	3.043	9.623	9.76	3.124	9.879
8.77	2.961	9.365	9.27	3.045	9.628	9.77	3.126	9.884
8.78	2.963	9.370	9.28	3.046	9.633	9.78	3.127	9.889
8.79	2.965	9.375	9.29	3.048	9.638	9.79	3.129	9.894
8.80	2.966	9.381	9.30	3.050	9.644	9.80	3.130	9.899
8.81	2.968	9.386	9.31	3.051	9.649	9.81	3.132	9.905
8.82	2.970	9.391	9.32	3.053	9.654	9.82	3.134	9.910
8.83	2.972	9.397	9.33	3.055	9.659	9.83	3.135	9.915
8.84	2.973	9.402	9.34	3.056	9.664	9.84	3.137	9.920
8.85	2.975	9.407	9.35	3.058	9.670	9.85	3.138	9.925
8.86	2.977	9.413	9.36	3.059	9.675	9.86	3.140	9.930
8.87	2.978	9.418	9.37	3.061	9.680	9.87	3.142	9.935
8.88	2.980	9.423	9.38	3.063	9.685	9.88	3.143	9.940
8.89	2.982	9.429	9.39	3.064	9.690	9.89	3.145	9.945
8.90	2.983	9.434	9.40	3.066	9.695	9.90	3.146	9.950
8.91	2.985	9.439	9.41	3.068	9.701	9.91	3.148	9.955
8.92	2.987	9.445	9.42	3.069	9.706	9.92	3.150	9.960
8.93	2.988	9.450	9.43	3.071	9.711	9.93	3.151	9.965
8.94	2.990	9.455	9.44	3.072	9.716	9.94	3.153	9.970
8.95	2.992	9.460	9.45	3.074	9.721	9.95	3.154	9.975
8.96	2.993	9.466	9.46	3.076	9.726	9.96	3.156	9.980
8.97	2.995	9.471	9.47	3.077	9.731	9.97	3.158	9.985
8.98	2.997	9.476	9.48	3.079	9.737	9.98	3.159	9.990
8.99	2.998	9.482	9.49	3.081	9.742	9.99	3.161	9.995
9.00	3.000	9.487	9.50	3.082	9.747			
9.01	3.002	9.492	9.51	3.084	9.752			
9.02	3.003	9.497	9.52	3.085	9.757			
9.03	3.005	9.503	9.53	3.087	9.762			
9.04	3.007	9.508	9.54	3.089	9.767			
9.05	3.008	9.513	9.55	3.090	9.772			
9.06	3.010	9.518	9.56	3.092	9.778			
9.07	3.012	9.524	9.57	3.094	9.783			
9.08	3.013	9.529	9.58	3.095	9.788			
9.09	3.015	9.534	9.59	3.097	9.793			

Answers to selected exercises

Answers to selected exercises

CHAPTER 2 SET THEORY

Page 13, The Language of Sets

1. a) well defined
   b) well defined
   c) not well defined
   d) well defined
   e) well defined

3. a) $\{1,7,3,2,6,8\}$
   b) $\{1,2,3, \ldots ,999\}$
   c) $\{x\colon x$ is a letter in the word *logic*$\}$
   d) $\{2,4,6, \ldots\}$
   e) $\varnothing$

5. a) $\in$ b) $\in$ c) $\notin$ d) $\notin$ e) $\in$

Page 24, Relations and Operations with Sets

2. a) equal
   b) equal
   c) not equal
   d) not equal

3. (From 1)
   a) $\{CBS,ABC\} \subset \{y\colon y$ is a major television broadcasting company$\}$
   b) $\{x\colon x$ is a season of the year$\} \subset \{$winter, spring, summer, fall, July$\}$
   c) Each set is a subset of the other.
   d) Each set is a subset of the other.

4. a) $A \cap P = \{x : x$ is an American political leader$\} = \{$Harry Truman$\}$

 b) $F \cup B = \{x : x$ is either French or British$\} = \{$Voltaire, Napoleon, Pasteur, Churchill, Shakespeare, Curie$\}$

 c) $B - D = \{x : x$ is British and not deceased$\} = \varnothing$

 d) $P \cap F' = \{x : x$ is a political leader and not French$\} = \{$Meir, Churchill, Caesar, Truman$\}$

 e) $(F \cup B)' = \{x : x$ is neither French nor British$\} = \{$Archimedes, Beethoven, Bach, da Vinci, Hemingway, Meir, Rembrandt, Caesar, Truman, Bernstein$\}$

6.

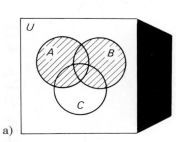
a)

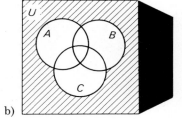
b)

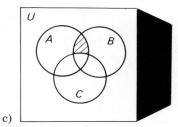
c)

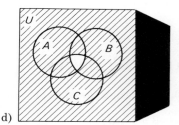
d)

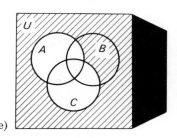
e)

7. a) $(A \cup B)'$ b) $(A \cup B) - (A \cap B)$

 c) $(A \cup B) - C$ d) $B - (A \cup C)$

9. a) 30   b) 7   c) 11   d) 2   e) 9

11. a) $A \cap B = A$ b) $A \cup B = B$

 c) $A - B = \varnothing$ d) $A' \cap B' = B'$

13. 7

Page 27, Mastery Test: Tools of Set Theory

1. a) The two sets are equal because each is a subset of the other.
   b) {x: x is a letter in the word *mild*}   {x: x is a letter in the word *middle*}
   c) {bread, cheese, ice cream} ⊂ {y: y is a manufactured food product}

2. a) $(A \cup C) - B$ b) $C - (A \cup B)$

3.

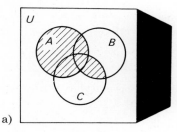

a)

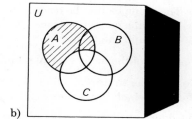

b)

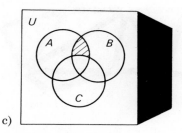

c)

4. a) {apple, banana}   b) ∅
   c) {apple, automobile, shirt, banana}   d) {apple, banana}

5. W and K are disjoint and also C and K are disjoint.

6. a) $2^5 = 32$ b) 31   7. a) 9   b) 15   c) 0   d) 3

Page 30, Part III

1. $n(A) = 18$, $n(B) = 15$, $n(C) = 14$

3. inconsistent; there are too many elements in $B \cup C$

5. $n(A) = 4$, $n(B) = 14$, $n(C) = 8$

7. 54; 14   9. 37

10. If we assume that no people use both the bus and the train, then we can easily account for more than 200 people.

CHAPTER 3 GRAPH THEORY

Page 49, Graphs, Puzzles and Map Coloring

1. a) The graph is connected; A and B are odd; C and D are even.
    b) The graph is not connected; all of the vertices are even.
    c) The graph is connected; A, B, E, and F are odd; C and D are even.
    d) The graph is connected; all of the vertices are even.

2. a) The graph can be traced.
    b) The graph cannot be traced; it is not connected.
    c) The graph cannot be traced; there are four odd vertices.
    d) The graph can be traced.

4. Yes. One of the many possible solutions is the sequence C, F$^\sharp$, Eb, C$^\sharp$, D, F, E, B.

5. No. In order to start and finish in the hallway, all areas must have an even number of doors, which is not the case here.

6. The door must be placed between B and the hall.

7.

a)

Washington Montana
Idaho
Oregon Wyoming

c)

Oklahoma Arkansas Tennessee North Carolina South Carolina
Texas Alabama Georgia
Louisiana Mississippi Florida

10. Four enclosures. One of the many possible solutions is:

I	II	III	IV
tiger	leopard	zebra	boar
heron	rhinoceros	crocodile	ostrich
		antelope	giraffe

Page 60, Directed Graphs

1.

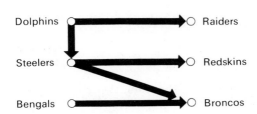

Dolphins — Raiders
Steelers — Redskins
Bengals — Broncos

4. a) *ABCE* and *ACDE*

    b) One answer is *CDAB*, length 3.

5.
$$\begin{array}{c c c c c c} & A & B & C & D & E \\ A & \begin{bmatrix} 1 & 1 & 0 & 0 & 1 \\ B & 1 & 1 & 0 & 0 & 1 \\ C & 1 & 2 & 1 & 1 & 1 \\ D & 2 & 1 & 1 & 1 & 2 \\ E & 0 & 0 & 0 & 0 & 0 \end{bmatrix} \end{array}$$

7. Bob, Harry, Ted, or Alice

8. Advisor, instructor, registrar, department chairperson, dean

9.

Ranking (highest to lowest)

Quigley
Thums
Ross
Stickle
Pratt

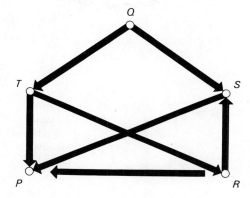

11

Ranking (highest to lowest)

C
D
A
E
B

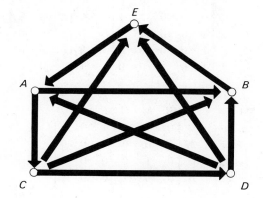

Page 71, Matrix Models for Directed Graphs

1. a)
$$\begin{array}{c c c c} & A & B & C \\ A & \begin{bmatrix} 0 & 1 & 0 \\ B & 0 & 0 & 1 \\ C & 1 & 0 & 0 \end{bmatrix} \end{array}$$

    b)
$$\begin{array}{c c c c c} & A & B & C & D \\ A & \begin{bmatrix} 0 & 1 & 0 & 0 \\ B & 0 & 0 & 0 & 0 \\ C & 0 & 1 & 0 & 1 \\ D & 1 & 1 & 0 & 0 \end{bmatrix} \end{array}$$

2.

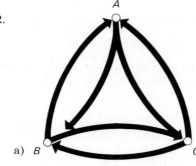
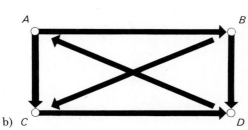

a) B b) C

3. a) $\begin{bmatrix} 3 & 1 \\ 3 & 7 \end{bmatrix}$ b) $\begin{bmatrix} 3 & 3 & 2 \\ 4 & 2 & 3 \\ 1 & 8 & 1 \end{bmatrix}$ c) $\begin{bmatrix} 3 & 0 & 6 \\ 6 & 0 & 18 \\ 3 & 0 & 0 \end{bmatrix}$ d) $\begin{bmatrix} 18 & 9 & 3 & 12 \\ 16 & 3 & 1 & 13 \\ 18 & 9 & 3 & 27 \\ 32 & 12 & 10 & 40 \end{bmatrix}$ e) $\begin{bmatrix} 0 & 0 & 0 \\ 0 & 0 & 0 \\ 0 & 0 & 0 \end{bmatrix}$

5. $a = 4$ and $b = 1$

9.

$$T = \begin{array}{c} \\ A \\ B \\ C \\ D \end{array} \begin{array}{c} \begin{array}{cccc} A & B & C & D \end{array} \\ \begin{bmatrix} 2 & 4 & 1 & 2 \\ 6 & 5 & 4 & 5 \\ 4 & 6 & 2 & 4 \\ 4 & 5 & 2 & 3 \end{bmatrix} \end{array}$$

The entries of T indicate the total number of paths of length 1, 2, or 3 between vertices.

10.

$$M = \begin{array}{c} \\ P \\ Q \\ R \\ S \\ T \end{array} \begin{array}{c} \begin{array}{ccccc} P & Q & R & S & T \end{array} \\ \begin{bmatrix} 0 & 0 & 0 & 0 & 0 \\ 0 & 0 & 0 & 1 & 1 \\ 1 & 0 & 0 & 1 & 0 \\ 1 & 0 & 0 & 0 & 0 \\ 1 & 0 & 1 & 0 & 0 \end{bmatrix} \end{array}$$

$$M^2 = \begin{array}{c} \\ P \\ Q \\ R \\ S \\ T \end{array} \begin{array}{c} \begin{array}{ccccc} P & Q & R & S & T \end{array} \\ \begin{bmatrix} 0 & 0 & 0 & 0 & 0 \\ 2 & 0 & 1 & 0 & 0 \\ 1 & 0 & 0 & 0 & 0 \\ 0 & 0 & 0 & 0 & 0 \\ 1 & 0 & 0 & 1 & 0 \end{bmatrix} \end{array}$$

$$M + M^2 = \begin{array}{c} \\ P \\ Q \\ R \\ S \\ T \end{array} \begin{array}{c} \begin{array}{ccccc} P & Q & R & S & T \end{array} \\ \begin{bmatrix} 0 & 0 & 0 & 0 & 0 \\ 2 & 0 & 1 & 1 & 1 \\ 2 & 0 & 0 & 1 & 0 \\ 1 & 0 & 0 & 0 & 0 \\ 2 & 0 & 1 & 1 & 0 \end{bmatrix} \end{array}$$

11.

$$M = \begin{array}{c} \\ B \\ M \\ C \\ R \\ T \\ J \end{array} \begin{array}{c} \begin{array}{cccccc} B & M & C & R & T & J \end{array} \\ \begin{bmatrix} 0 & 1 & 1 & 1 & 0 & 1 \\ 0 & 0 & 0 & 1 & 0 & 0 \\ 0 & 1 & 0 & 0 & 0 & 1 \\ 0 & 0 & 1 & 0 & 0 & 1 \\ 1 & 1 & 1 & 1 & 0 & 0 \\ 0 & 1 & 0 & 0 & 1 & 0 \end{bmatrix} \end{array}$$

$$M^2 = \begin{array}{c} \\ B \\ M \\ C \\ R \\ T \\ J \end{array} \begin{array}{c} \begin{array}{cccccc} B & M & C & R & T & J \end{array} \\ \begin{bmatrix} 0 & 2 & 1 & 1 & 1 & 2 \\ 0 & 0 & 1 & 0 & 0 & 1 \\ 0 & 1 & 0 & 1 & 1 & 0 \\ 0 & 2 & 0 & 0 & 1 & 1 \\ 0 & 2 & 2 & 2 & 0 & 3 \\ 1 & 1 & 1 & 2 & 0 & 0 \end{bmatrix} \end{array}$$

$$M + M^2 = \begin{array}{c} \\ B \\ M \\ C \\ R \\ T \\ J \end{array} \begin{array}{c} \begin{array}{cccccc} B & M & C & R & T & J \end{array} \\ \begin{bmatrix} 0 & 3 & 2 & 2 & 1 & 3 \\ 0 & 0 & 1 & 1 & 0 & 1 \\ 0 & 2 & 0 & 1 & 1 & 1 \\ 0 & 2 & 1 & 0 & 1 & 2 \\ 1 & 3 & 3 & 3 & 0 & 3 \\ 1 & 2 & 1 & 2 & 1 & 0 \end{bmatrix} \end{array}$$

Ranking (highest to lowest)
Tom
Bruce
John
Ron
Charlie
Max

Page 74, Mastery Test: Tools of Graph Theory

1. Both graphs can be traced.

2. One of the many possible solutions is:
   Color 1: California and Utah
   Color 2: Nevada and New Mexico
   Color 3: Arizona and Colorado

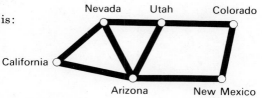

3. 3

4. Alice, Frank, or Carol can initiate a message.

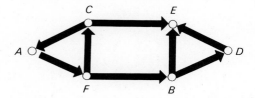

5. a)

$$M = \begin{array}{c} \\ S \\ C \\ R \\ A \end{array} \overset{\begin{array}{cccc} S & C & R & A \end{array}}{\begin{bmatrix} 0 & 1 & 1 & 0 \\ 0 & 0 & 1 & 0 \\ 0 & 0 & 0 & 0 \\ 1 & 0 & 1 & 0 \end{bmatrix}} \qquad M + M^2 = \begin{array}{c} \\ S \\ C \\ R \\ A \end{array} \overset{\begin{array}{cccc} S & C & R & A \end{array}}{\begin{bmatrix} 0 & 1 & 2 & 0 \\ 0 & 0 & 1 & 0 \\ 0 & 0 & 0 & 0 \\ 1 & 1 & 2 & 0 \end{bmatrix}}$$

   Row A of the matrix $M + M^2$ indicates that there is at least one directed path of length 1 or 2 from auto to each of the other three vertices.

6. a)

$$M = \begin{array}{c} \\ A \\ B \\ C \\ D \end{array} \overset{\begin{array}{cccc} A & B & C & D \end{array}}{\begin{bmatrix} 0 & 1 & 1 & 0 \\ 1 & 0 & 0 & 1 \\ 0 & 1 & 0 & 1 \\ 0 & 0 & 1 & 0 \end{bmatrix}}$$

   b)

$$M^2 = \begin{array}{c} \\ A \\ B \\ C \\ D \end{array} \overset{\begin{array}{cccc} A & B & C & D \end{array}}{\begin{bmatrix} 1 & 1 & 0 & 2 \\ 0 & 1 & 2 & 0 \\ 1 & 0 & 1 & 1 \\ 0 & 1 & 0 & 1 \end{bmatrix}}$$

   c) 1; 2

Page 83, Part III

1. a) begin, A, B, E, I
   c) day 14
   e) 17 days

   b) begin, A, B, E
   d) after 16 days
   f) 2

3.

Task	Day it begins
A	1
B	1
C	3
D	3
E	6
F	8
G	8
H	12

The project takes 17 days to complete.

4.

Task	Day it begins
A	1
B	1
C	1
D	3
E	4
F	5
G	7
H	7
I	11
J	11

The project takes 13 days to complete.

5.

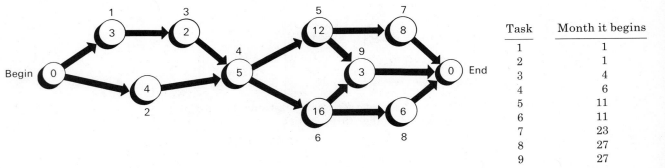

Task	Month it begins
1	1
2	1
3	4
4	6
5	11
6	11
7	23
8	27
9	27

The project takes 32 months to complete.

7.

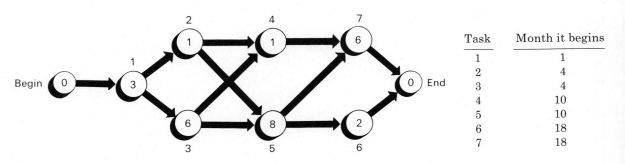

Task	Month it begins
1	1
2	4
3	4
4	10
5	10
6	18
7	18

The project takes 23 months to complete.

CHAPTER 4   LEGISLATIVE APPORTIONMENT AND INEQUALITIES

Page 98, Measures of the Unfairness of an Apportionment

1. .2; .0194

3. 2.0; .2353

5. .0333; .0714

6. a) .4391   b) .0588

Page 102, Inequalities

1. a) $x = -\frac{10}{6}$ b) $t = 6$ c) $q = -\frac{1}{3}$ d) $y = -\frac{8}{3}$
   e) $n = \frac{1}{2}$ f) $m = -1$

2. a) equivalent   b) not equivalent   c) equivalent   d) not equivalent

5. $w \leqslant 5$ 7. $q \geqslant 9$ 9. $x > -5$ 11. $x \geqslant 7$ or $-3 < x < 2$

Page 107, An Apportionment Principle

1. Delaware   3. Vermont   4. Vermont then New Hampshire

6. a) dorm A b) dorm C 7. dorm G

Page 108, Mastery Test

1. a) not equivalent   b) equivalent   c) not equivalent   d) equivalent

2. $x \leqslant -\frac{27}{8}$ 3. a) .1   b) .2857   c) Arizona   4. Alabama

Page 114, Part III

1. 2 representatives go to A, 3 to B, and 5 to C

3. 17 seats

4.	Humanities	Science	Education
2	450	800.00	50.00
3	150	266.67	16.67
4	75	133.33	8.33
5	45	80.00	

3 fellowships go to humanities, 4 to science, and 1 to education

6.	X	Y	Z
2	4.50	8.00	12.50
3	1.50	2.67	4.17
4	0.75	1.33	2.08
5	0.45	0.80	1.25

3 representatives go to state X, 3 to Y, and 4 to Z

CHAPTER 5   LINEAR EQUATIONS IN THE PLANE

Page 125, Linear Relations

1. x	$\frac{1}{3}$	0	3	-1	2	$\frac{4}{3}$
y	-3	-4	5	-7	2	0

3. a) equivalent
   c) equivalent

   b) not equivalent
   d) equivalent

4. a) $2x + 4y = 1$

   b) $4x - 3y = 12$

5. a) belongs b) does not belong c) belongs

6. a) $x = \frac{1}{3}y$
   c) $x = 2y - 5$

   b) $x - y = 10$
   d) $2x + 3y = 5$

7. a) $\{(P,I): I = .08P, P \geqslant 0\}$
   b) $\{(P,N): N = P + .07P, P \geqslant 0\}$
   c) $\{(I,T): T = 145 + .16(I - 1000), 1000 < I \leqslant 1500\}$
   d) $\{(C,F): F = \frac{9}{5}C + 32\}$

9. $T = 20U + 1000;$ \$2000; \$3000

11. $V = 350 - 70y;$ \$280; \$105; \$70

12. $\{(I,T): T = 3260 + .28(I - 16,000), 16,000 \leqslant I \leqslant 20,000\}$

Page 138, Graph of a Linear Relation

1. a) to the right of the y-axis
   c) on the x-axis
   e) in quadrant IV, including the positive x-axis
   f) in quadrant I, including the positive x- and y-axes

   b) in quadrant II
   d) on the negative y-axis

2. a) not on the same line
   c) on the same line

   b) not on the same line
   d) on the same line

3. a)

   b) same graph as (a)

   c)

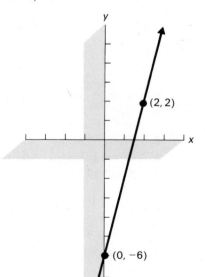

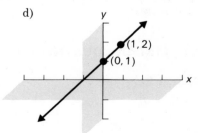

5. a)

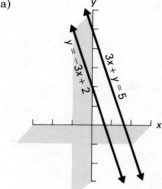

b)

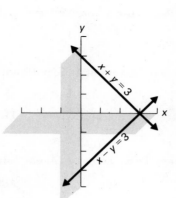

c)

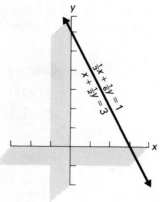

7. a)

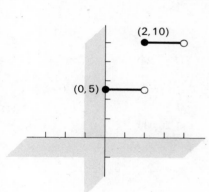

b)

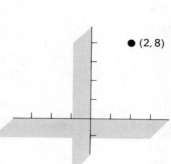

c)

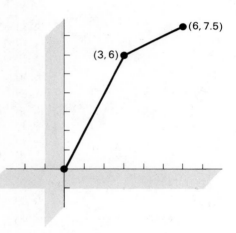

9.

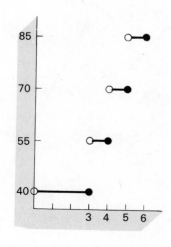

10.

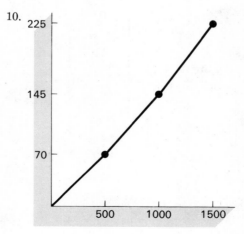

11. It is reasonable to use a linear relation in (a) and (e); in all of the others it is not reasonable.

Page 143, Solving a System of Linear Equations

1. (2,8) 3. no solution 5. equivalent

7. $(\frac{7}{3}, \frac{16}{3})$ 9. no solution 11. (0,0)

Page 144, Mastery Test: Tools of Linear Equations

1. a) true b) true c) false d) true

2.

x	0	$\frac{1}{3}$	1	3	-1	$\frac{5}{3}$	$\frac{1}{2}$
y	$\frac{1}{2}$	0	-1	-4	2	-2	$-\frac{1}{4}$

3. a) $y = \frac{2}{5}x + \frac{1}{5}$ b) $y = 3x - 6$
c) $y = -2x - 1$ d) $y = x - 3$
e) $y = \frac{3}{2}x - 2$

4. a) $y = 16x$ b) $M = .62K$
c) $T = .023I$ d) $C = \frac{5}{9}(F - 32)$

5. a) $\{(y,V): V = 200 - 50y, 0 \leqslant y \leqslant 4\}$
b) $\{(V,T): T = .081V, V \geqslant 0\}$
c) $\{U,C): C = 12U + 1500, U \geqslant 0\}$
d) $\{(I,T): T = 3260 + .28(I - 16,000), 16,000 \leqslant I \leqslant 20,000\}$

6. $A(0,0)$ $B(6,0)$ $C(6,-4)$ $D(0,-6)$ $E(-4,-4)$ $F(-4,6)$ $G(4,4)$

7. a) -2 b) $\frac{1}{2}$ c) $\frac{7}{5}$ d) $\frac{11}{6}$

8. a) b)

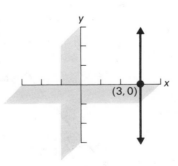

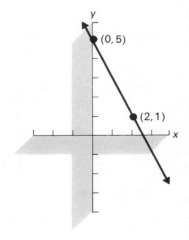

c)

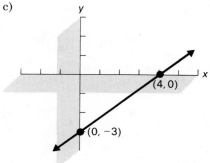

d)

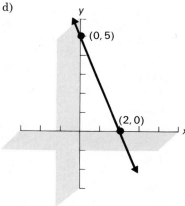

9. a) $(-1,\frac{1}{4})$ b) $(2,-2)$ c) $(\frac{2}{3},\frac{5}{3})$ d) $(-1,2)$

Page 152 Part III

1. Let x = number of \$40 hides
   y = number of \$25 hides
   Solve: $3x + 2y = 9$
   $ 2x + 3y = 11$

   Solution: $x = 1$ and $y = 3$;
   \$115

2. Let x = price per set
   y = number of sets
   Solve: $y = .06x$
   $ y = -.09x + 20$

   Solution: $x = \$133.33$

3. Let x = price per pound
   y = number of head
   Solve: $y = \frac{10}{5}x$
   $ y = -\frac{25}{6}x + 325$

   Solution: $x = 52.7$

5. Let x = number of harnesses
   Cost $= 2x + 116,\quad x \leqslant 20$
   Cost $= 1.5(x - 20) + 156,\quad x > 20$
   Receipts $= 6x$

   Solution: 28 harnesses;
   \$4.50

6. Let x = amount invested at 13%
   y = amount invested at 8%
   Solve: $\quad x + y = 10,000$
   $ .13x + .08y = 1000$

   Solution: $y = 6000$ which is the maximum
   amount that can be invested at 8%

8. Let x = grams of oatmeal
   y = grams of dates
   Solve: $3.9x + 2.74y = 3000$
   $ .142x + .022y = 70$

   Solution: $x = 414.8$ and $y = 504.5$

9. Let x = gallons of acidic water
   y = gallons of river water
   Solve: $x + y = 150{,}000$ Solution: $x = 5000$ and $y = 145{,}000$
   $\quad\quad\;.03x = .001 \cdot 150{,}000$

10. Let x = number of undergraduates
    y = number of graduate students
    Solve: $\dfrac{x}{15} + \dfrac{y}{5} = 350$ Solution: $x = 4500$ and $y = 250$

    $\quad\quad 200x + 400y = 1{,}000{,}000$

11. Let x = the value of agricultural goods produced
    y = the value of industrial goods produced
    Solve: $x - (.25x + .15y) = 15$ Solution: $x = \$58.5$ billion
    $\quad\quad y - (.10x + .45y) = 100$ $y = \$192.5$ billion

13. Let x = the amount of gold held by Us
    y = the amount of gold held by Them
    Solve: $\frac{3}{4}x + \frac{3}{5}y = x$
    $\quad\quad \frac{1}{4}x + \frac{2}{5}y = y$ Solution: $\dfrac{y}{x} = \dfrac{5}{12}$

15. Let x = the number of people living in urban areas
    y = the number of people living in nonurban areas
    Solve: $\frac{2}{5}x + \frac{1}{7}y = x$
    $\quad\quad \frac{3}{5}x + \frac{6}{7}y = y$ Solution: $\dfrac{y}{x} = \dfrac{21}{5}$

CHAPTER 6   LINEAR PROGRAMMING

Page 168, Linear Inequalities in the Plane

1. a) 11   b) 10   c) $4x - y < 12$ d) $4x - y < 12$

3. a)   b)

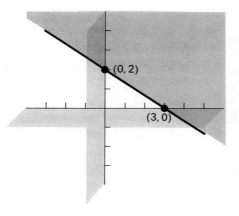

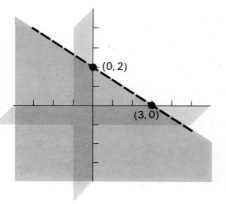

5. a)

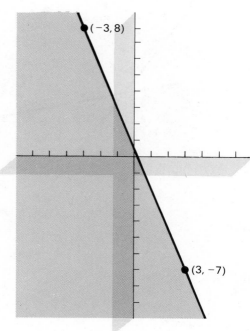

6.

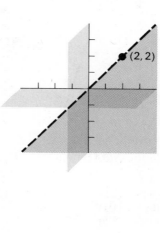

7. $x + y \leqslant 10$

9. a)

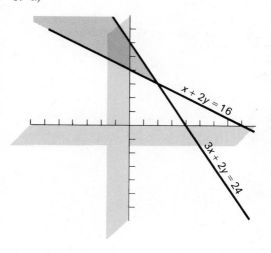

b)

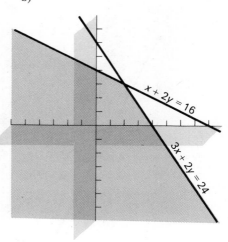

11.

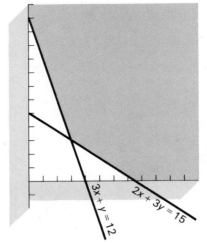

$3x + y = 12$
$2x + 3y = 15$

12.

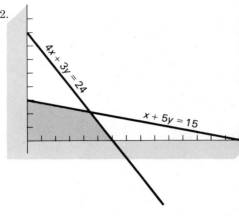

$4x + 3y = 24$
$x + 5y = 15$

13. the empty set

15. (0,10), (0,14), (5,12), (15,0), (10,0)

17. (0,3)
    (0,2) (1,2) (2,2) (3,2) (4,2)
    (0,1) (1,1) (2,1) (3,1) (4,1) (5,1)
    (0,0) (1,0) (2,0) (3,0) (4,0) (5,0) (6,0)

19. a) 15   b) 17   c) 3

Page 179, The Linear Programming Problem

1. a) 10   b) 12   c) 5

2. a) (9,9); 45   b) (9,9)

5.

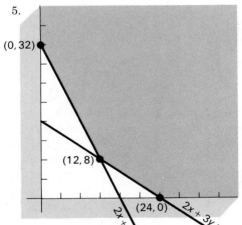

(0, 32)
(12, 8)
(24, 0)
$2x + y = 32$
$2x + 3y = 48$

a) (24,0)
b) all points on the line $2x + 3y = 48$
   between (12,8) and (24,0)
c) (12,8)

7. Solution: The optimal solution is $x = 4$ and $y = 6$ with the maximum value of P being 36.

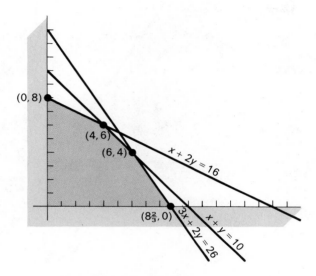

9. Solution: The optimal solution is $x = 5$ and $y = 5.5$; the minimum value of C is 10.5.

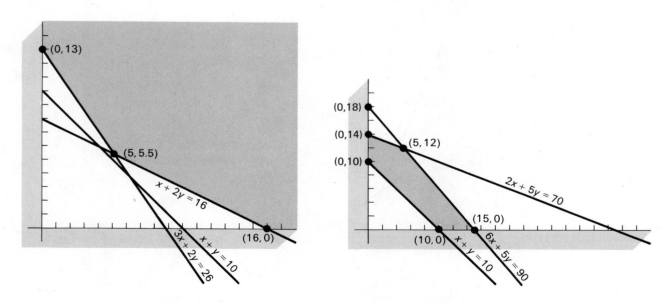

11. Solution: The maximum value of D is 80, which occurs at (5,12); the minimum value of D is 40, which occurs at (10,0).

Page 182, Mastery Test: Tools of Linear Programming

1.

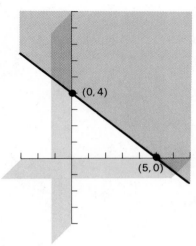

(0, 4)

(5, 0)

2.

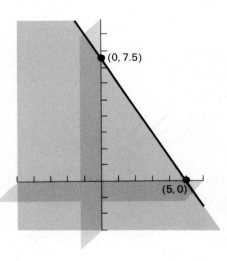

(0, 7.5)

(5, 0)

3.

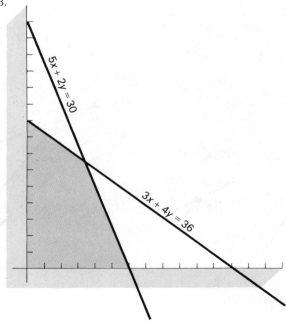

5x + 2y = 30

3x + 4y = 36

4.

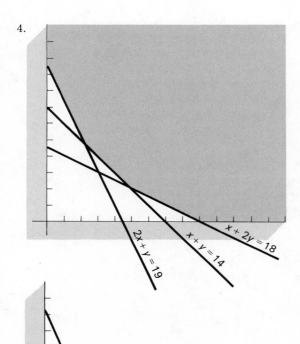

2x + y = 19

x + y = 14

x + 2y = 18

5. Solution: The maximum value of P is 43, which occurs at $x = 4$ and $y = 9$.

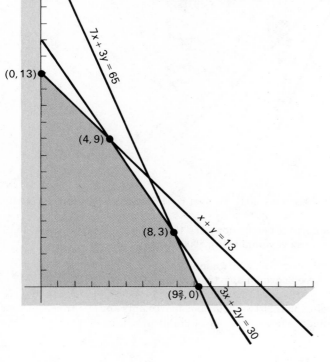

$(0, 13)$

$(4, 9)$

$7x + 3y = 65$

$(8, 3)$

$x + y = 13$

$(9\frac{2}{7}, 0)$

$3x + 2y = 30$

6. Solution: The optimal solution is $x = 13$ and $y = 0$ with the minimum value of C being 39.

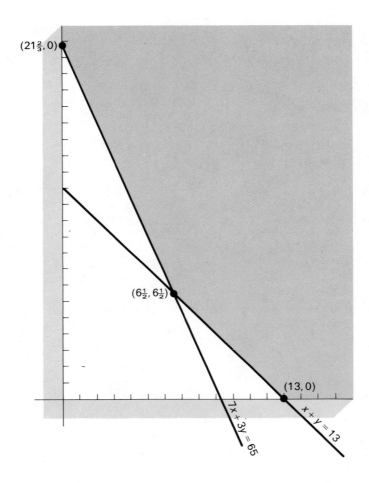

Page 188, Part III

1. Solve: Maximize $T = 60{,}000x + 20{,}000y$, subject to the same restrictions as in the example.
    Solution: The optimal solutions occur along the line $6x + 2y = 90$ between $(15,0)$ and $(9,18)$ with the largest tax revenue being \$900,000.

3. Let x = the number of two-bedroom apartments
       y = the number of three-bedroom apartments
    Solve: Minimize $C = x + 2y$, subject to $600x + 700y \geqslant 26{,}000$, $y \geqslant x$, $x \geqslant 0$
    Solution: The optimal solution is 20 two-bedroom apartments and 20 three-bedroom apartments; the minimum predicted number of children will be 60.

5. Let x = units of food A to be in the diet

$\quad\quad y$ = units of food B to be in the diet

Solve: Maximize $F = .3x + .4y$, subject to $2000 \leqslant 300x + 200y \leqslant 3000$, $60 \leqslant 5x + 10y \leqslant 80$

Solution: The combination which gives the maximum amount of fiber, 3.9 units, is 7 units of food A and 4.5 units of food B.

7. Let x = the number of acres of corn to be planted

$\quad\quad y$ = the number of acres of wheat to be planted

Solve: Maximize $P = 375x + 200y$, subject to $x + y \leqslant 150$, $100x + 30y \leqslant 9000$, $x \geqslant 0$, $y \geqslant 0$

Solution: The best program is to plant 64.3 acres of corn and 85.7 acres of wheat; this will give the highest profit, \$41,252.50.

9. Let x = number of men's suits to be made

$\quad\quad y$ = number of women's suits to be made

Solve: Maximize $P = 12x + 16y$, subject to $.5x + .5y \leqslant 15$, $1.5x + y \leqslant 38$, $x + 1.5y \leqslant 40$, $x \geqslant 0$, $y \geqslant 0$

Solution: The best program is to make 10 men's suits and 20 women's suits with the largest profit then being \$440.

10. Let x = the number of units of item A to be made

$\quad\quad y$ = the number of units of item B to be made

Solve: Maximize $L = 2x + 3y$, subject to $4x + 5y \leqslant 1000$, $81x + 90y \geqslant 18{,}900$

Solution: The optimal solution is to produce 100 units of item A and 120 units of item B. The amount of hours spent in this production program is 560, which will be the largest.

CHAPTER 7   COUNTING

Page 206, Counting Principles and Permutations

1. $4! = 24$ 2. $4^4 = 256$ 3. $4 \cdot 3 \cdot 3 \cdot 3 = 108$

5. $14 \cdot 2 \cdot 3 \cdot 3 = 252$ 7. $4 \cdot 3 \cdot 3 \cdot 2 \cdot 2 = 144$ 9. $10! = 3{,}628{,}800$

11. $P[15,3] = 2730$ 13. $P[7,3] + 1 = 211$

14. $5! = 120$; $4 \cdot 3! = 24$ 15. $k = 5$

16. $6! + P[6,5] + 5 \cdot P[5,3] = 1740$ 17. $10 \cdot 10 \cdot 2 - 1 = 199$

Page 210, Combinations

1. 4   3. $C[50,5] = 2{,}118{,}760$

5. $C[13,5] = 1287$; $4 \cdot C[13,5] = 5148$

7. $C[3,1] \cdot C[7,2] \cdot C[11,3] = 10{,}395$

9. $C[12,5] \cdot C[7,3] \cdot C[4,4] = 27{,}720$; $C[11,4] \cdot C[7,3] \cdot C[4,4] = 11{,}550$

10. $C[5,3] \cdot C[7,3] + C[5,4] \cdot C[7,2] + C[5,5] \cdot C[7,1] = 462$

11. $C[4,1] + C[4,2] + C[4,3] + C[4,4] = 2^4 - C[4,0] = 15$

Page 211, Mastery Test: Tools of Counting

1. a) 60   b) 10   c) 10   d) 20
   e) 210   f) 5040   g) 1   h) 1

2. $3! = 6$; $4! = 24$ 3. $26^3 \cdot 9^3 = 12{,}812{,}904$

4. $4! = 24$ 5. $P[7,4] = 840$ 6. $2^4 = 16$;   6

7. $C[15,3] = 455$ 8. $C[100,10] \cdot C[150,15]$ 9. $5! = 120$

10. Without area codes there are at most $8 \cdot 10^6 = 8{,}000{,}000$ telephone numbers available and there are more telephones in the country than this.

Page 215, Part III

1. $C[13,5] = 1287$ 2. $C[13,5] - C[11,4] = 957$

3. $C[13,5] - C[7,2] \cdot C[5,2] = 1077$

4. $C[5,2] = 10$ 5. $C[4,1] + C[4,2] = 10$

6. $C[10,2] = 45$; $C[10,2] + C[10,1] + C[10,0] = 56$

7. $C[10,7] + C[10,8] + C[10,9] + C[10,10] = 176$

8. $C[5,4] + C[5,5] = 6$

CHAPTER 8   PROBABILITY

Page 230, Probabilistic Models

1. $S = \{r,b,w\}$; $Pr(r) = \frac{2}{6}$; $Pr(b) = \frac{1}{6}$; $Pr(w) = \frac{3}{6}$

3. $S = \{$candy only, both, nothing$\}$; $Pr($candy only$) = \frac{34}{50}$, $Pr($both$) = \frac{5}{50}$, $Pr($nothing$) = \frac{11}{50}$; the outcome will be one of those in S; $\frac{39}{50}$

5. $\frac{1}{500}$; $\frac{5}{500}$ 7. $2 \cdot \dfrac{1}{2^4} = \dfrac{1}{8}$

8. a) For example, there is only one way to obtain a 2, namely (1,1), while there are two ways to obtain a 3, (2,1) and (1,2).
   b) $S = \{(1,1),(1,2), \ldots (1,6),(2,1), \ldots (6,1),(6,2) \ldots (6,6)\}$

9. $\frac{6}{36}$ 10. $C[7,3]/C[8,4] = \frac{1}{2}$

12. $(C[5,4] + C[5,5])/2^5 = \frac{6}{32}$

14. $\dfrac{4 \cdot C[13,5]}{C[52,5]} = .002$ 15. $1 - \dfrac{C[48,5]}{C[52,5]} = .341$

Page 238, Expected Value

1. $6.50   3. $35,000   5. $57.14

7. The insurance company should sell the policy for more than $900.

9. 70.5

10. Order either 1300 or 1200 copies; the expected value in either case is the highest, $46.00.

Page 246, Multistage Experiments

1.

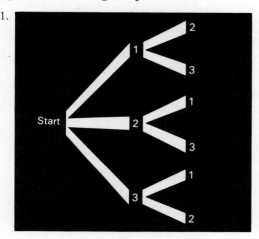

The probability of any one of these outcomes is $\frac{1}{6}$; $\frac{1}{2}$.

3. $\frac{3}{8} \cdot \frac{5}{7} = \frac{15}{56}$; $1 - \frac{5}{8} = \frac{3}{8}$

5. $Pr(\text{wins}) = \frac{18}{37}$, $Pr(\text{loses}) = \frac{37}{74}$
$Pr(\text{breaks even}) = \frac{1}{74}$

6. .35;   .56

7. $-\frac{1}{74}$; $-\frac{1}{74}$

8. $.655; $2.28

Page 253, Binomial Experiments

2. .1916;   .6079

4. a) .2461   b) .1762;   .7368

5. There must be an even number of questions on the test.

7. .1663

Page 254, Mastery Test: Tools of Probability

1. $Pr(O_1) = Pr(O_3) = \frac{1}{7}$, $Pr(O_2) = \frac{2}{7}$, $Pr(O_4) = \frac{3}{7}$

2. a) $\frac{15}{50}$ b) $\frac{36}{50}$ c) $\frac{9}{50}$

3. $S = \{(r,b),(r,w),(b,r),(b,b),(b,w),(w,r),(w,b),(w,w)\}$
    Each of the outcomes (b,w) and (w,b) have probability $\frac{1}{5}$, all other outcomes have a probability of $\frac{1}{10}$.

4. The possibility set is that of exercise 3 with the additional outcome (r,r). The outcome (r,r) has probability $\frac{1}{25}$; the outcomes (r,w), (r,b), (b,r), and (w,r) each have probability $\frac{2}{25}$; all other outcomes have probability $\frac{4}{25}$.

5. a) $\dfrac{C[80,10]}{C[100,10]}$ b) $1 - \dfrac{C[80,10]}{C[100,10]}$

6. \$2.00;   loss of \$.50   7. \$700

8. The slower machine is the better buy because the expected profit per item is higher for it than for the faster machine; namely, .3984 as compared to .3968.

9. .855   10. .2581;   .6840   11. 17

Page 257, Part III

1. .2173   2. .1643   3. No   4. the true-false test

6. .1331   8. .2824

CHAPTER 9   DESCRIPTIVE STATISTICS

Page 275, Distributions and Measures of Central Tendency

1. a)   b)

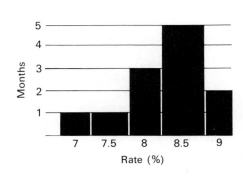

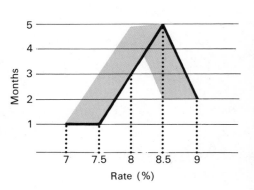

3. Mean = 8.25, median = 8.5, and mode = 8.5

5.

Score	Frequency
2	1
3	0
4	0
5	3
6	2
7	4
8	4
9	3
10	3

mean = 7.3
median = 7.5
mode = 7 and 8

7. Mean = 15.43, median = 15, and mode = 15

Page 284, Measures of Dispersion

1. a) 2.236   b) 7.071   c) .7071
   d) 4.796   e) 14.14

3. a) mean = 7   b) mean = 7
      standard deviation = 2.074   standard deviation = 2.898

5. The variance of A is bigger than B; 5.17 as compared to 2.42

7. The family was better off in 1960, since $10,000 was then 4 standard deviations above the mean as opposed to 1970 when $14,000 was only $1\frac{1}{2}$ standard deviations above the mean.

10. c) The mean is increased or decreased by the number which is added to or subtracted from each score; the standard deviation does not change.

11. Mean = 599.5; standard deviation = 2.291

12. Both the mean and the standard deviation is multiplied by the number.

13. Mean = 20; standard deviation = 10

Page 288, Chebyshev's Inequality

1. .75   3. $A = 40$ and $B = 60$ 5. .84

7. No. The probability of the tire wearing out between 30,000 and 32,000 is only .36.

10. The probability of a tube in model A burning out between 1825 and 1975 days is at least .982, whereas for model B it is .96; a tube in model A has a better chance of lasting 5 years.

Page 289, Mastery Test: Tools of Descriptive Statistics

1. a) 1 inch which occurred 6 times   b) 9 inches which occurred 2 times
   c) 5 inches   d) 35 months
   e) $\frac{10}{35}$

2. a) 5.4   b) 4   c) 5

3. a) 5   b) 76   c) 3.564

4. There are many possible answers.

5. Mean = 8, variance = 44, and standard deviation = 6.633

6. The family was better off in 1965, since their income was then 3.33 standard deviations above the mean as compared to 1975 when their income was only 2.14 standard deviations above the mean.

7. .9375   8. 2600 days

Page 295, Part III

1. *B*   2. *F*   3. *C*   4. *B*   6. *A*

8. The grade would be a *D* with the 100 included but it would be an *F* with the 100 replaced by a score of 60.

10.

a) Score	Grade	b) Score	Grade
56, 57	*D*	56	*F*
58, 59, 60,		57, 58	*D*
61, 62, 63, 64	*C*	59, 60, 61	*C*
100	*A*	62, 63	*B*
		64	*A*

11. 5.01

12. Mean = 65, standard deviation = 4.43

13. To be combined with class *B* so as to increase the standard deviation, in which case your grade might be included in the *C* range.

CHAPTER 10   THE COMPUTER: A TOOL OF MAN

Page 315, Computers and Algorithms

1. Input: one pair of numbers
   Output: one number, which is the product of the two numbers used as input

2. Input: a positive integer followed by as many pairs of numbers
   Output: for each *pair* of numbers used as input, the product will be displayed

3. Input: a pair of numbers
   Output: one number, which is 2 times the first number of the pair plus 3 times the second number of the pair

5. Input: a positive integer followed by as many groups of 4 numbers
   Output: for each group of 4 numbers put into the computer, the mean will be computed and displayed

6. Input: one pair of numbers
   Output: the larger of the two numbers used as input

8. The algorithm as given in the flowchart will display only the last number put
   into the computer. In order to correct the flowchart, place the step "$S \leftarrow 0$"
   before the step "$K \leftarrow 1$".

10. For the flowchart as given, the output will be the numbers $\frac{1}{2}$, $\frac{1}{2} + \frac{1}{3}$, $\frac{1}{2} + \frac{1}{3} + \frac{1}{4}$,
    up to $\frac{1}{2} + \frac{1}{3} + \frac{1}{4} + \cdots + \frac{1}{19} + \frac{1}{20}$. In order to correct the flowchart to obtain the
    desired output, replace the step "$K \leftarrow K + 1$" by "$K \leftarrow K + 2$".

12. The procedure will not terminate because N will always have the value 1. The
    correction that must be made is to place the step "$N \leftarrow N + 1$" after the decision
    "$N = K$?" and before returning to the input step "Obtain A". Also, we need the
    output step "Display L" on the Yes branch of the test "$N = K$?".

Page 340, Writing Flowcharts

1. | Box number | Instruction needed |
   |:---:|---|
   | 1 | "$S \leftarrow 0$", "$N \leftarrow 1$" |
   | 2 | "Display S" |
   | 3 | "$N \leftarrow N + 1$" |
   | 4 | "Stop" |

4. | Box number | Instruction needed |
   |:---:|---|
   | 1 | "$K \leftarrow 1$" |
   | 2 | "$N \leftarrow N - 1$" |
   | 3 | "$P \leftarrow P \cdot N$" |

3.
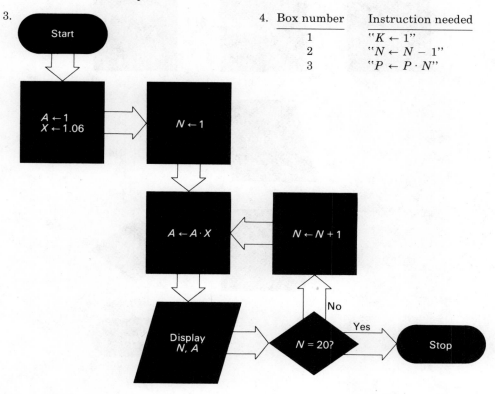

5.

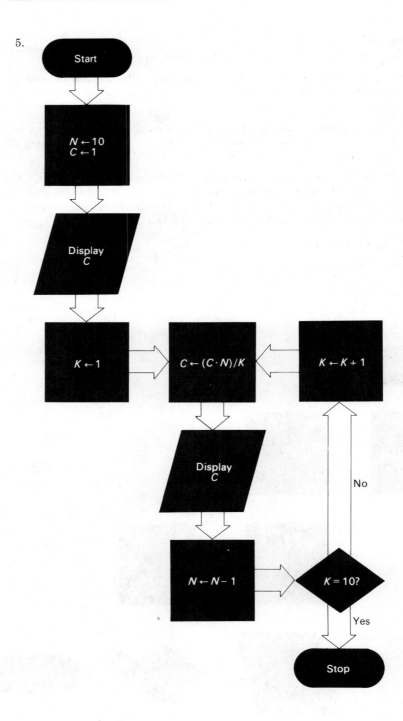

7.

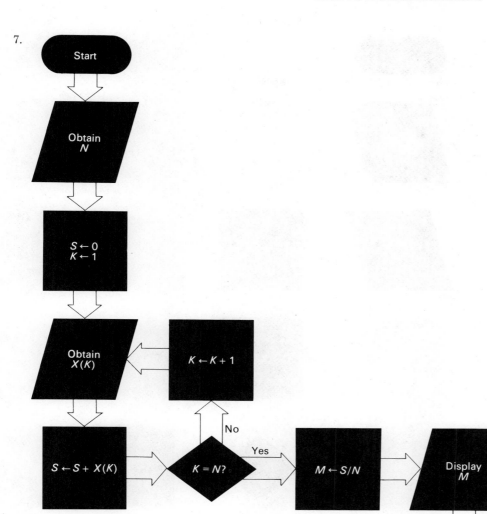

9.

Box number	Instruction needed
1	"Obtain BP, BR"
2	"$X \leftarrow -X/(AP/AR)$"
3	"$X \leftarrow X/(BP/BR)$"

11.

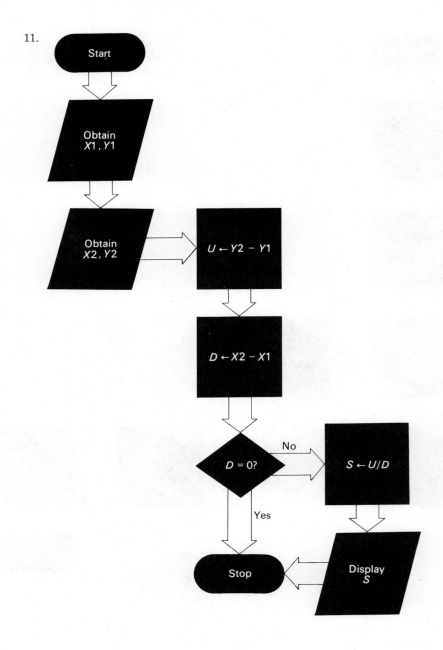

14. a)

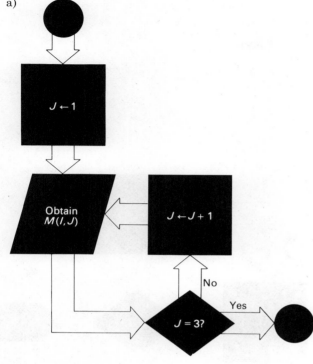

b)

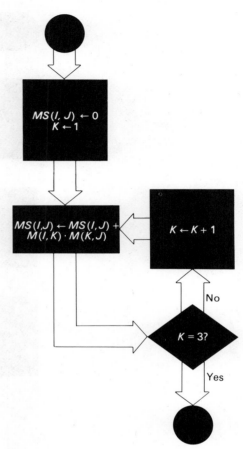

16.

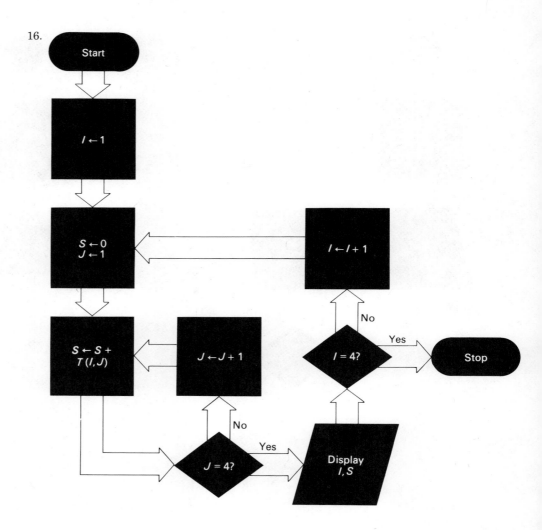

Index

Index